RELUCTANT NATION

To DKF

RELUCTANT NATION

AUSTRALIA AND THE ALLIED DEFEAT OF JAPAN 1942–45

DAVID DAY

OXFORD

OXFORD UNIVERSITY PRESS AUSTRALIA

Oxford New York Toronto
Delhi Bombay Calcutta Madras Karachi
Petaling Jaya Singapore Hong Kong Tokyo
Nairobi Dar es Salaam Cape Town
Melbourne Auckland

and associated companies in
Berlin Ibadan

OXFORD is a trade mark of Oxford University Press

First published 1992

National Library of Australia
Cataloguing-in-Publication data:

Day, David, 1949- .
Reluctant nation: Australia and the allied defeat of Japan 1942–45.

Bibliography.
Includes index.
ISBN 0 19 553242 2.

1. World War, 1939–1945—Australia. 2. World War, 1939–1945 Japan. 3. World War, 1939–1945—Campaigns —Pacific Ocean. 4. World War, 1939–1945—Campaigns —Japan. 5. Japan—History—1926–1945. I. Title.

940.540994

Cover design by Sarn Potter
Typeset by Syarikat Seng Teik Sdn. Bhd., Malaysia
Printed by Fong & Sons Printers, Singapore
Published by Oxford University Press,
253 Normanby Road, South Melbourne, Australia

Contents

Preface

Reluctant Nation is the second part of a two part study of Australian defence and foreign policy during the Second World War. The first part, *The Great Betrayal*, published in 1988, examined Australia's predicament during the first three years of the war as her traditional confidence in Imperial defence faltered in the face of diminished British power. As Britain grappled with threats to her existence and to that of her Empire, she made choices that left the vital Singapore naval base practically undefended, and Australia and New Zealand, as a direct consequence, vulnerable to Japanese invasion.

This second part examines Australia's reaction to her plight in 1942, how she met the Japanese challenge under the leadership of the American Supreme Commander, Douglas MacArthur, and then began to prepare for the post-war world. The lesson of the war was clear—Australia could not rely for her defence on the beneficence of great power allies with world-wide responsibilities and interests. Greater self-reliance, combined with a web of alliances and meaningful treaties of friendship and defence, together with a restored world organization, would offer Australia the prospect of a secure and independent future.

During 1943–4 Australia attempted to lay the foundations for a new order by seeking to construct a post-war economic and military sub-empire in the South Pacific. The over-arching British and American empires would guarantee its existence from external threat while the expanded Australian defence forces would enforce internal control. Huge Lancaster bombers assembled in local aircraft factories would cast a shadow of terror from the air while battle-hardened troops maintained order on the ground. As it happened, defence self-reliance and dreams of a sub-empire were crushed by persistent colonial attitudes, forceful Anglo-American opposition, and an appreciation of the political and economic costs of such a course to Australia.

All these studies arose from a question I posed myself when leaving for Cambridge to begin a Ph.D.: to what extent was the Second World War a turning point in modern Australian history? The answer appears to be that it was not as great a turning point as most people think. Australia emerged from the war anxious to strengthen her discredited defence links with Britain, abandoning wartime hopes of greater self-reliance. Hence the title of this volume.

For permission to quote from various private and official papers, I would like to thank the following: Rupert Wilkinson for the diary of Gerald Wilkinson; D. C. Edwards for the papers of Sir Ralph Edwards; the librarian of the British Library of Political and Economic Science for the papers of Hugh Dalton; the archivist of the House of Lords Record Office for the papers of Lord Beaverbrook; the Master, Fellows and Scholars of Churchill College in the University of Cambridge for the papers of Lord Hankey; Frank Strahan, archivist of the University of Melbourne for the papers of W. S. Robinson; Leo Cooper Ltd for extracts from *Chief of Staff*; Collins Publishers for extracts from the *Harold Nicolson diaries*; Constable Publishers for extracts from *Winston Churchill: the struggle for survival*. Extracts from Crown copyright documents appear with the permission of the Public Record Office. To any copyright owners I have been unable to contact, I sincerely apologize.

I am grateful to the dedicated staff at the following institutions who facilitated the research that made this book possible: Australian Archives; Manuscript Room of the British Library; Cambridge University Library; Churchill College Archives Centre; Flinders University Library; House of Lords Record Office; Imperial War Museum; India Office Library; Liddell Hart Centre for Military Archives at

King's College in London; London School of Economics; Melbourne University Archives; National Library of Australia; National Maritime Museum; Public Record Office; and Reading University Library.

This work would not have been possible without financial assistance from several sources. The Nuffield Foundation and the Smuts Fund assisted several research visits to Australia, as well as to various archives in London. This study was begun at Churchill College, written at Clare College, and completed at Bond University. Both Churchill and Clare Colleges were most generous in their support of my work, providing a physical and intellectual environment in which minds could range at will and work could prosper. I am most grateful to the Masters and Fellows for their assistance and their friendship. The Australian Studies Centre in London and the Imperial History Seminar in Cambridge provided convivial surroundings in which parts of this book were originally presented. I have benefited also from discussions with, among others, Carl Bridge (University of New England), Warren Kimball (Rutgers), Peter Lyon (Institute of Commonwealth Studies), John McCarthy (Australian Defence Force Academy), Tom Millar (London School of Economics), David Reynolds (Christ College) and Claude Wischik (St John's College).

I thank my wife, Silvia, for patient support and comfort, and for numerous valuable suggestions regarding the style and content of this work. Nan McNab did a sterling job of editing, overseen by Peter Rose of Oxford University Press who has guided this book into print.

Throughout my time in Cambridge, I was privileged to enjoy the friendship, support and stimulation of Professor David Fieldhouse to whom this volume is most gratefully dedicated.

Abbreviations

AA	Australian Archives, Canberra
AIF	Australian Imperial Force
AMF	Australian Military Forces
ANZAC	Australian and New Zealand Army Corps
BBC	British Broadcasting Corporation
BCOF	British Commonwealth Occupation Force
BL	British Library, London
BOAC	British Overseas Airways Corporation
BPF	British Pacific Fleet
CC	Churchill College Archives Centre, Cambridge
COS	Chief(s) of Staff
DAFP	Documents on Australian Foreign Policy 1939–49, R. G. Neal et al., vols 1–8, Canberra
D DAY	6 June 1944, Allied landing in France
FDR	Franklin D. Roosevelt
FUL	Flinders University Library, Adelaide
GG	Governor General
HLRO	House of Lords Record Office, London
IOL	India Office Library, London
IWM	Imperial War Museum, London
KC	Liddell Hart Centre for Military Archives, King's College, London
LSE	London School of Economics
NATO	North Atlantic Treaty Organization
NLA	National Library of Australia, Canberra
NMM	National Maritime Museum, Greenwich
PBY	Flying boat
PRO	Public Records Office, London
RAAF	Royal Australian Air Force
RAF	Royal Air Force
RUL	Reading University Library, Reading
SEAC	South East Asia Command
SWPA	South West Pacific Area
UAP	United Australia Party
UMA	University of Melbourne Archives
USAAF	United States Army Air Force
USCOS	United States Chief(s) of Staff
VE DAY	8 May 1945, Official end of hostilities in Europe
VLR	Very Long Range (aircraft)
W/T	Wireless Traffic

1

The Isolation of Australia

June 1942

BY JUNE 1942 the forces of Emperor Hirohito had carved out a glorious Asian empire for their diminutive and bespectacled monarch. A succession of European outposts had fallen with startling speed to the Japanese. The Americans in the Philippines, the British in Hong Kong, Malaya, Singapore and Burma, and the Dutch in the East Indies proved no match for the imperial forces of the rising sun. Almost as surprised as the Allies by the extent of their victories, the Japanese paused to secure their new empire from counter-attack. The isolated British Dominion of Australia remained as a potential threat to the enjoyment of their victories. The Japanese adopted a strategy aimed at neutralizing this threat before an Allied force could assemble.

Two Allied naval victories removed the immediate prospect of a Japanese invasion of Australia. In the Battle of the Coral Sea from 5 to 8 May 1942, the Japanese were forced to turn back from a seaward attack on Port Moresby, the torpid administrative capital of Australia's Papuan colony. One month later, a more decisive defeat was inflicted on the Japanese Fleet during the battle for the central Pacific island of Midway, the final stepping stone between Tokyo and Hawaii. Forewarned by intercepted radio messages, the American Fleet was able to ambush the Japanese and sink four of their aircraft-carriers. The Japanese were forced onto the defensive for the remainder of the war. The huge risk that they had taken with their surprise attack on Pearl Harbor the previous December now seemed very ill-advised.

Despite these setbacks, Japan remained in a strong defensive position. Their naval force was the most powerful in the Pacific in terms of battleships and cruisers, and her land-based air strength compensated in part for the loss of the four fleet carriers. The decisive difference after Midway was that Japan could no longer afford to gamble her fleet on offensive actions that might, if they failed, lay Japan itself open to Allied attack. They husbanded their strength for a long war, while they tried to wear down the American will to avenge the humiliation of Hawaii. Although Japanese victory was less likely after the Battle of Midway, Japanese defeat remained a formidable and costly prospect for the Allies who had to strike across the immense distances of the Pacific.

The Americans could adopt two obvious strategies. One was to strike across the central Pacific, leap-frogging from one distant island to another until Japan was within reach. This could not be attempted until the American fleet had been able to make good its own losses. The alternative strategy was to use Australia as a springboard from which to strike at Japan from the south. The Japanese tried to defeat both these strategies by strengthening their Pacific island bases and by capturing a further string of islands off the east coast of Australia, thereby driving a deep wedge between Australia and the United States. The reinforcement route to Australia was threatened and the country's vital coastal cities were brought within the range of Japanese bombers. This was the challenge facing the Allies following the naval victory of Midway Island.

Australia's defensive position in mid-1942 had certainly improved since the dark days of Pearl Harbor when she possessed neither air, sea nor land defences sufficient to meet a serious invasion. Even so, there was little cause for complacency. America's defeated General Douglas MacArthur, ordered out of the Philippines by Roosevelt before it fell to the Japanese, boosted Australian morale when he arrived but provided little real security. American forces were a long time coming and did not begin to outnumber the Australian forces till 1943. However, modern American fighter aircraft did arrive to form the beginnings of a defensive barrier for Australia.

Australia's traditional defender, Britain, retained its forces far from the loyal Dominion's shoreline. The Royal Navy fleet that was meant to spring to Australia's defence had sunk without trace in the South China Sea where the two battleships, the *Prince of Wales* and the *Repulse*, rested on the sea-bed, dispatched by Japanese torpedo bombers. More British ships shared their fate, overwhelmed by the wave of Japanese successes. The partially reinforced remnants of Britain's Far Eastern naval forces were fixed in a triangle of the Indian Ocean between Aden, Cape Town and Ceylon, protecting the supply convoys which travelled from Britain and India to the Middle East.

Australia's own war effort was split between the war in the Pacific and the struggle against Germany. Before Pearl Harbor, three divisions of Australian troops had been committed to the British cause in the Middle East. Thousands of airmen were based in Britain under the Empire Air Training Scheme and the Australian naval forces, such as they were, had been scattered as far as the Mediterranean. Even when Australia was clearly threatened by Japan, it was slow to disengage its forces from the European war. By June 1942 less than half the troops had returned from the Middle East. Two brigades had been diverted to the symbolic defence of Ceylon while the 9th Division remained in Egypt. Australia used this Division at various times, to bargain for British or American reinforcements, but the attempts were totally unsuccessful.

The problem faced by Australia was that the Allies were committed to finishing the European war before they turned against Japan. Germany posed a greater threat than Japan to both Britain and America, and Churchill was determined to outmanoeuvre those American naval strategists keen to avenge Pearl Harbor and build on the successes of the Coral Sea and Midway Battles. Initially Churchill suggested launching a second front in Europe during 1942. But, as he well knew, a cross-Channel invasion was not an immediate possibility. Still, it attracted American attention as Churchill intended, and when the time came to plan the invasion in detail, he convinced the Americans that the only feasible option was for Allied forces to invade North Africa.[1]

In principle the Americans were responsible for the naval defence of Australia's eastern seaboard, whereas the British were responsible for the Indian Ocean. But, as we have seen, the diminished Eastern Fleet was stationed far from Australia. Nevertheless Churchill knew he must prosecute the war against Japan with some semblance of vigour. 'Pacific first' elements in the United States, as well as the Chinese and the Australians, comprised part of a powerful group calling for greater efforts. The Chinese Ambassador in London, Dr Wellington Koo, had tackled the British Foreign Secretary, Anthony Eden, on British 'ineptitude' in the Far East. As a result, Eden had warned Churchill in early June that the Chinese attitude towards Britain had 'developed from hostility to contempt' and that China was hoping to assume post-war leadership in Asia and 'help their neighbours to throw off the yoke of western imperialism'.[2]

There was a deep-seated fear in Whitehall of an Asiatic bloc expelling Western influence from the region. Churchill had referred to such a possibility in the past[3], and now quickly assured the Chinese ambassador of Britain's aggressive intentions against the Japanese while simultaneously ordering his army chief in the Far East, Field Marshal Wavell, to take the offensive in Burma. In his meeting with Dr Koo at Downing Street on 3 June, Churchill outlined a plan to cut Japanese communication lines in Burma with southern China. Although he was careful not to promise huge reinforcements for such an offensive, Churchill did suggest that the forces in the Middle East might be drawn upon to help in the task. At the same time, he stressed the overriding importance of the war against Germany, promising to bring the full weight of British forces against Japan when victory in Europe had been won.[4]

Meanwhile, Churchill assured his Chiefs of Staff that no further troops would have to be commited against the Japanese. As he informed them, he had 'never suggested sending any further troops to the East than those now on the sea or under orders'.[5] The problem was that, as Wavell quickly advised him, no offensive could be mounted without further troops.[6] Desperate to keep up the appearance of action, Churchill continued to pressure his chiefs to plan for an early offensive.[7] As the naval commander of the Eastern Fleet, Admiral Somerville, caustically commented, Churchill is 'cut out as a figurehead but *not* as a military war winner . . . His strategical intuition is disastrous'.[8]

Churchill certainly did not want his strategic daydreams spoilt by any of Wavell's facts of military life. In the afterglow of the Midway victory he disparaged Wavell's Burma plans as 'very nice and useful nibbling'; but he wanted Burma recaptured and a thrust made towards Bangkok, for which, he hinted, British reserves might be provided. Mindful of his Chiefs' opposition, Churchill described his message to Wavell as a personal suggestion, enjoining him to make 'war on a large scale' and seize the initiative from Japan 'instead of being through no fault of your own like clay in the hands of the potter'.[9] Wavell was unmoved by Churchill's enthusiasm, not least because of trouble in the Middle East where Rommel's *Afrika Korps* were threatening to recapture the strategic town of Tobruk. Pointing again to the lack of air and sea power in his Far Eastern command and the poor condition of his Indian troops, Wavell offered instead to transfer some of his troops to the Middle East if the situation required it.[10] As if the deteriorating position in the Middle East were not enough, Gandhi added to the pressure on Britain by launching his disruptive 'Quit India' movement.

There was still no sign that American involvement in the war would tip the balance, and with the news of further setbacks, a mood of pessimism began to spread through Westminster. Churchill's political critics whispered their opposition, increasingly confident that military events might bring political change. As one critic, Sir Edward Grigg, privately opined, 'Winston . . . has lost the power to give us moral strength—he just stimulates and titivates, and makes everyone feel that things are going better than they really are'. He judged that Winston was safe while the 'country is at present in the clouds' but that it will soon 'have to be steered through another patch of great disillusionment' which might well provoke a political crisis.[11]

Churchill had narrowly escaped public censure earlier in 1942 and he knew the political risks associated with further military defeats. This would have spurred him on to take the offensive anywhere, whether it be Burma, North Africa or even Norway. All these options were proposed by Churchill at the beginning of June 1942. In order to settle on some plan and keep the Americans with him in Europe, Churchill signalled to Roosevelt on 13 June that he wished to visit Washington.[12] With a fresh disaster looming in the Middle East, it would not do any harm to be away from knife-sharpening MPs in Westminster. Roosevelt had made the disturbing suggestion that the skeletal Eastern Fleet should join the American Pacific Fleet in an attack on the islands off Burma or against Timor, an island off the north coast of Australia. Such an action, the Americans suggested, might ease the Japanese pressure on New Guinea that was threatening northern Australia.[13]

Urged on by his naval chief, Admiral Pound, Churchill resisted American pressure and firmly rejected Admiral Somerville's suggestion that British forces used for the invasion of the Vichy French island of Madagascar should proceed eastward to help defend Australia.[14] Although the Australian Government no longer feared imminent invasion and occupation of the Australian continent by the Japanese, they were desperate to capitalize on the Midway victory and expel the enemy from the adjacent islands, particularly New Guinea and New Britain where the Japanese were establishing a major base at Rabaul that would allow their bombers to strike at northern Australia.

At the same time as the Japanese Fleet had been massing for its descent on Midway, five Japanese submarines arrived off Sydney Heads on 29 May and dispatched midget submarines into the harbour to attack what they believed to be battleships and cruisers. There were in fact only two cruisers at anchor, one Australian and one American, and the midgets missed both. Had they succeeded (and their torpedoes missed one cruiser by only a few feet), Australia would have been practically devoid of naval defence.

After this close shave, and the Allied victory at Midway, MacArthur was keen to attack New Britain while the Japanese were still reeling from their humiliating naval defeat. On 11 June he met the Australian Prime Minister, John Curtin, in Melbourne and warned him against simply calling for reinforcements to defend Australia, the security of which 'had been assured'. It would, MacArthur claimed, be 'merely interpreted [in London and Washington] as a timid cry for help'. Instead, Australia had to press for a supply of modern aircraft for the Royal Australian Air Force (RAAF) and munitions for the Army.[15]

MacArthur and Curtin should have been able to influence London and Washington and raise the status of the South West Pacific Area (SWPA) for which MacArthur was the Allied Supreme Commander. However, their potential influence was considerably diminished, and not simply because of the Anglo–American commitment to the 'Germany first' strategy. MacArthur was regarded with suspicion and even enmity in Washington where his Republican credentials were not appreciated by the Democratic administration of Franklin Roosevelt. He was also disliked within the American Navy, particularly by the new naval chief, Admiral 'Ernie' King, who freely described him to his British counterparts as a prima donna.[16]

The problem with MacArthur's proposed strategy of striking north from Australia was that it cut across the preferred naval option of a central Pacific thrust, hopping from Hawaii to Midway and onward through the enemy islands of the Marianas and Guam. Although the Army supported MacArthur's strategy, it could not proceed without naval support and this was firmly refused by King.

Curtin fared just as badly with his requests to London. Churchill's view of the Dominion was jaundiced following a series of heated disputes over strategy, and he was almost at the stage of refusing to deal with Australia at all, leaving most of the communicating to his Dominions Secretary. Whereas during the first three months of 1942 Churchill sent forty-three cables to Curtin, he sent only thirty-three during the next nine months.[17] The withdrawal of Australian forces from Tobruk in 1941 and the behaviour of some Australian troops during the fall of Singapore combined to leave a sour taste in the mouths of British Government Ministers.[18]

To make matters worse, refugees from Singapore arrived in Britain and embellished the stories of Australian cowardice with first-hand details. So worried was the Australian High Commissioner in London, the former Prime Minister Stanley Melbourne Bruce, that he visited the War Office about the matter. As he feared, the War Minister, Sir James Grigg, was well aware of the reports but assured Bruce that it was 'all hearsay' and applied to British and Indian troops as well. Nevertheless, Grigg also mentioned that he had thought it sufficiently important to take up with Churchill who had decided that 'it was better not to approach the Australian Government on the matter'.[19]

Stories of Australian cowardice at Singapore continued to corrode the previously high opinion of Australian military valour. Lord Croft, an under secretary at the War Office, described the Australians as being 'terribly disappointing—I won't say more about them'.[20] From Canberra, Cross confided to Attlee in London about widespread stories of Australian troops 'having "broken" in Greece, Crete, Libya and, above all, at Singapore'.[21] Further reports filtered through to officials in Australia House, but there was little that could be done about it. As Bruce's secretary noted, it required delicate handling.[22] British officials were unwilling to admit that these stories influenced their treatment of Australia and Bruce's attempts may only have lent credence to the stories and widened the circle of people privy to the details. Although he could do little to stop the insidious spread of gossip, Bruce did warn Curtin of the 'undercurrent of resentment' against Australia and of the 'feeling in some quarters [in London] that Australia is entirely selfish and out to get what she can for herself irrespective of the common interest and the wide strategical necessities of the war'.[23]

This simmering resentment may have affected a series of decisions that damaged Australia's defence position. First came a proposal to divert to Ceylon eight British motor torpedo boats that had originally been destined for Australia.[24] At the same time, Churchill tried to retain the two Australian brigades in Ceylon despite Australia's urgent need for them.[25] He also ignored various Australian suggestions for the return of the 9th Division from the Middle East or, alternatively, the provision of equivalent forces.[26] In fact, Britain's Joint Planning Staff recommended in mid-June that, in the case of conflicting demands for the war against Japan, 'priority should be given to supplies to India and Siberian Russia rather than to Australia'.[27]

Britain then diverted to the Middle East Spitfire aircraft *en route* by ship to Australia. Churchill had personally promised these aircraft to the Australian Minister for External Affairs, Dr H. V. Evatt, when he was in London in May on an urgent mission to secure supplies. The Spitfires comprised part of a commitment of three squadrons, two of them Australian and one British, that were meant to buttress the air defence of Australia. However, on 21 June, the day after a ship carrying forty-two of the aircraft left Britain, the town of Tobruk surrendered to Rommel, and 33 000 British servicemen were captured. Apart from the fall of Singapore, it was the most ignominious British defeat during the war. Worse still, German tanks were now pushing on towards the Egyptian frontier and the valuable British naval base at Alexandria. Australia's shipload of Spitfires was seized for the defence of Egypt.

The seizure of the Australia-bound Spitfires as part of a rescue package for the Middle East was yet another case of competing priorities in which Australia was put last. Once again it was done with the active connivance of her High Commissioner. Bruce had just been admitted to meetings of the British War Cabinet as a special representative of the Australian Government, but he was not averse to advising Britain on how she could impose her views on the Dominion.

Bruce accepted absolutely the British military opinion that Australia was not seriously threatened, and supported Britain's policy of concentrating on the Middle East. Despite this, Churchill never trusted the Australian; condemning him for his pre-war association with Neville Chamberlain's appeasement policies, disliking him for his occasional straight talking and his continuing support of Churchill's political opponents.[28]

On 24 June Curtin received notice of Britain's unilateral diversion of the aircraft.[29] Only the previous day he had congratulated Evatt on the success of his mission: 'Australia is assured of the certainty that all our needs are known to our allies and will be afforded greatly increased strength of armaments'.[30] Both men knew that Evatt's success was more illusory than real,[31] but they could at least point to the commitment of Spitfires; now even that had disappeared. Evatt was furious. He refused to accept Britain's *fait accompli*.[32] Bursting into a meeting at Parliament House between Curtin and the British High Commissioner, Sir Ronald Cross, Evatt reversed his Prime Minister's initial acceptance of the British decision, suggesting instead that Britain should consider evacuating the Middle East.[33]

Evatt absolved Churchill of personal blame for the diversion, accusing instead the British Air Ministry and Bruce, whom he denounced as being ready to 'let us down at the slightest pretext'. In a private cable to Brendan Bracken, the British Information Minister and close confidant of Churchill, Evatt threatened 'considerable trouble' if the Spitfires were diverted and pleaded with Bracken to

> intervene at once with the Prime Minister so as the bargain made with Australia shall be implemented and not defeated. Even putting up such a proposal is calculated to open old wounds. I am sure you will pardon the directness of my protest but I am utterly disgusted to find that people can be so forgetful of their obligations as to rob us of aid which is *en route* solely because of our foresight.[34]

Evatt was mistaken in believing that Churchill would lend a sympathetic ear to his plea. From London, the Dominions Secretary and Deputy Prime Minister, Clement Attlee, cabled to Churchill his opinion that the dispute was a storm in an Australian beer glass, that the politically ambitious Evatt was 'trying to undermine Curtin's position' and was 'violently upset' because the Spitfire diversion 'delays his presenting to Australia the fruits of his visit'.[35] There was some truth in Attlee's assessment, but he failed to properly appreciate that Evatt was also angry because he had been let down once too often by Britain. Evatt had braved the perils of air travel, of which he was, like Curtin, desperately afraid, to travel to the Allied capitals to put the Australian case. Now his principal achievement was threatened. Acting on Attlee's advice, Churchill disregarded Evatt's importuning and stuck by the diversion, arguing that it would not involve 'any substantial departure from what was agreed with Evatt'.[36]

Faced with intransigence where he had expected sympathy, Evatt dropped his opposition 'with a heavy heart', informing Churchill that it was 'solely as a result of his personal intervention that we are consenting to one month's postponement of delivery of [the] first instalment of Spitfires'.[37] The dispute fizzled out for the time being, but there was a residue of bitterness in London and Canberra. Churchill revenged himself on Evatt when he later refused to intercede with the Americans to have the Spitfires treated as an extra allocation of aircraft. The Americans simply reduced by an equivalent number the aircraft they promised to send to Australia, thereby destroying the defence advantage for Australia and some of the political advantage for Evatt.[38]

Evatt's foreign policy failures, of which this was only one, were not solely due to the lack of British or American sympathy for Australia. It also had much to do with Evatt's personality. Many of those with whom he believed he had developed a close relationship actually despised this restless and ambitious politician. Although they could not help admiring his obvious intelligence and ability, they were repelled by his character and most kept him at arm's length. More than most politicians, he was regarded warily as a solitary tiger in the political jungle.

Evatt counted as one of his allies in Washington the American Supreme Court Judge, Felix Frankfurter, one of Roosevelt's trusted advisers. However, privately, Frankfurter admitted to being 'a little disillusioned about his [Evatt's] character'.[39] The Australian academic and one-time diplomat, W. Macmahon Ball, wrote politely that Evatt's 'temperament and disposition made it hard for him to work with others in lasting mutual confidence'.[40] Others were less circumspect. Churchill was reported to have said that he 'simply can't *bear* this fellow', while Bracken described Evatt as 'dreadful'.[41] Another British Minister, who had extensive dealings with Evatt during the war, was relieved to lose political office in 1945 because he would not have to deal any more with Evatt, whom he described as 'particularly repulsive'.[42] As political head of Australia's fledgling diplomatic service, there was obviously little that Evatt could teach his officers about the finer points of diplomacy.

Evatt's task was not made any easier by the common knowledge of his ambition. His rivalry with Curtin created the appearance overseas of Australia speaking with two voices. The British government exploited this division in Canberra and sometimes circumvented the hostility they anticipated from Evatt by approaching Curtin directly. The rivalry was exacerbated by the fact that all top-level overseas cables passed through Evatt's department, allowing him to oversee them and even intervene to change their contents.[43] This included cables to and from Bruce in London which had traditionally been kept within the purview of the Prime Minister alone.[44]

While Evatt was worrying about his Spitfires, MacArthur was pressing for more activity against Japan. Meeting with Australia's all-party Advisory War Council in Melbourne on 17 June, MacArthur argued forcefully against the strategy of defeating Germany first. He pointed out that there was no possibility of a second European front in the near future. According to MacArthur, the Allies could best help Russia in her European struggle by knocking Japan out of the war and freeing Russia's Siberian forces to fight Germany instead of being tied up with the Japanese threat in Manchuria. He warned that, despite the recent naval successes, the situation in the Pacific was still unstable and that Japan might be able to win the Pacific war even if the Allies won in Europe. Moreover, an early offensive in the Pacific was vital for the 'prestige of the white races' and to prevent the 'coloured races' from consolidating behind Japan and making her position 'unassailable'.[45]

The idea of building up reinforcements in Australia before pushing back the Japanese, found favour even in some British official circles. As commander of the Eastern Fleet, Admiral Somerville had already unsuccessfully suggested sending British forces to Australia. A report by his deputy, Vice Admiral Willis, on 19 June supported the suggestion. Willis urged that, once the situation had stabilized in the SWPA, a joint Anglo–American fleet, equal in strength to the Japanese main fleet, should concentrate in Australian waters. This would leave the Indian Ocean weakened but, Willis argued, the risk was worth taking since British interests would not be seriously damaged.

Although the South African Prime Minister, Field Marshal Smuts, supported Willis's report, it made little headway in London.[46] This was despite a chorus of calls for some concrete expression of British concern for Australian security. In the anxious days of early 1942 there had been much angry talk in Australia of being abandoned by Britain. This had been strenuously contested by the many Anglophiles as well as by those commercial and industrial groups with strong ties to Britain. Oxford University Press released in May 1942 a small book by the Archbishop of Brisbane, entitled *Has Britain let us down?* It defended Britain's actions in the Far East and rallied 'the vast bulk of loyal opinion . . . [to] make it clear that we are proud of our membership in the Commonwealth of Nations, and that we reckon ourselves, come weal or come woe, as for ever a British people.'[47] It became a best-seller.

As the danger receded and the Americans started to arrive, the criticism of Britain abated somewhat.[48] In fact Britain was in danger of being supplanted. On 16 June the Governor of New South Wales, Lord Wakehurst, informed a friend in the Dominions Office of the excellent impression the Americans had made in Australia:

> They are well behaved and friendly and most of them look well turned out. It has come as a surprise to some Australians to find that you needn't put on bad manners and a slovenly appearance to show what a fine democratic fellow you are. It is curious how much more mature and grown up the Americans seem than the Australians.[49]

Just four days later, the leader of the opposition United Australia Party (UAP) and former Prime Minister, the venerable Billy Hughes, echoed Wakehurst's view but recognized the threat to British influence posed by the American 'invasion'. Cabling direct to Churchill, Hughes warned that the position in Australia was

> not good. Nearly 90,000 American troops . . . crowd the streets producing great impression on the public mind. MacArthur's confidential report to the Government strongly anti-British.[50] He is highly thought of, is Government adviser, his soldiers are here, the Government leans naturally to America. Strangely forget what they owe to Britain.

To counter this, Hughes urged Churchill to send a 'substantial number' of British troops to the Dominion.[51] Apart from the single British Spitfire squadron, Churchill refused to do any such thing.

The problem originated in part from the division of the Pacific and Indian Oceans into spheres of American and British control. Britain now tended to view the Pacific as an area solely of American interest, an attitude that American officials actively encouraged. Britain's Admiral Cunningham, who was sent to Washington to deal with the 'rather truculent and didactic' Ernie King, was 'horrified by all the jealousy and suspicion' between the British and American naval staffs. He confided to Willis that

> nowhere are our two staff sitting together and studying war problems on a broad basis. We have divided the world into spheres of influence and each country is fighting its own war in its own sphere and resents the other poking his nose into or even examining the problems in the sphere in which he is predominant.[52]

This division certainly seems to have permitted Churchill to wipe his hands of Australia and the Pacific. Even after the European war ended in 1945, he was committed to limiting severely the British contribution to the war against Japan.[53]

The mounting military disasters in the Middle East did embolden Churchill's political critics to renew their attacks on his handling of the war. Since he filled the dual role of Prime Minister and Defence Minister, it was his head on the block whenever Britain's war took a turn for the worse. When Tobruk fell to Rommel while Churchill was visiting Washington, the American press on 22 June talked of his imminent political downfall. Eden, woken by an anxious call from Churchill in Washington, assured his leader that he 'had heard not one word of this'.[54] This was most unlikely since Eden was often mentioned as the most suitable successor for Churchill, either as Minister for Defence or Prime Minister, or both.[55]

In fact, the movement for political change found ready supporters in Westminster but no real leaders. When it blossomed into a motion of no confidence in the House of Commons, it was found, to Churchill's relief, that his critics could not agree on the sort of changes they wanted. Those critics of stature, such as Stafford Cripps and Anthony Eden, stood by the Government and once more saved the day for Churchill. On 2 July the motion of no confidence was defeated by 475 votes to twenty-five.[56] This massive majority was comforting for Churchill but he would have noted that the number of MPs prepared to oppose his leadership had risen from three in a similar debate in May 1941 to twenty-five. Churchill had gained the time to put matters right, but the spectre of losing Egypt haunted him.

On 30 June he sent a stirring message to Richard Casey, the former Australian politician and diplomat whom he had appointed resident British Minister in the Middle East and member of the War Cabinet. At his palatial villa among the pyramids outside Cairo, Casey read Churchill's instructions to organize the 'defence to the death of every fortified area or strong building, making every post a winning post and every ditch a last ditch. This is the spirit you have to inculcate. No general evacuation, no playing for safety. Egypt must be held at all costs.'[57] Churchill had sent a similar message to Wavell just before the fall of Singapore and the prospects for Cairo looked similarly grim. According to stories put around Whitehall by his enemies in the Foreign Office, Casey panicked and 'blotted his copy book badly in Cairo by completely getting the wind up when Rommel was advancing, and issuing all sorts of notices advising flight'. According to the gossip, Mrs Casey had 'flown away in an aeroplane' but the wife of the British ambassador had 'angrily refused to go with her'.[58]

These stories, and Casey's constant illnesses, destroyed his career in Cairo. After Churchill visited Cairo in August he appointed a former Colonial Secretary, Lord Moyne, as Casey's deputy with a view to Moyne becoming his successor. But he kept Casey in Cairo for just over a year, perhaps because his membership of the War Cabinet deflected Australian calls for their own full membership of the British War Cabinet. Bruce had been pushing hard in London for an expanded role within the War Cabinet. In late 1943, when Churchill was much stronger politically and Bruce much more quiescent, Casey was finally shunted off to the Governor's residence in Bengal, and Moyne took

Casey's chair in Cairo.[59]

Curtin viewed the situation in the Middle East with as much alarm as Churchill. Australia's 9th Division was about to be thrown into the defence of Egypt and Curtin was aware that his own political fate might depend upon the fate of that Division. If it met with disaster like the 8th Division in Singapore, Curtin might be criticized by his own party for allowing it to remain in the Middle East. On his visit to London in May, Evatt had refused to press for the Division's return to Australia. Curtin suspected that Evatt had anticipated a military reverse and was now waiting to profit by it.[60]

In confidential briefings to prepare the press, Curtin warned that the situation in the Middle East might 'entail some sacrifices by Australia, and the taking of some more risks by Australia'. Hunched down in his leather chair among the huddle of scribbling journalists, the haggard leader of the embattled Dominion warned that the country must expect a 'harder and longer war' and that 'he would have to commence his fight for strength in the Pacific all over again'.[61] Even then, the Japanese Army was planning to attack Port Moresby across the jungle-clad Owen Stanley Range while the Japanese navy prepared to reach out from its base at Rabaul and extend its control along the length of the Solomon Islands.

2

A Question of Priorities

July 1942

THE FIRST WEEK of July 1942 was a time of great political crisis for Churchill and of military crisis for the Allies. Churchill survived the crucial vote of no confidence on 2 July not because the House had any real confidence in his leadership but because his critics were divided. Nevertheless, Churchill was quick to claim victory, maintaining that his opponents were eccentric rather than dangerous.[1]

Much of the parliamentary disquiet was centred on the deteriorating military situation in the Middle East. After Rommel's *Afrika Korps* had overrun Tobruk they pushed Britain's Eighth Army back almost four hundred miles into Egypt. In Cairo, smoke hung over the British headquarters as soldiers hurriedly burnt secret files in anticipation of the city's fall. Hitler's panzer divisions in Russia were taking advantage of the late spring weather to mount a massive offensive towards the south-east, attempting to strike out for the oil supplies of the Caucasus with one arm of their advance while the other arm struck at Stalingrad before turning to capture Moscow from the rear. That at least was the plan, and it looked like it might succeed as spring turned into summer and the initial attacks brushed aside all Russian resistance.

The German offensive threatened to create a pincer movement on the embattled British Middle East position. If the

Germans succeeded in their rush to the Caucasus, there was little to stop them moving on into Persia and capturing the oil supplies upon which Britain's war effort was largely based. The Germans would have almost encircled the Turks, making it difficult for them to maintain their neutrality. The Middle East and the Mediterranean would have been untenable for Britain. The Allied strategy of backing Stalin was dealt a grievous blow in the icy reaches of the Arctic Ocean, where an Allied convoy to Russia was almost totally destroyed by German aircraft and submarines. In the holds of the twenty-three sunken ships were some 500 tanks and 260 aircraft. Future convoys were temporarily suspended.

Australia was caught up in the deteriorating situation on several fronts. The 9th Division was threatened by the German onslaught in the Middle East, and it would now be more difficult to extricate it from Churchill's grasp. The Spitfire aircraft being shipped to Australia from Britain had been diverted to the Middle East, and Britain's fleet in the Indian Ocean was drawn upon to restore the position in the Mediterranean rather than to buttress the Americans in the Pacific.

Although the prospect of an imminent Japanese invasion had receded, Australia remained in considerable peril. The Japanese were planning to attack Port Moresby and the Solomons to the north-east of Australia, and they had established two secure bases on the north New Guinea coast from which they threatened the remainder of that country.

It was a time for taking stock in Canberra. In a report drawn up for Prime Minister Curtin, his Defence Department Secretary, Frederick Shedden, claimed that the setbacks suffered by Japan assured Australian security for the time being. Nevertheless, he warned that Japan was still very dangerous. Like MacArthur, Shedden argued that, while the Japanese were temporarily off-balance, they should be dealt a series of heavy blows that would drive them from the Australian coastline and demolish the hastily constructed defences of the new Japanese empire before they could be consolidated. But no early offensive could be launched against Japan while the 'Germany first' strategy dominated Allied thinking. This was the problem Shedden proceeded to confront.[2]

The usually mild and retiring civil servant berated the British for expecting Australia to defend British possessions in Malaya, Burma and Ceylon, areas from which, he caustically observed, 'large revenues had been derived by British trade and commerce, but the defence of which had been largely neglected'. Shedden urged Australia to shout even louder, making the most 'vigorous presentation of the Australian viewpoint by its representatives on the United Kingdom War Cabinet and Pacific War Councils'.[3] It is difficult to imagine how Australia could have done more than it had done in early 1942 when its appeals for assistance had led to much anti-Australian feeling within Whitehall and had little effect on the supply of reinforcements and munitions to the Australian theatre.[4]

Nevertheless, there was good sense in Shedden's recommendation to hit the Japanese hard. The Pacific was a much more decisive theatre than the Middle East and the effect of knocking Japan out of the war at an early stage would have been of immense assistance to the Russian fight against Germany. So long as Japan remained in the war, Russia was forced to fight Germany with one hand tied behind her back, maintaining considerable forces in Siberia to guard against a Japanese attack from the rear.

Shedden's report also had important implications for Australia's attitude towards her own security. Although he failed to recognize that Britain's refusal to provide for Australian defence was a result of her declining imperial power and a calculated decision to concentrate that power in the Middle East, Shedden was mindful that America offered Australia assistance from self-interest rather than from any innate sense of protectiveness. Australia must therefore continue to build up her own land and air forces to the 'highest degree possible'. As he reminded Curtin, it remains the 'obligation of every sovereign State to provide for its national defence to the maximum degree possible'.[5]

Despite this reminder, there were worrying signs that the lessons of the war had not been learnt. Already the Government had approved the formation of a cabinet committee to plan post-war demobilization, and General Blamey was reorganizing the Army to reduce the number of fighting units.[6] The United States, too, was moving to prevent

Australia from playing a full part in its own defence. In early 1942 the RAAF, which was being amalgamated with the USAAF squadrons as they arrived in Australia, had been placed under the control of an American, General Brett. However, in July, Brett was replaced by another American, General Kenney, who proceeded to divide the force along national lines, reserving for the Americans the role of pursuing the offensive against Japan. Australia's Air Force became a second-line, and very much second-rate, defence force.[7]

Australia was aware of the problem but unable to influence events. By welcoming MacArthur as the saviour of the nation and conceding to him the supreme command, Australia lost an opportunity to develop a sense of greater self-reliance. For the duration of the war she became a *de facto* American colony, just as she had been a British colony since 1788 and a *de facto* colony of London since federation in 1901. The dependent mentality nurtured by a century and a half of colonialism, and buttressed by the difficulties of defending the island continent with a population of just seven million, remained (and still remains to some extent) a barrier to the full exercise of her nationhood.

On 2 July Australia's Defence Committee agreed to give MacArthur power to allocate all supplies within the SWPA. The implications of that decision were potentially very serious for Australia, as the Chief of the Air Staff, Air Vice Marshal George Jones, pointed out to Shedden a few days later. For example, MacArthur could allocate the British Spitfires to American rather than Australian squadrons; only MacArthur's sense of fair play would ensure that Australian defence forces were given a proper role in the defence of their homeland and the wherewithal to perform that role.

This was to prove a slender reed upon which to rely, but there was little Australia could do about it without offending MacArthur. As Jones pointed out, although the issues were crucial to Australia, they would have to be circumspect in raising the matter with MacArthur 'lest any impression be created that we feared his powers would not be exercised in a manner which would give the R.A.A.F. a fair deal'.[8] In fact the problem was insoluble once the supreme command for the country's defence had passed out of the hands of her political leaders.

Naturally MacArthur hailed Shedden's report to Curtin, but there was little chance that it would be accepted in London or Washington. In fact, Churchill continued his campaign to fix any wavering American eyes upon the struggle against Germany. The destruction of the convoy to Russia spurred him to repeat his suggestion that Britain invade Norway and neutralize the German bases there to protect the northern convoy route. He had bungled such an invasion in 1940 and had risen to the prime ministership from the ashes of that disaster. According to Churchill, the operation presented 'such attractive possibilities from so many points of view',[9] even though his Chiefs of Staff advised against it.

No doubt one of the attractions for Churchill was that an invasion of Norway would soak up some of the forces accumulating in Britain for the invasion of Europe, which had been put off at least until 1943. It would also maintain the offensive spirit among the Allies, not least within the increasingly critical House of Commons, and deflect criticism from the dire situation in Egypt. Churchill was also anxious to kill Admiral King's plan for the Eastern Fleet to attack Timor, at the same time as an American attack on Japanese positions in New Guinea and Rabaul.[10]

King led the group in Washington keen to increase pressure on Japan and critical of Britain's wish to divert Anglo–American forces to the Middle East. Even the American Army supported King as it became clear that Britain was misleading the Americans about the chances of a cross-Channel operation in 1942, and perhaps even in 1943. However, the Army and Navy were hopelessly divided on how to tackle the Japanese. King, jealous of any threat to naval power, favoured a central Pacific thrust rather than a major offensive from Australia which would reduce his Pacific Fleet to an appendage of MacArthur's force.

Despite inter-service rivalry, many military leaders now argued that their best chance of getting to grips with the enemy at an early stage was in the Pacific. The American Secretary of State for War, Henry Stimson, shared their exasperation, venting his anger in his diary on 10 July when he railed about 'the British [who] won't go through with what they agreed to' regarding a cross-Channel invasion. In retaliation, Stimson suggested that the Americans should 'turn our backs on them and take up war with Japan'. A showdown with Churchill seemed imminent, but fortunately for Churchill, Roosevelt held the line for Britain, maintaining his support for the 'Germany first' strategy.[11]

Churchill's opposition to the Timor operation was difficult to sustain while the Eastern Fleet was relatively unemployed in the Indian Ocean. His solution was to suggest sailing its heavy units through the Suez Canal into the eastern Mediterranean to interpose itself between Britain's embattled forces in Egypt and any German thrust from Greece across the Mediterranean. Given the poor state of Axis sea power in the Mediterranean, such a German thrust was most unlikely, but Churchill persisted, arguing that it would also provide an 'invaluable distraction' for the convoy that the Admiralty was planning to force through the western Mediterranean to her island fortress of Malta. In addition, Churchill suggested, it would 'steady things in Turkey' and, if Rommel's attack on Egypt was successfully repelled, the simultaneous return of the Fleet would 'very nearly restore our prestige in the Eastern Mediterranean'.[12]

The Admiralty joined Churchill in opposing the Timor operation but it rejected his plan to shift the Eastern Fleet to the Mediterranean, arguing that it would subject them to 'an unjustifiable risk.'[13] The Admiralty pointed out quite rightly that, given Axis air superiority, it was air power rather than sea power that was needed in the Mediterranean. To soothe the Americans, Admiral Pound cabled to Washington on 12 July with instructions for his naval representative to inform King that the Royal Navy was fully occupied with its present commitments and unable to consider the Timor operation within the time frame suggested by Washington.[14]

Pound's case was undermined somewhat by Churchill who, on the following day, partly relented from his opposition, reminding Pound that he (Churchill) had previously organized such diversionary operations. He observed that Admiral Somerville, whose Eastern Fleet included two modern aircraft-carriers and a powerful battleship, had been 'doing nothing for several months and we cannot really keep this fleet idle indefinitely'.[15] However, instead of backing King's plan, Churchill pressured Wavell to do the impossible—to use his limited resources in India in a combined operation against the Japanese forces in Burma. In various forms, this would be Churchill's grand plan for the Far East over the next couple of years as he pressed his generals in India onto the offensive in Burma or Sumatra without providing the means for them to do so. In this instance, the competing demands of the Mediterranean removed the landing ships from the Indian Ocean. These ships were vital for the success of Wavell's plans, which were consequently put back to 1943.[16]

When a top level American delegation arrived in London on 19 July to confirm Allied military plans for 1942, British opinion was against the favoured American plan for a cross-Channel invasion in 1942 and proposed instead a campaign to clean up North Africa. There was a chance, as General Dill had warned Churchill four days earlier, that the Americans might be so disappointed that they would withdraw to 'a war of [their] own in the Pacific, leaving us with limited American assistance to make out as best we can against Germany'.[17] Fortunately for Churchill, Roosevelt instructed his delegation to fall in with the British plan to take North Africa, thereby securing British interests in the Middle East and foregoing the chance to strike a fatal blow at the temporarily enfeebled Japanese. As a result, the Pacific war was needlessly prolonged, and Australia remained under the shadow of the strengthening Japanese empire.

Meanwhile, Curtin put the question of the 9th Division's return from the Middle East on the agenda again. Australia was struggling to provide manpower for her own national defence and to service the growing American forces. Retrieving the 9th Division would relieve part of this problem and stop the flow of reinforcements demanded by the 9th. On 16 July, three days after the two AIF brigades in Ceylon boarded their eleven ships to return to Australia, Curtin cabled to Churchill warning that the return of the remaining AIF division would have to be considered in the near future.

Curtin insisted that it was possible for Japan to invade Australia in force, although he acknowledged that Churchill believed it was improbable. He reminded Churchill that, apart from the 9th Division, Australia still had a hefty commitment to the European war—some eight thousand air personnel in Britain and elsewhere under the Empire Air Training Scheme and nearly three thousand sailors attached to the Royal Navy. It was time, Curtin concluded, for the AIF to be concentrated in Australia 'for home defence and for participation with our American allies in offensive action against Japan when this becomes possible'. He warned also of 'considerable difficulties in the despatch of further reinforcements from Australia'.[18] Not least among these difficulties for Curtin, was the potential criticism from his more radical colleagues. Curtin asked Churchill for his views before he brought it before the Australian War Cabinet.

Churchill's view was clear, as it always was when the question arose of subtracting forces from the effort against Germany. 'We must try to stop this',[19] he told his Chiefs of Staff, ordering that they prepare a 'short, definitive opinion . . . setting forth the dangers of withdrawing this division at the present time or, indeed, during this year'.[20] With such a clear brief, the Chiefs submitted a suitable minute setting out the dangers of a premature withdrawal. Their advice was not without foundation.

The huge German offensive sweeping through southern Russia was dangerously close to Persia from where Australia as well as Britain drew much of her oil supplies. As the Chiefs warned ominously, the shortage of tankers meant that such supplies 'could not be brought from elsewhere without the greatest difficulty, if at all'. They also claimed that the withdrawal of the Division, 'coupled with the transportation of a division to replace it, would involve an unjustifiable and dangerous shipping commitment'. Churchill sent these views on to Curtin.[21]

On paper, Churchill's arguments looked impressive. Ships were in short supply, with all available ships being used to carry American forces across the Atlantic or reinforcements to the Middle East. The German capture of the Persian refinery at Abadan certainly would be a terrible blow to the Allied war effort. However, if the Division stayed in the Middle East, ships would have to maintain the flow of reinforcements. The dispatch of a 'green' American division to Australia was hardly equivalent to the return of her seasoned troops, and there was also the matter of morale among the troops of the 9th Division, fighting so far from home when their country was itself under threat.

It is quite possible that the cable from Churchill contained precisely the battery of arguments Curtin needed to demolish the criticisms of colleagues anxious to see the return of all Australian forces. There was also pressure on the Government to rationalize its manpower resources, both to maximize the war effort and to gain the best starting position in the race for post-war trade.

In early July Curtin had suggested to his Cabinet that the country should have a period of one hundred days' austerity which would be 'carried out in association with rationing and with the making of a maximum effort to put Australia on a war footing'.[22] When his Ministers submitted plans for increasing war expenditure from £320 million to £500 million, Curtin demanded that they be cut in line with the nation's capacity to fulfil them and that the war effort and civilian economy be kept in balance.[23] On 23 July the Tariff Board submitted a report urging the establishment of a reconstruction authority, noting that 'some effort must be diverted from waging the war if reconstruction planning is to be effective'.[24]

While Curtin was mindful of these problems, he was also anxious to foster goodwill in Whitehall and seems to have been prepared to leave the 9th Division in the Middle East to obtain it. Bruce had notified him of the rumours of Australian cowardice at Singapore and of the antipathy in Whitehall towards Australia.[25] Curtin was also concerned about the dangerous voyage the Division would have to make across the Japanese-controlled reaches of the Indian Ocean. Delaying the return of the troops might add to the security of their passage.

At about this time Curtin cabled to the King requesting that the term of office of the Governor-General, Lord Gowrie, be extended.[26] It was established Labor Party policy for an Australian to fill this office, but the appointment of Sir Isaac Isaacs in the early 1930s, who was doubly damned in many conservative Australian eyes for being both Australian and Jewish, had caused political divisions. By extending

Gowrie's term, Curtin was postponing the problem of choosing a replacement and perhaps earning additional credit in London by retaining a Briton as the King's representative.

Evatt was also keen to repair relations with Britain. Since his return from London in June, his harsh views of Whitehall seemed to have softened. The Parliamentary Under-Secretary at the Dominions Office, Paul Emrys-Evans, claimed that Britain's relations with Australia, the 'most difficult' of all the Dominions,[27]

> seem to be a little happier than they were and I think the Evatt visit was useful, not that he created an altogether happy impression here, but he learned something . . . Brendan Bracken said to me one day, 'I practically live with Evatt, and he is a very rude man'. On the whole, however, the bulletin might be 'better than could have been expected'.[28]

When Evatt reported to Australia's Advisory War Council upon the results of his trip, the tenor of his talk was much more muted than that of Menzies when he made a similar report following his trip to London in 1941. Evatt even claimed that Whitehall was 'coming round to the view . . . that there should be a concentration of naval forces against Japan'. This was an optimistic gloss not justified by his talks in London but it reflected political credit upon himself.[29]

Evatt was a victim of his own vanity, and the Australian businessman, W. S. Robinson, played on this in a devious double game with Evatt and the Australian government. Robinson was a leading figure in Australia's metal industry and had been an important adviser to Prime Minister Billy Hughes during the First World War. He had built up an impressive array of business and political contacts across the world but his loyalties were to Britain and the Empire rather than to Australia. As late as 1963 he wrote of being unable to 'think or refer to Britain as other than "Home"'.[30] The metal industry boomed during the war, and Robinson was quick to capitalize on his many contacts and build his industrial base. He had accompanied Evatt to London and had helped to moderate Evatt's calls for reinforcements.[31]

On 20 July Robinson wrote to Evatt from his Collins Street office seeking permission to return to London by way of the United States. He claimed that he

> could be of use in Britain on Australian account without, in any way, interfering with S. M. B[ruce]'s outfit. I might, on occasions, be able to help his officials through my personal relations with the various ministers and departmental heads . . .
> In general, my voice might add a little weight to the claims of the Pacific Front.[32]

Evatt was completely captivated by Robinson and was even eager to have him replace Bruce as High Commissioner. But Bruce's post was not within the gift of the External Affairs Minister.

So Evatt used Robinson as a *de facto* ambassador in London, providing an alternative line of communication with Downing Street. Robinson was on close terms with Brendan Bracken and Oliver Lyttelton, respectively Ministers for Information and Production, and was often invited for weekends with Churchill. There is no doubt that Robinson sometimes served Australia well, helping to smooth over rough patches in the often stormy relationship.[33] In contrast to their hostile attitude towards Evatt, Bracken and other British Ministers had a deep respect for Robinson.[34]

It is difficult to assess the true worth of Robinson's role in London. There were certainly negative aspects. His imperial perspective no doubt affected the manner in which he presented the Australian case and probably caused him to play down Australian demands of a more nationalistic nature. In addition, the knowledge in London that Bruce did not enjoy Evatt's confidence must have undercut Bruce's effectiveness. Certainly, Bruce resented Robinson immensely, observing that he 'would not trust him a yard' and counselling visiting Australian politicians and British officials to stay well clear of him.[35]

One of the proposals Robinson pushed hard was the planned survey of an alternative aerial route across the south Pacific suggested by the Australian aviator, P. G. Taylor. Taylor had run foul of MacArthur and authorities in Washington who regarded the route as a threat to America's trans-Pacific monopoly. By using Clipperton Island off the west coast of Mexico, Taylor planned to avoid refuelling in Hawaii, thereby circumventing the American monopoly and creating a secure air route far from the reach of the Japanese.[36]

There was much to commend the Clipperton route as far as Australia was concerned and Evatt and Robinson combined, behind MacArthur's back, to support Taylor's endeavour. On 13 July Robinson wrote to Evatt urging that 'when the necessary machine and other equipment, plus personnel, can be released [the survey of the Clipperton route] should be done and let us pray that it will be possible to do it quickly.'[37]

With both Churchill and Roosevelt agreed on the North African operation, Australia's prospects were bleak. MacArthur's hopes of tackling the rapidly developing base at Rabaul before the Japanese recovered after Midway were stone dead. Instead, he informed a disappointed Curtin on 17 July that Washington would only approve a limited offensive against the Solomons rather than the simultaneous attack against Japanese bases in the Solomons, New Guinea and Rabaul, as he had urged. MacArthur warned that, as a result, the outlook was very grave and it had become a question 'as to how we could save the war'. He complained that the low priority accorded the SWPA meant that inferior aircraft and tanks were being dumped on his Command.[38] Curtin kept up a brave face, publicly maintaining the need to hit the Japanese hard and confiding to journalists on 23 July that he 'would be willing to take risks with Australia in order to drive the Japanese out of the war'.[39]

The Japanese were already creeping towards Australian positions in New Guinea. A small company of Australians had managed to traverse the rugged trail over the Owen Stanley mountains to the outpost at Kokoda in central New Guinea. On 17 July Blamey's headquarters ordered the occupation of Buna on the north coast of New Guinea. Four days later, the Japanese struck first when they landed troops at nearby Gona and pushed south towards Kokoda. A seesaw battle developed for control of this strategic point, the capture of which was necessary if Japan were to attack Port Moresby and expel the Australian forces from New Guinea. By 27 July the Australians were forced to withdraw from Kokoda and retreat back along the mountain trail to Moresby.

As his troops struggled back from Kokoda, Curtin learnt

from Churchill on 29 July that the delivery of Spitfires had been further delayed by an administrative mix-up; only twelve out of the fifty promised for dispatch during July had actually been sent. It was a sign of strained relations that Churchill felt it necessary to cable the news direct to Curtin, promising that forty-six aircraft were being loaded onto a 'special ship' that would weigh anchor for Australia on 5 August.[40] Curtin took the news calmly and, despite this further small setback, also agreed to the temporary retention of the 9th Division in the Middle East and the dispatch of a further draft of reinforcements.

When tackled by members of the Advisory War Council, Curtin assured them that he would not allow the dispatch of the reinforcements to

> turn temporary retention of the Division in the Middle East into permanent retention. In view of the manpower situation and the danger threatening Australia, there was to be no departure from the principle that all the A.I.F. Forces abroad should ultimately return to Australia to fight in this theatre.[41]

Furthermore, in his message to Churchill, Curtin attempted to link his agreement with the Division's retention to an agreement on Churchill's part to ensure that Australia received sufficient aircraft to build the RAAF up to seventy-three squadrons. He also requested the 'regular allotment' to the SWPA of Australian squadrons with operational experience in Europe.[42] The reply from London contained no joy for Australia. There were the usual assurances about Britain's protectiveness towards Australia but no commitment about the strength of the RAAF.[43]

Churchill had more important things on his mind. On 1 August he had left London for Cairo to meet with his military commander, General Auchinleck. The general had incurred Churchill's ire by digging his forces into a defensive line at Alamein and delaying plans for a counter-attack until at least September. Churchill needed a victory in the Middle East to compensate for the humiliation of Tobruk's surrender. He was aware of Rommel's supply problems and feared that any delay would allow the 'desert fox' to regain the initiative and move on to overwhelm Egypt.'[44]

More importantly, Churchill was planning his first meeting with Stalin in Moscow. There he would try to sooth the Russian leader's anger at the British refusal to create a second front in Europe in 1942 to relieve the enormous German pressure bearing down on the Russian front line. This was one of the most crucial meetings of the war. Upon it might depend the continued existence of that most peculiar triple alliance between Russia, Britain and America. Churchill's own political future was also bound up in the talks with Stalin. If he could cement a working relationship with Stalin to complement the one he had developed with Roosevelt, it would eclipse the rash of military defeats and make his political position practically unassailable.

3

Retreat From Kokoda

August 1942

AT THE BEGINNING of August 1942, the tenuous Allied hold on New Guinea was threatened by a Japanese thrust from the north coast. The small Australian garrison holding the strategic airfield of Kokoda in central New Guinea fell back towards the defensive bulk of the Owen Stanley mountains which provided a considerable physical obstacle, protecting the vital supply point of Port Moresby on the southern coast. If Port Moresby fell to the Japanese, the Allies would have to retreat from their scattered outposts to the Australian mainland. Such a defeat would have allowed Japan to stiffen the perimeter of her Pacific empire and create a launching pad for a possible attack on Australia itself. The retreat from Kokoda alarmed Canberra where the conduct of the Allied campaign had already caused political rumblings.

In the Middle East, other Australian troops were facing a decisive battle as Rommel's *Afrika Korps* struck deep into Egypt threatening to destroy the huge British army that had been assembled to hold this Imperial linchpin. The potential military disaster grabbed the attention of political leaders whose hold on power might well depend upon the British Army's continuing hold on Egypt. By allowing Australia's 9th Division to remain in the region rather than withdrawing it for the crucial struggle against Japan, Curtin's personal political position, and perhaps that of his Labor government, might well depend on those thousands of Australian soldiers not falling into Rommel's hands.

Churchill was similarly embattled. Although he had achieved his major strategic object of drawing the United States into the war against Germany, there had been no significant victory in the European theatre. Instead, the nine months of American involvement had coincided with a run of disastrous defeats for Britain from Singapore to Tobruk. In Westminster MPs grumbled about Churchill's leadership and scoured their ranks for a suitable alternative. Eden, Cripps and the ageing Lloyd George were all mentioned as likely contenders.

Churchill had the measure of his opponents. Eden, the highly-strung Foreign Secretary, was very much a puppet on Churchill's string, kept dangling by repeated assurances that he was the heir apparent and that he would not have long to wait. Why then should Eden draw his dagger on a failing leader? He regarded the talk of himself deposing Churchill as part of a conspiracy orchestrated by the press baron, Lord Beaverbrook, and designed to 'upset Winston, put in a weak govt under himself (A.E.) or Cripps as P.M. which Govt. he (B) would then also knock out in order to come in himself. He [Beaverbrook] would then make a negotiated peace.'[1]

Certainly Beaverbrook was an inveterate conspirator who had been instrumental in the fall of Asquith in 1916 and had encouraged Menzies to oppose Churchill in 1941. Nevertheless, Eden's fear that Beaverbrook would occupy Downing Street in 1942 was rather far-fetched. More importantly, Eden would have realized that the 'conservative caucus [was] against him because of his progressive views'.[2] This feeling of relative political isolation, combined with his indecisiveness, would have stayed Eden's hand.

While criticisms of Churchill swirled through the carpeted corridors of Whitehall, Curtin was becoming alarmed at the Australian casualties during the retreat into Egypt. In a briefing with Canberra journalists, Curtin distanced himself from his decision, made just five days earlier, to allow the 9th Division to remain in the Middle East. He now spoke bitterly of the Australians being 'butchered' when they should have been returned to the Pacific theatre with the other two divisions. 'He was angry', he said, that the Australians 'had been again made the "chopping block" although there were 900 000 other troops in the Middle East'.[3] This was a familiar and not unjustified complaint, but it was also true that Australian governments had made clear to Whitehall on several occasions that they wished their troops to be in the front line rather than in reserve where they believed there would be discipline problems.[4]

On 6 August Churchill thanked Curtin for sending reinforcements for the 9th Division, and cited as 'practical proof' of their concern for Australian defence the 'agreement to despatch, at real sacrifice to ourselves, the three Spitfire squadrons'. Curtin was also informed that the Combined Chiefs of Staff in Washington would be deciding shortly on the aircraft strength to be allotted to the SWPA. Churchill assured Curtin that the American Chiefs of Staff had agreed that 'adequate measures would be taken to ensure the safety of Australia and the provision of the necessary equipment'.[5]

By agreeing to further reinforcements for the 9th Division, Curtin had committed it to the offensive that Churchill would soon be urging upon his new commander of the Eighth Army, General Montgomery. The life of Rommel's army depended on its line of communication and supply which was savaged by British submarines across the Mediterranean, and stretched to breaking point across hundreds of miles of desert. As Churchill confided in a letter to his wife, 'The more I study the situation on the spot the more sure I am that a decisive victory can be won if only the leadership is equal to the opportunity'. His new leader, Montgomery, was 'disagreeable to those about him', but, as Churchill observed, Montgomery would 'also [be] disagreeable to the enemy'.[6]

Although Churchill concentrated on the European theatre, a watching brief was also kept on the Pacific, partly to ensure that excessive Allied resources were not diverted from the war in Europe. As part of this brief Whitehall requested Australia to allow a military mission to be attached to the office of the British High Commissioner in Canberra so that London could be kept informed of military developments in the SWPA. This seemed an innocuous request that could be to Australia's advantage since it would provide a direct channel to the British War Office. Australia had already conceded supreme control of its forces to the Americans, so no one in London thought there would be any objection to the appointment of a military mission, particularly as Australia enjoyed such facilities in London.

Objections there certainly were. Curtin worried that the proposal would interfere with his relationship with MacArthur. General Blamey, the Australian army chief, had had a difficult time in the Middle East where he had operated under the control of a British commander; he was keen to preserve his relative independence in Australia, and urged that the Government simply strengthen its representation in London so that Whitehall could be provided with all the information it needed through Australian channels.[7]

Blamey's objections caused dismay and disbelief in the War Office; it was difficult to understand why facilities extended to Australian officers in London could not be reciprocated to British officers in Australia. In fact, as Bruce informed Curtin on 20 August, Britain had expected Canberra to welcome her proposal as an indication of increased British interest and concern in the region. The matter was taken up at cabinet level where it was seen as a further sign of Australian bloody-mindedness. Bruce pressed Curtin to withdraw his objections while there was still time and 'put an end to an episode which is tending to engender a quite unnecessary amount of friction'. In the face of Bruce's pleading, and armed with assurances that the British officers would not intrude on the triangular relationship between London, Washington and Canberra, Curtin finally consented to the plan in mid-September, although MacArthur still strenuously opposed it and granted the officers only the most limited access to his headquarters.[8]

The British proposal was not as innocent as it seemed. Its timing gave some indication of its intent. There had been no proposal for such a mission to Australia during the dangerous months leading up to the attack on Pearl Harbor, or during the first half of 1942 when Australia's survival was at stake. Curtin and MacArthur had appealed for greater resources to be allocated to the SWPA, but such appeals, if they were successful, would threaten the British effort against Germany. London could best guard against this with a military mission which would provide an independent, authoritative assessment of the needs of the SWPA and could undermine further appeals. The British proposal could perhaps mute Australia's voice in the international councils of war.

At the end of November General Blamey discovered that the original proposal, for three British officers headed by a brigadier, had expanded to a full-blown mission led by a major-general. Blamey denounced this as a British trick designed to 'prevent as much as possible our direct dealing with America and to ensure British control of Australian requirements'. He urged the Government to ban the mission from Australia's 'inner councils' and warned that it was capable of becoming a great 'thorn in our side'. But Blamey was too late. Curtin had taken Bruce's warning about British displeasure seriously.[9]

Unknown to Curtin, Bruce's attitude to the mission was affected by his continuing involvement in two struggles in London. One was to give greater substance to his role as Australian representative to the British War Cabinet, while the other involved him in a backstairs conspiracy against Churchill's leadership. Throughout June and July Bruce had been struggling to gain enough power to prevent his being excluded from the War Cabinet when matters of interest to Australia were being decided. In fact, Bruce was keen to be involved in all questions coming before the War Cabinet other than those of purely British domestic concern.

Just as Bruce thought he had a workable understanding with Attlee, the dispute over the military mission threatened to increase British official hostility towards Australia.[10] It was important for Bruce that Australia be seen as a loyal and supportive Dominion whose representative's voice deserved to be heard on the question of a replacement for Churchill.

With Churchill away in Cairo and Moscow, criticism of his leadership increased. Lord Astor, the influential newspaper owner, dispatched an emissary to Wales to sound out Lloyd George and check on the rumours of his increasing senility. The elderly statesman declared himself ready to join in a move against Churchill and willing to serve under Eden, but not Cripps.[11] Bruce, however, was using his position at Australia House to back Cripps. Part of the strategy was to attack Churchill where he was most vulnerable, that is, over his concentration of air resources on the bombing of Germany, to the detriment of the anti-U-boat war in the Atlantic and the defence of Ceylon.

On 12 August, in Churchill's absence, Bruce submitted a memorandum to the War Cabinet urging that aircraft be diverted from the bombing of Germany to the war at sea. The defence of Ceylon, he claimed, was given high priority on paper but was not accorded sufficient means. Bruce did not mention diverting part of the British bomber force to the defence of Australia where it was sorely needed. Had his call been successful, it would have seriously undermined Churchill's position as war leader since he had staked so much on the bombing offensive and had already deflected moves for some of the aircraft to be used as long-range patrols over Britain's north-west approaches. But Churchill's colleagues were unwilling to challenge his strategy, apart from requesting details of the present allocation of aircraft.[12]

This setback did not deter Bruce from his plan to depose Churchill. Five days later he met Cripps, ostensibly to discuss the diversion of aircraft to the war at sea, but the meeting soon degenerated into a diatribe against Churchill's leadership. There was more than a hint of irony in the former conservative Prime Minister of Australia meeting with a British Labour Party Minister to urge the ousting of the Conservative Prime Minister of Britain. According to Bruce's account, they 'both agreed that while the Prime Minister had been an invaluable asset in what might be described as the emotional period following on Dunkirk he was not temperamentally suited to running the war in the planning and organising period in which we now are.' However, soon after being ushered from Cripps's presence, Bruce was handed various cables from Moscow reporting on the progress of Churchill's difficult talks with Stalin.[13]

Churchill had staked his political future on the outcome of his Moscow trip and he was quick to take advantage of the relationship that he claimed to have developed with Stalin. In a cable to Attlee, while resting in Teheran on the return journey, Churchill claimed his trip had been a personal success: 'I am sure that the disappointing news I brought could not have been imparted except by me personally without leading to really serious drifting apart. It was my duty to go. Now they know the worst, and having made their protest are entirely friendly. . .'[14]

In one of the fastest about-turns in political history, Bruce immediately realized that his commitment to Cripps would have to be withdrawn. After reading the reports from Moscow, he described Churchill not as a liability but as 'an asset of incalculable value (who) has established his position with Stalin in an amazing way and in doing so has re-affirmed his position, vis-a-vis the U.S.A.' In view of Churchill's prestige in Russia and America, Bruce concluded that 'any idea of action which might lead to his displacement as Prime Minister is quite out of the question'.[15]

Bruce's attitude towards Churchill changed from outright opposition to gentle persuasion. But it was too late. Bruce had destroyed his chances of wielding influence in Downing Street and nothing could convince Churchill that the High Commissioner was not seeking his political demise.[16] In order to smooth his way in London Bruce moved to over-turn Australian objections to the military mission and re-direct Anglo–Australian relations onto a path of co-operation rather than conflict.[17] Curtin's willingness to concede reflected Australia's continuing hesitancy about its national interests and destiny.

Despite Australia's greater confidence in the outcome of the war and in her chances of survival as a nation, this hesitancy persisted. In fact the population seemed ready to return to the lackadaisical pre-war attitudes, a formidable barrier to the sort of all-out war effort that ultimate victory demanded. On 3 August, as the battle for Kokoda hung in the balance, Australian Ministers met in Canberra to decry the apathy of their fellow citizens. Curtin, the ex-alcoholic union boss and anti-war activist, angrily pointed to the prevalence of drinking, gambling and strikes, and urged that the laws 'be invoked to put down these national abuses'. 'War', he said, 'could not be waged without ruthlessness.'[18]

But Curtin's attitude was softening too. With American forces steadily accumulating in Australia—some 90 000 men had arrived by August—his fears for his country's survival were largely allayed. On Sunday 10 August he gathered his

Chiefs of Staff together in Melbourne to address the state premiers. Blamey dismissed the fighting in the jungle around Kokoda, which every day was splashed across the front pages of the nation's newspapers, as being 'not of great importance'. As for the nation's security, Blamey claimed that 'there had been an immense improvement in Australia's position in recent months' and that, 'while one could not say definitely that Australia would not be attacked, he did not think it would be very easy for the enemy to attack on a large scale at present'.[19]

Much of Blamey's confidence was misplaced since it was based on a false impression of Allied naval supremacy. Thus, he claimed that Western Australia was secure because its two infantry divisions at Perth were augmented by the 'protection of the fleet stationed in the Indian Ocean'. Australia's navy chief, Admiral Sir Guy Royle, also spoke of the Allied 'preponderance of naval power' including 'a substantial force in the Indian Ocean' that could reinforce those in the Pacific.[20]

Unknown to Blamey, and perhaps also to Royle, the major ships of the British Fleet had been transferred to the Mediterranean to sustain the British bastion of Malta and for the forthcoming invasion of North Africa. There were just two modernized battleships and an aircraft-carrier of any power and they were concentrated in the Cape Town–Ceylon–Aden triangle where they protected the reinforcement route to the Middle East and the British hold on India. The sea that washed the shores of Western Australia supported a skeleton fleet incapable of preventing a Japanese invasion. Fortunately for the people of Perth, the Japanese were occupied with their conquest of the Solomon Islands off Australia's north-eastern coast. They spread steadily through New Guinea, gradually encircling the lightly-defended base of Port Moresby. Meanwhile, the Government was ignorant of their exposed west coast and remarkably complacent about the deteriorating position in New Guinea.

The position in the Solomons would have destroyed Australian complacency had they known of it. On 7 August American forces tried to pre-empt the Japanese move along the island chain by landing on Guadalcanal and Tulagi. One of Australia's few remaining heavy ships, the cruiser *Can-*

berra, was taken from the SWPA without the Government's knowledge, to support the South Pacific Command's landings. The following night Japanese ships torpedoed four of the Allied cruisers, including the *Canberra*. The knowledge of this grievous blow to Allied naval strength took some days to reach Australia. Despite the sinkings, American troops were ashore and in a position to contest the Japanese hold over the Solomons. The outcome of this struggle would depend on the ability of the opposing navies to keep their men supplied.

Following his meeting in Melbourne with state premiers, Curtin addressed the annual dinner of the Commercial Travellers' Association. The 'war is one which the men of our race are going to wage to a victorious end', he claimed, confident that his fledgling nation would maintain its disputed grip on the continent. Secure in this belief, Curtin called on Australia to 'share the perils of a common cause'.[21] This was a far cry from his calls during April and May for Australia to be accorded special treatment as the last bastion of the white man in the southern hemisphere.[22]

Curtin's comments were partly intended to rebut criticism from the Opposition leader, the elderly Billy Hughes, who had led the country during the First World War and who had deposed Robert Menzies as conservative leader twelve months previously. Curtin was on a political knife-edge in Canberra, relying on the vote of an independent MP who had only backed Curtin because of the likely beneficial effect on the war effort. If Hughes could show that Curtin was neglecting the country's defence, his chance might come to reassume the mantle of war leadership. The Japanese moves in New Guinea gave Hughes his chance. MacArthur's own plans to occupy the north coast of New Guinea were preempted, and the army appeared to have been caught wrongfooted. Australia seemed set to repeat the disastrous experience of Malaya. Roundly blasting the generals for 'a lamentable lack of vision', Hughes publicly called for the Government not to wait for the 'death stroke' but 'to go out and sweep the enemy from his vantage ground'.[23]

In order to do just that Curtin pushed forward with plans to boost the size of the RAAF so that it would be strong enough both to ensure Australia's security and to provide a striking force for use against the Japanese. He had learnt from Bruce that 25 per cent of the American air strength intended for Britain was now available for use in the Pacific following the decision not to mount an invasion of France in 1942.[24] Curtin appealed for a fair share of the aircraft for Australia. On 12 August he briefed the Australian Minister in Washington, the eminent jurist Sir Owen Dixon. Curtin conceded that his ideal of seventy-one squadrons could be achieved in stages, the first being the supply of aircraft to the seventeen existing squadrons operating with obsolete or insufficient aircraft. After that, aircraft to allow the formation of 44 squadrons could be provided by January 1943. Curtin suggested that Dixon enlist the assistance of British officials in his appeal to Roosevelt and link the request for aircraft to Australia's agreement to retain the 9th Division in the Middle East and her commitment to supply Britain with aircrew under the Empire Air Training Scheme.[25]

Characteristically, Evatt backed up Curtin's move by circumventing official channels with a personal cable to the American navy chief, Ernie King, in which he lauded the American attack on the Solomons but warned that the Japanese would probably 'soon attempt some spectacular counter blow, and probably at the mainland of Australia'. Evatt urged that Australia receive its share of the aircraft being diverted to the Pacific so that she might be 'secure against large scale invasion'. Given such security, Britain would not have to abandon the Mediterranean if Australia were invaded.[26] Evatt was right to stress Australia's vulnerability but King knew enough about Japan's increasing weakness to realise that she was unlikely to attack Australia. The risks to Australia were becoming more theoretical than real.

Neither Britain nor America was willing to respond positively to Australia's pleading. Building up Australia's air power would be a useful adjunct to the post-war power of the Empire, but London was obsessed with keeping America interested in the fight against Germany and allowing to the Pacific only enough munitions to conduct a holding war against Japan. So British officials might make sympathetic noises in response to the Australian appeal but their aims were diametrically opposed.[27]

Curtin was aghast when he learnt that the American Chiefs of Staff proposed to allocate sufficient aircraft to Australia for just thirty squadrons instead of forty-four, and then only by April 1943 instead of January. To sweeten the pill, they proposed that aircraft for an additional ten squadrons could be transferred to the RAAF from the United States Army Air Force (USAAF) squadrons already in Australia. This was a neat piece of accounting that would profit Australia little since it would gain ten RAAF squadrons at the expense of ten American ones. To Evatt's horror, the three Spitfire squadrons that he had won were subtracted from the total allocation of American aircraft. Existing Australian squadrons would have to subsist with their out-moded types, for there would be few aircraft to replace them. Curtin also discovered at that time that MacArthur intended to assign only secondary defensive roles to those squadrons.[28]

On 25 August Evatt dispatched an angry cable to Dixon questioning how a proposal to increase the strength of the RAAF in return for leaving the 9th Division in the Middle East had been transformed into a proposal simply to transfer existing aircraft from American to Australian forces, thus making no overall improvement in the Dominion's defence position. Evatt argued that the 'proposal to count in the special Spitfires which I obtained is quite opposed to the express agreement I made with Churchill' and that the 'incident is most distressing'.[29]

Evatt was particularly angry about the Spitfires. He had touted them as the major achievement of his long journey to Washington and London and it was galling to find that they had become a symbolic addition to Australia's defences. When the Spitfires were diverted to the Middle East, Hughes had publicly taunted him: '[Where is] the vaunted success of Dr. Evatt's mission, if as it now appears, we are not going to get the goods after all?' On the same day as his caustic cable to Dixon, Evatt gave an off-the-record briefing to journalists in which he tried to resurrect the Spitfires as a great personal triumph.[30] But instead of seeking to have Evatt's points redressed, Dixon simply reiterated the American argument about shortages of aircraft and limply replied that the American Chiefs of Staff 'appear to regard the result as defensively adequate'.[31]

Although both Evatt and Dixon were lawyers, they were complete opposites in personality. Evatt was ambitious, thrusting and forthright in his opinions. He favoured America where he believed he had a friend in Roosevelt, rather than Britain where he saw enemies behind every oak door. Downing Street was the exception; he seems to have genuinely believed that Churchill was 'showing a disposition to help us'.[32] By contrast, Dixon was an Anglophile, with a reserved and reflective nature that suited the lawyer but hampered him as a diplomat in Washington.[33]

While Evatt was shooting off cables to Washington, the Japanese were tightening their hold on New Guinea and continuing their steady push southwards towards Port Moresby from Kokoda. On 25 August they landed at the important staging post of Milne Bay. Most of the Australian troops in New Guinea were drawn from the largely conscripted militia and their performance under fire was heavily criticized by the commander of Allied forces in New Guinea, General Rowell. At the end of August, he reported to Blamey's Chief of Staff in Australia, General Vasey, that one militia company 'broke and scattered' in the face of the Japanese attack. Rowell wrote that he was 'beginning to wonder whether any of these A.M.F. units can be relied on . . . and I really shudder to think what the position might be if the Wog elects to come here.'[34] The Japanese moves also shook the confidence of MacArthur and of Australia's leaders, who began to fear that the smaller Japanese force might overwhelm the New Guinea force as surely as they had the British forces in Malaya.

On 25 August, as Japanese ships were moving their invasion force towards Milne Bay, Curtin warned Churchill that events were moving towards a 'naval clash which may well decide the course of the conflict in this theatre'. He asked again when part of Britain's Eastern Fleet might be transferred to the Pacific to create a naval force capable of defeating the Japanese. He also confronted Churchill with the aircraft situation in Australia and the miserly plan by the American Chiefs of Staff to equip just thirty squadrons of the RAAF.

Considering Australian sacrifices in the Allied cause,

Curtin argued it was only right that the Allies should provide a proper 'degree of reinsurance'. He reminded Churchill that the 9th Division had only been retained in the Middle East on condition that British representatives in Washington 'would be instructed to do their utmost' to ensure the adequate supply of aircraft to Australia. He claimed that the country was 'more vulnerable to invasion than any other part of the Empire' and that the proposed allocation would 'not assure even the defence of Australia as a base'. He asked Churchill to intercede immediately with Roosevelt while there was still time.[35]

Churchill would have none of it. He was willing to offer Australia the gift of a British cruiser to replace the *Canberra*,[36] but he refused to intercede with Roosevelt. Meanwhile, Japan strengthened the defences of her newly-won empire and spread consternation throughout Australia. MacArthur enlisted Curtin's help to co-ordinate a two-pronged attack on the American Chiefs of Staff to pressure them into increasing the forces allotted to the SWPA. In his message to Washington on 30 August MacArthur warned of the crisis confronting his command and predicted 'the development within a reasonable period of time of a situation similar to those which produced the disasters that have successively overwhelmed our forces in the Pacific since the opening of the war'.[37] For a time, it seemed as though the Japanese pincer movement towards Port Moresby and through the Solomons might cut the Allied lifeline to Australia and isolate the continent in a Japanese sea.

4

The Battle for Moresby

September 1942

BY THE BEGINNING of September the Australian forces defending Port Moresby were being pushed back towards the town. The militia troops proved no match for either the Japanese or the punishing terrain. Their commander, General Rowell, appealed to Australia for reinforcements of regular troops that would stand steady in the face of the Japanese assaults.[1] But even the troops of the battle-tested 7th Division were unable to stem the remorseless advance of the lightly-equipped Japanese. When the Australians established fixed positions blocking the tortuous muddy track, the Japanese would simply melt into the jungle, to reappear behind the Australians in successive outflanking movements that sent the defenders rushing backwards in a desperate attempt to maintain their grip on the tenuous supply line to Port Moresby.

Although the situation was certainly critical, every retreat by the Australians lengthened the lines of communication for the Japanese and reduced their advantage. A five mile retreat added five miles of mud over which native porters in single-file had to transport Japanese ammunition and supplies from Buna on the north coast. For the Australians, it was five miles less from Moresby to the front line.

Meanwhile, on Guadalcanal in the Solomon Islands, there were pitched battles between American and Japanese troops for this potential launching pad for air attacks upon Australia. Both sides were entrenched in strong positions and neither had the ability or the numbers to overwhelm the other. Unknown to the Japanese, the Americans had some 20 000 men dug in a protective circle around Henderson Field, the rudimentary airfield from which they were struggling to establish air superiority. The Japanese repeatedly sent too few men to subdue the Americans. Defeat followed defeat and still the Japanese commanders refused to recognize their folly. Their superiority in the air and surrounding seas was gradually forfeited by their inferiority on the ground.

Following the Japanese naval success in the Solomons and the steady advance of their troops towards Port Moresby, Allied assurances of Australian inviolability were called into question. There was little to prevent the Japanese doing what they pleased in the south-west Pacific. As the former chief of the Committee of Imperial Defence, Lord Hankey, observed gloomily from distant London: 'the loss of the pivotal base at Singapore, as was always foreseen, has enabled the Japanese to sprawl all over the Far East. They have an embarrassing choice of attractive offensives open to them—China, Eastern Siberia, Australia, New Zealand and India.'[2]

The situation was not quite as dark as Hankey was inclined to paint it. The Japanese had postponed plans for further offensives, except those that secured their hastily-won empire. Mounting a scaled-down offensive to cut Australia's lifeline to America remained a top priority for Tokyo. Although Australia was secure from invasion in the medium term, her links to America were of the most vital importance. If they could be cut, Australia would again be vulnerable to conquest. And if Port Moresby fell, a door into northern Australia would be forced wide open.

From his newly-established headquarters at Brisbane, MacArthur spread alarm among the complacent Australians as he enlisted Curtin's support in his campaign for further reinforcements.[3] As part of this campaign Curtin dispatched an urgent cable to Roosevelt on 31 August, expressing concern about the Allies' ability to defend Australia and emphatically rejecting the plan to withdraw American airmen from Australia and transfer their aircraft to the RAAF. Once again he pointed 'respectfully' to the Australian contribution, some 48 000 men, to the Allied war effort in Europe and the Middle East. If Australian casualties in those distant theatres were added to this, the number was around 85 000, not much less than the total American force then in Australia—some 98 000 men. The implication was clear: the Allies so far had contributed little more to Australia's security in the Pacific war than Australia had contributed to the Allied effort in the European war. Lastly, Curtin claimed that intelligence reports indicated an imminent and intense Japanese offensive in the SWPA. He called for Japan's defeat to be the Allies' 'first priority'.[4]

Curtin sent a copy of this cable to Churchill asking that he back Australia's case. Instead, Churchill challenged the basis of Australia's fears and refused to intercede with Roosevelt to increase the supply of American aircraft or to have the Spitfires counted as a separate allocation. While expressing his 'entire sympathy and anxiety' to help in strengthening Australia's defences, Churchill claimed that the Americans were not neglecting their responsibilities and that Curtin must trust that their plans would be 'adequate to ensure Australia's defence though admittedly not enough for a strategic offensive'.[5]

Churchill's dismissive cable had been drafted by the Air Ministry and approved by Churchill without reference to the Chiefs of Staff. Despite Australia's wholehearted involvement in the Empire Air Training Scheme, and the service of so many Australian airmen in the European theatre, the Air Ministry was most possessive about releasing aircraft to the distant Dominion. It approved of a greater proportion of Australia's aircraft allocation from America going to the RAAF rather than the USAAF, but opposed any increase in the total allocation. As the Chief of the Air Staff confided to Churchill, 'the danger . . . is not that the forces sent to that area will not be large enough, but that they will be greater than is necessary' and that they would be provided at the cost of operations in the European theatre.[6]

Bruce had been taken unawares when the draft reply to Curtin's cable was approved and dispatched during one of Churchill's late night sessions. In an attempt to soothe Canberra he claimed that the cable 'was not really considered by the Prime Minister who at the moment is almost entirely preoccupied with the battle in the Middle East . . .'[7] So concerned was Churchill with that battle, and with the German thrust through the Caucasus and round the eastern shore of the Black Sea, that he proposed, on 28 August, sending two hundred tanks to the Turks, who were stubbornly neutral despite constant British blandishments.[8] At the same time, he was shipping a huge convoy of munitions through the Arctic Ocean to Russia. Despite the protection of a full fleet of seventy-seven warships, thirteen of the forty-four heavily-laden cargo ships were sunk by the German aircraft swarming above, or the submarines lurking below the icy waters.

Bruce promised to try and enlist Churchill's 'personal interest' in making a fraction of this effort for Australia, although Bruce himself had earlier called on Curtin to drop the demand about the Spitfires being considered as an additional allocation of aircraft.[9] This Curtin readily agreed to do. By now, the issue of the Spitfires had become of political, rather than practical, importance.[10] For reasons of maintenance and national prestige, MacArthur's command was not enthusiastic about having Spitfires in the SWPA, suggesting that they be stationed in Perth, a proposal described as 'absurd' by the chief of the RAAF.[11] In the event, they were sent to Darwin where they performed useful duty for a while, then were left to languish when the front line moved away from Australia's northern shores.[12]

The Spitfires could still be of political benefit to the Government, and of personal political benefit to Evatt who had arranged for their transfer. On 3 September he touted the Spitfires to the House of Representatives as 'another splendid contribution, which Britain agreed to, [that] will be of inestimable value to our defence'.[13] However within Government circles privy to the facts about the Spitfires, Evatt's reputation must have suffered.

Churchill was aware that Rommel's strike into Egypt, launched on 30 August, was a do-or-die attempt by an army running critically short of fuel and ammunition. Within three days, Churchill had the success he was seeking when the German attack was broken off to conserve fuel. The British forces re-grouped for an offensive that Churchill was keen to begin almost immediately. However, Montgomery counselled patience, mindful of previous hasty offensives mounted at Churchill's behest and ending in defeat for the army and dismissal for the commander. Despite Churchill's badgering during September, his commanders carefully gathered their forces for a counterstroke that would repel Rommel once and for all.[14] The Middle East campaign had been see-sawing for two years. Now, with the Anglo–American landings planned for early November along the North African coast to Rommel's rear, the time for a final decision was approaching.

In 1941, when Singapore had been left with inadequate protection, it was clear that British priorities had changed and that the Middle East enjoyed a higher priority than the Far East. In rejecting Curtin's pleas, Churchill was simply reconfirming that in his estimation, no defeat in the Far East could compare with the loss of the Middle East. The continual pledges to protect Australia were made with fingers crossed, in hopeful expectation that they would never be tested. In any event, they were so hedged about with conditions that only an invasion in force by the Japanese would have triggered a British response.

As Curtin discovered, the American response was no more sympathetic to Australia. Roosevelt backed Churchill's strategy to fight Germany first and risk leaving the Japanese in the Pacific. Roosevelt was considering the American Navy's alternative island-hopping strategy which would enable the United States to sidestep Australia in their march towards Japan. So there was no joy for Australia when Roosevelt replied, two weeks after receiving Curtin's request, that the planned aircraft allocation would not be altered. It was another finger-crossing exercise, with Australia as the potential loser.

Australia received another body blow on 8 September when Churchill finally informed Curtin that he could not arrange for the British Eastern Fleet to join the American Pacific Fleet to concentrate Allied naval power in the Pacific. As he now admitted, the Eastern Fleet existed in little more than name only. On 30 August, an Australian intelligence report claimed that the Eastern Fleet comprised five battleships, three aircraft-carriers, three heavy cruisers, sixteen light cruisers, nine armed merchant cruisers and about twenty-five destroyers. In fact the backbone of the Fleet consisted of just two battleships and an aircraft-carrier and they were tied up in the protection of the Middle East reinforcement route along the east coast of Africa.[15] Australia's western shoreline had been protected for the past three months by a figment of Curtin's imagination.

That same day, depressing news was received from Port Moresby, where the Japanese were pushing steadily south. Blamey's headquarters reported 'heavy and confused fighting' in which one AIF battalion was outflanked by the Japanese and cut off from retreat. It was three weeks before the exhausted troops stumbled into the comparative safety of Port Moresby after an epic and circuitous trek across forbidding terrain.[16] Curtin despondently announced to journalists that Australians faced 'a long struggle . . . to hold this place' and suggested that they 'might have a 100 years war'. He railed against Australia's allies, who were only too willing to provide 'plenty of ships to transport reinforcements to the 9th Division in the Middle East' but not 'to reinforce our men in Port Moresby'.[17]

Curtin was depressed by his continuing failure to extract defence resources from either London or Washington, and by the deteriorating situation in New Guinea. On 7 September in Canberra he had met General Blamey who reported yet more bad news. Not only was Australia being denied aircraft and ships but the Americans did not intend to increase their existing commitment of two divisions until at least 1 April 1943. Moreover, these two divisions were 'green' and still training for active operations. Blamey reported that he had only two divisions in his command that were capable of being used in offensive operations against the Japanese, while the division defending the Townsville area of north Queensland was under strength by two thousand men who had been released to cut sugar-cane. Although Australia would soon have eleven divisions, Blamey calculated that at least fourteen were needed to ensure the security of the settlements scattered round the coast of the huge continent. But when Curtin suggested withdrawing the 9th Division from the Middle East, Blamey argued for its retention if the United States could be persuaded to provide three more divisions.

Blamey understood the difficulties of withdrawing the Division when it was committed to the offensive against Rommel. However, his suggestion that it could remain in the Middle East indefinitely was foolish considering the drain on Australia's military manpower. Blamey may have been prompted in part by a desire to keep the Division's commander, the highly-regarded General Morshead, occupied in a distant theatre and unable to threaten his supremacy in Australia.[18] Blamey faced a tough time as Land Forces Commander and the fewer close rivals the better for his survival at the pinnacle of the Australian Army. Apart from the military defeats that beset him, his Falstaffian figure was assailed by the press, with pointed barbs about his private life. So far, Curtin had helped to shield him from attack, pointing out to journalists that, in originally appointing Blamey, he 'was seeking a military leader not a Sunday School teacher'.[19] But Curtin's support could not be vouchsafed for ever, particularly when proposals began to circulate about the possibility of a military dictatorship with Blamey at its head.[20]

In the midst of these military disasters, Australia was being prepared for the post-war world. With his interests in Australia's mining and manufacturing industry and his political links, the globe-trotting Australian businessman, W. S. Robinson, hoped the war would boost the Dominion's post-war industrial expansion. In September, with the connivance of the Government, he dispatched a mission of business colleagues to Britain and the United States, with secret orders to seek out new industrial processes that could benefit the Australian war effort and, in the longer term, the companies over which Robinson ruled. In a memo to Evatt, he offered to finance the investigative trip out of his companies' funds, provided the Government advanced him the necessary American dollars. It was, he wrote, 'an urgent necessity if Australian industrial progress is to be maintained'.[21]

Robinson arranged visits for his colleagues at aircraft plants and aluminium smelters, but warned them not to send any information they gleaned to Australia by ordinary means. It was illegal to send such information outside the United States. Instead, Robinson would process their reports through official channels at the Australian legation in Washington, where they would not be subject to American inspection. He told his gentlemen 'spies' that they should put aside any outdated ideas and 'think in terms of the new world now being created—one in which the light metals—aluminium, magnesium, beryllium and their alloys will reign supreme with the aid of the internal combustion engine, oil and aircraft'.[22]

Robinson conjured up an Australian economic empire that could replace Japan's co-prosperity sphere. He suggested to his 'spies'

> a new map of the possible 'sphere of influence' of Australia in the new world—a sphere in which, if Australians care to work and progress, they must play the most important part in developing . . . Define it in arcs from the big centres of population on the east coast of Australia. Let these arcs be of a diameter no greater than the distances today between Melbourne and Perth or Melbourne and Townsville. Within their circumference there are unlimited resources. If they are not developed by us—are not under the protection of Britain and the United States and their Allies—if they are controlled by those antagonistic to Australia—the chances are Australia will not long survive.[23]

Robinson's vision of an Australian economic empire inspired Evatt who, in his report to the House of Representatives on 3 September, had claimed for Australia a 'particular interest in closer economic relations with its nearer neighbours'. However, since most of these neighbours were colonies, Evatt pushed for a doctrine of trusteeship that would guarantee them economic and political freedom. What he meant was, that these colonies would be either ceded to Australian control or otherwise opened up for Australian businessmen. He cited the Australian mandate territory of New Guinea as an example of the doctrine of trusteeship in practice.[24] In fact, New Guinea was a plantation society where native land had been expropriated and native labour ruthlessly exploited for the profit of a few Australian and British companies.[25] That example aside, Evatt's seemingly idealistic doctrine threatened Britain's hold over its colonial interests, and enhanced Australia's position in the Pacific. Evatt's statement carried the seeds of a conflict that would increasingly trouble relations between

Canberra and London.[26] However, an Australian empire required an Allied victory and, even as Evatt and Robinson were writing, Australia's survival was not guaranteed.

In the face of Japan's continuing success and the Allies' refusal to base substantial naval forces in Australia, Curtin took up Blamey's call for land forces and continued to demand aircraft for the RAAF. He asked Roosevelt for three divisions of American troops in return for the continued retention of the 9th Division in the Middle East.[27] At the same time, Curtin sent a strong cable to Churchill demanding for Australia the same protection that Britain provided for India and her other Indian Ocean possessions. He pointed out that, without the necessary naval forces to intercept and destroy a Japanese invasion force, Australia was reliant on her air and land defences. Since these were also seriously deficient, he warned that as soon as Japan had captured Port Moresby, the way would be 'open to a direct attack on the mainland under cover of land-based aircraft, which may well defy all naval attempts to interrupt their line of communication and dislodge them'.[28]

Churchill passed Curtin's cable to the Chiefs of Staff, noting that they could consider it 'at leisure'.[29] From Washington, Britain's Field Marshal Dill reported that Roosevelt had referred Curtin's cables to the Combined Chiefs of Staff and asked for guidance from London to ensure that the replies of Roosevelt and Churchill were along 'parallel lines'. Churchill now requested his Chiefs of Staff to prepare a reply for him along the lines that 'we do not think there is danger to Australia and we do not intend to move the Eastern Fleet there'.[30] Meanwhile, the Japanese pushed on to within fifty miles of Moresby, brushing aside the ineffectual Australian resistance.

MacArthur refused to accept Blamey's assurances about New Guinea and confided to Curtin that he had lost confidence in the ability of the local commanders. He was planning to send American troops there as fast as possible to stiffen the defence and he urged that Blamey be sent to Port Moresby to take personal command of the defence so as to save both the strategic town and his position as Land Forces Commander. On a more confident note, MacArthur said he had detected a change in the American Service chiefs' attitude towards the Pacific war which would eventually increase the flow of supplies and reinforcements for the SWPA. In the meantime, said MacArthur, 'the problem is one of fending the enemy off for some months. Support is coming, but the query is will it be too late?'[31]

Among the many appeals for assistance was one from the Opposition MP, Sir Earle Page, recently returned from Britain where he had served briefly and ignominiously as Australia's special representative. On 14 September, in his capacity as a member of the Advisory War Council and with the full support of Curtin, he sent an urgent, and almost hysterical message to Bruce to exert maximum pressure on Churchill for the sake of Australian security. Britain, he wrote, risked 'precipitating an Empire political crisis that will leave indelible scars for generations'. Even his own 'intense Empire patriotism' would be tested to the utmost. If the Dominion's appeals to the Mother Country went unanswered, Page dramatically predicted that 'the heart will largely have been taken out of Australia'.[32] As an indication of his anxiety, Page joined with a colleague on the Advisory War Council to urge that the Australian Army meet the military emergency by making greater use of horses. The suggestion was not adopted.[33]

Surprisingly, Page's cable stirred Bruce in a way that Curtin had been unable to do. Bruce requested an immediate meeting with Churchill but his request was dismissed with the excuse that Churchill had a heavy programme of engagements and was suffering from a sore throat. He suggested that Bruce see the acting Dominions Secretary, Lord Cranborne. Bruce was not put off so easily. He agreed to meet Cranborne out of deference to Churchill's 'slight throat trouble', but repeated his request, as Australia's representative in the War Cabinet, for 'an opportunity to discuss with you at the earliest possible moment a matter of the greatest consequence'.[34] Cranborne supported Bruce's request and assured Churchill that the distinguished Australian was not out to make trouble. In fact, wrote Cranborne, Bruce accepts that 'there is no immediate danger of invasion to Australia' but needs to be able to tell Curtin and Evatt that 'he has put their views to you personally . . . He is, I believe, really anxious to help'.[35]

The two men finally met after a War Cabinet meeting on 21 September. Bruce told Churchill that he realised no one in London and Washington believed Australia was facing a full-scale invasion. Nevertheless, he continued, it was his 'duty to put the case as we saw it'. Such an approach left Churchill in little doubt that Bruce did not share Canberra's fears. In fact, he assured Churchill that there was no need to strengthen the forces in the Pacific until after the Anglo–American invasion of North Africa. When Churchill began to dispute some of the points being made on Curtin's behalf, Bruce refused to take up the challenge, telling Churchill that 'nothing was to be gained by pursuing the argument at this stage' but warning that he 'would have to continue to press him upon the question'.[36]

Bruce devoted much of his meeting with Churchill to arguing the case for diverting aircraft from the bombing of Germany into the war in the Atlantic. Australia could have used such aircraft in bombing Rabaul, the strategic centre of Japanese power that was fast becoming a potent threat to Australian security, but on this Bruce was silent. He was more concerned to assure Churchill of his new-found loyalty following the successful talks with Stalin. According to Bruce, it was now 'unthinkable that he should go' from Downing Street.[37] This did not mean that Bruce supported the way Churchill was organizing the war, but that he recognized that Churchill was the inspirational figurehead under which a reorganized government should be established. There was widespread support for such a reorganization throughout Westminster, and Bruce himself was approached by one dissident Conservative who wished to promote him as First Lord of the Admiralty.[38] At the close of their long discussion, Churchill conceded to Bruce that, as High Commissioner, he had been 'discreet and . . . had sent a number of most helpful telegrams to Australia'. This, Bruce confidently felt, marked the beginning of a new and promising chapter in their previously stormy relationship.[39] For Australia, though, it meant little.

On 21 September Curtin gave a solemn background briefing to the press gallery. He was, he said, profoundly disturbed by the replies from Churchill and Roosevelt, which meant that 'it is vain to appeal for these places to be made a major theatre. I am not surprised. You were told all this when I was in Opposition. The bloody country was told what would happen long before the war came.' Australia therefore had a 'six months' menace to survive' which she had to do with 'blood and sweat and hard work'.[40] As part of the all-out effort, Curtin introduced 'raceless Saturdays', one day in each month when Australians could not gamble on the horses. However, an enterprising Sydney promoter announced a boxing match for the raceless Saturdays. Curtin expressed his shock and anger at such 'alternative assemblages for the purpose of fun [being held] while the enemy thunders at the gates'.[41] The troops in New Guinea had no time to comment as they prepared for the battle that would determine the fate of Port Moresby and perhaps of Australia itself.

The battle for Moresby had reached its nadir on 17 September when the defending troops had been forced to pull back to Imita Ridge, one of the last strong points protecting the township. Unlike desert warfare in the Middle East, where some of the troops had first seen battle, the war in New Guinea was a war in slow motion, where men might slog through the mud for days to gain an advantage of yards. The enemy was the terrain, the weather and the mosquitoes as much as the Japanese. More men fell victim to malarial mosquitoes than to enemy bullets. Such was the rate of sickness that it was necessary to have three men in reserve to keep one in the field.

At Moresby a few thousand Australians were dug in to make a last ditch stand, with orders from General Rowell to 'fight it out at all costs'.[42] For the first time, artillery pounded the Japanese positions, but the climactic battle never came. Instead, Tokyo ordered the Japanese to relinquish their hard-won gains and retreat northwards to the coast. They had failed to occupy the airfield at Milne Bay and were encountering unexpectedly stiff opposition from the Americans on Guadalcanal. Suddenly, Moresby became one commitment too many for the overstretched Japanese.

Just when victory was so close they could taste it, it was whipped away. Instead of a cleansing bath in Moresby, the Japanese troops faced many more nights in the jungle, fighting a difficult delaying action against an enemy now enjoying the advantage of artillery and air support. When Blamey arrived in Moresby on 23 September to accept responsibility for the retreat, Rowell's troops were ready to advance. On 28 September they recaptured Ioribaiwa Ridge, the final redoubt of the Japanese offensive which now became the starting point for their long retreat.

5

Victory at Alamein

October 1942

WHEN ASKED IN retrospect to nominate the most difficult months of the war, Churchill immediately chose September and October 1942.[1] The fact that Australia's security was safeguarded during this period did not figure in his assessment, only events in the European theatre. The German offensive in southern Russia seemed set to overwhelm Stalingrad and spread through the Caucasus to threaten the eastern flank of Britain's Middle Eastern fortress. Montgomery planned to move westward from Egypt to catch Rommel while his *Africa Korps* were ill-prepared, and an Anglo–American force was assembling secretly for a large-scale invasion of French North Africa. With Montgomery's forces, it would drive Rommel back to Europe and provide a springboard for an Allied invasion of Italy. Upon the success of these various enterprises might depend the fate of the entire Allied war effort as well as the political fate of the Allied war leaders.

Stalin's personal prestige was locked into the battle for Stalingrad; Roosevelt was facing an important congressional election that would test public support for his handling of the war; and Churchill's political position hinged on the forthcoming battle for control of North Africa. As he told Eden on 1 October, if the Allied operation failed, he would be 'done for' and would have to 'hand over to one of you'.[2] This was not the rambling of a depressive, but an assessment of political reality confirmed by other observers in Westminster. On 2 October the Conservative MP, Victor Cazalet, informed the former Foreign Secretary and now British Ambassador in Washington, Lord Halifax, that 'There will come a day, unless a victory intervenes, in which it will not be a question of who is to succeed Winston, but the conviction that with him as P.M. we cannot win the war'.[3]

Churchill threw everything into the struggle. To sustain the Russian resistance, he ordered ships full of munitions across the Arctic Ocean before winter set in. They would sail independently, unprotected by the huge fleet that had been assembled for the previous convoy. Of the thirteen ships that set out during October, only five reached Russia.[4] Despite this loss, the Russian resistance did not crumble, but recoiled before the German offensive, only to spring back in mid-November when German strength was spent and cold was strangling Hitler's attenuated lines of supply and communication. Hundreds of tanks, aircraft and armaments poured from Russian factories deep in the distant Ural Mountains.[5] Allied efforts to sustain Russia achieved little and simply reduced the assistance available for the struggle against Japan.

Despite the temporary relief for Port Moresby, Curtin's political security rested on the quixotic support of an independent MP and divisions amongst his rivals within the Labor Party. Curtin's political problems threatened to multiply when Australia learnt that the 9th Division was committed to the battle for El Alamein, with Evatt preparing for a political offensive against Curtin, should the military offensive in the Middle East end in disaster for the Australians.[6]

Although Australia's defence position had deteriorated with the Japanese thrust at Moresby, it did not provoke resentment against Britain. In mid-September, as Moresby seemed set to fall, the British High Commissioner in Canberra, Sir Ronald Cross, reported to Clement Attlee, Britain's Dominions Secretary, that there had been

> a steady improvement in the feelings towards the United Kingdom. A stranger landing here would not, I think, notice

> any ill-feeling. We, who have been pricked into sensitiveness on the subject and who are consequently not inclined to take things for granted, have a pleasurable sensation of basking in the sunshine of restored amity.[7]

With nearly 100 000 Americans scattered round the continent, and with most of the Australian forces returned home, Australians looked not to Britain but to Curtin, Blamey and most particularly, MacArthur for salvation from the Japanese. This was a relief for hard-pressed British representatives in Canberra but it posed serious questions about the future of Anglo–Australian relations, which for so long had rested on the defence guarantee.

The Government, however, still deeply resented British perfidiousness. At the beginning of October, the Defence Department compiled a report for Shedden listing the number of British aircraft received during the first eight critical months of 1942. Of the 449 aircraft, 366 were for training under the Empire Air Training Scheme and of no use for defence against the Japanese. There were only eighty-three operational aircraft and seventy-seven of those were from previous contracts and not in response to Australia's plight. The only special British contribution was the six Spitfires that had so far arrived, and these had been subtracted from the American allocation.[8] These revealing figures said much about the Imperial defence of Australia and put Churchill's oft-repeated pledges in proper perspective. As a point of reference, in the first nine months of 1942 Britain produced some 13 500 operational aircraft and America over 18 000.[9]

It was presumably no coincidence that Evatt moved at this time to ratify the Statute of Westminster, an act the British Parliament passed in 1931 that acknowledged the Dominions as nations largely independent from Britain. Australia had so far steadfastly refused to ratify it. Even now, Evatt denied that the move indicated any break with Britain, far from it, he claimed. It was simply a technical measure 'to remove threats of invalidity which now hang over important Commonwealth laws and regulations'.[10]

Menzies, who had tried unsuccessfully when he was Attorney General in the 1930s to ratify the statute, agreed with Evatt that it was of 'relatively minor importance' and 'derives its chief interest from the fact that it refers us back to the controversies of 1931 and reminds us, if we are given to thinking about such problems, that in our Empire relations we have by no means reached either finality or certainty'.[11] However, Menzies was no longer the Opposition Leader; Billy Hughes seized on the act, decrying Labor's disloyalty to the Mother Country. He ignored appeals from Evatt not to make it 'an occasion for raising any issue such as Australia v. Britain' and strenuously, but unsuccessfully opposed the bill's passage.[12]

Apart from the defence guarantee tying a dependent Australia to Britain, economic and defence links had intertwined to produce a seemingly unbreakable bond. Now American economic interests were following their troops into Australia. Many Australians welcomed this for a variety of reasons. Some acknowledged America's economic strength and Britain's relative weakness since the First World War, and wished to link Australia to the rising star rather than to the dying one. Others recognized that the pre-war policy of Imperial preference was bankrupt, that Britain was unable to absorb Australia's agricultural surpluses, and Australia was unwilling to accept sometimes second-rate British industrial goods in preference to those from the United States.

Even the arch-Anglophile, S. M. Bruce, saw that Australia's post-war economic revival would depend on expanding world trade generally, rather than returning to the restraints of the economic Empire of the 1930s. In October 1942, when a British Cabinet Minister warned him that 'Australia would have great difficulty in getting rid of the Americans after the war', Bruce retorted that Australia would 'welcome the maximum co-operation by the Americans both financially and by their exports in developing industry in Australia'.[13]

It was a conundrum that London was unable to solve. By concentrating on the defeat of Germany, she was forced to rely on the Americans to save the Empire, but feared that in saving it, Washington would either keep it or, perhaps even worse, set it free. Australia had been opened up to the Americans with hardly a second thought, but India was a different matter. It was only in September 1942, after the start of Gandhi's 'Quit India' movement and the realization that no British offensive would be possible in Burma without American assistance, that the British War Cabinet discussed encouraging the Americans to build up their air forces in India.[14]

The British attitude towards Australia remained deeply critical within both Westminster and Whitehall, tainted as it was with several major disputes during the previous two years and by continuing Australian attempts to dilute the Allied commitment to the 'Germany first' strategy. At the end of October, Oliver Harvey, private secretary to Anthony Eden at the Foreign Office, wrote in his diary of the Australians 'running screaming to America' and having 'behaved like sillies, if not worse. Witness the withdrawal [in 1941] of the Australian Division from Tobruk on the eve of the battle'.[15]

During the defence of Port Moresby, not only the commander lost confidence in his troops. In Washington, Roosevelt asked why Australian troops could not prevail against a smaller force operating at the end of an extended line of communication that stretched over some of the most impossible terrain in the world.[16] MacArthur was the source of much of this criticism. When Curtin challenged him in September about the number of troops defending Port Moresby (some 40 000), he replied that even '140 000 men would be no good unless they'd fight'.[17]

MacArthur had complained to Wilkinson that the 'whole senior officer class in Australia [was] too complacent and easily pleased with work $\frac{1}{2}$ done. [They] Live off but not up to the name of Anzac'. General Gordon Bennett, the Australian commander who had fled Singapore leaving his troops to face years of captivity, was considered by MacArthur to be a 'cantankerous crab—a low Australian type'. Only Blamey received praise as a 'tough commander', though he was also 'sensual, slothful and [of] doubtful moral character'.[18] Australia's reputation only began to improve when the Japanese were repulsed from Moresby and the 9th Division shared in the resounding victory against Rommel,[19] but it never recovered its former shine. Instead, Australians would be kept from the truth by carefully censored British official histories, designed not to offend the sensibilities of a former colony.[20]

The image of Australia in Downing Street was further

distorted by Bruce's political meddling. He still sought to change the political composition of the British War Cabinet so that Churchill's power over policy would be curbed and his own role enhanced. Following his amicable meeting with Churchill in September, he naively believed that he could put such a proposal for change directly to Churchill. It was an amazing political *faux pas* at a time when Churchill was already being beset by Cripps, and it left him wildly angry with the arrogant Australian. As Bruce discovered on 6 October, Churchill 'became somewhat violent on the subject' and dictated a 'fairly acid' letter pointing out that it was up to the House of Commons and not the Australian High Commissioner to decide on the composition of the British War Cabinet. Ironically, it was only the timely arrival of W. S. Robinson at Chequers that dissuaded Churchill from firing his poisoned barb at Bruce.[21]

Instead, Churchill simply excluded Bruce from War Cabinet meetings and ignored his repeated protests via Attlee. By the end of October, Bruce threatened to resign over the issue, but eventually he learned to live with the situation he had created. He had gone out on a limb only to see it cut off by Churchill and to find, with the collapse of Cripps' challenge, that there was no way back up Britain's political tree.[22]

Meanwhile, London had passed the Dominion's appeals for assistance to Washington. When the Dominions Office asked Churchill, on 3 October, whether he was planning to respond to the series of cables from Curtin, his secretary advised him that Dill had urged Britain to keep a low profile and allow any Australian criticism to be directed at Washington. That being so, he continued, 'you may feel the less said the better'. Curtin had asked, among other things, how Britain's much-depleted Eastern Fleet intended to deter a possible Japanese attack upon Western Australia, or deal with it if it eventuated. Although the Admiralty was preparing a reply to this question, Churchill thought they had 'better leave it for a while'.[23]

Churchill had earlier instructed the Chiefs of Staff to draw up a draft reply to Curtin's urgent questions at their leisure, which they certainly did. It was not until 9 October, almost a month after receiving Curtin's cable, that the Chiefs approved the Admiralty reaction, which was to make no reply at all. They claimed that Roosevelt had effectively answered the cables in the negative, and an attack on Western Australia was 'most unlikely to take place' and was anyway 'an entirely hypothetical question'.[24] It was, however, a hypothetical question to which Curtin was keen to know the answer.

Curtin's electorate and home were in Western Australia and he was well aware of its isolation from the populated south-east, the source of help should an invasion occur. Curtin often retreated to his home at Fremantle to recuperate from the rough and tumble of Australian politics, travelling for two to three days by train on a single line track, a vivid reminder of Perth's exposed position.

Despite the Japanese setback at Port Moresby, Australia was still going through 'pretty hectic and anxious times', as Lord Gowrie observed on 20 October. All eyes were now on the 'vitally important' battle in the Solomons which would not be finally decided until early 1943. Until that time, Australia's security could not be assured. Even Roosevelt was concerned, confiding to Churchill that there was 'no use blinking at the fact that [in the SWPA] we are greatly outnumbered'.[25]

Curtin cast around for a solution to Australia's immediate security problem and its looming manpower problem, as his defence forces were sucked into the struggle for New Guinea. Every man sent out of Australia was one less for the nation's defence. Curtin's nightmare was that the Japanese might succeed in the Solomons and leap-frog to the north Australian coastline, leaving the Allied forces in New Guinea isolated and doomed to defeat. By early October the troops faced the arduous task of pushing the Japanese back from Moresby. With tremendous supply problems across mountainous terrain, there would be no easy victory for the Australians but rather a slow, hard slog. Curtin rejected the calls for more speed, telling journalists that he

> was not going to be driven by public demand to send men to form suicide squads in New Guinea . . . New Guinea wasn't Australia. It was only a place of military strategy and if we took it entirely it would not affect the war greatly because the situation was dependent on factors far removed from New Guinea.[26]

Despite Curtin's reluctance to commit more men to New Guinea, any prolonged battle under such debilitating conditions would soak up further precious manpower. Again the question arose of retrieving the 9th Division from the Middle East. After first assuring himself of the support of MacArthur, the Chiefs of Staff, the Advisory War Council (except for Menzies) and his own War Cabinet, Curtin called, on 17 October, for the return of the 9th Division. Australia's contribution to the Empire Air Training Scheme would continue much as before. He informed Churchill that, due to the looming manpower crisis and the operations in New Guinea and the Solomons, Australia could no longer supply reinforcements for the Division. Since the Government would not countenance it being broken up into smaller units and distributed among British formations, the only alternative was for it to be returned to Australia as soon as practicable.[27]

Curtin's cable was sent just six days before the planned British counter-attack against Rommel's forces at Alamein. The 9th Division had been assigned a front line position, and there was no question of withdrawing the Australians before the battle, but Churchill did feel that it 'seems important' to release the troops after the battle. He even suggested to the Chiefs of Staff that the RAF could surely release the Australian airmen within its ranks.[28] Churchill was not being solicitous of Australian security but responded more in exasperation and from a desire to remove this continuing bone of contention. Besides, the flow of Americans was making the Australians less necessary. In any event, Churchill decided to make no immediate reply to Curtin for fear of jeopardizing the security of the Middle Eastern operations.

Churchill's brief show of concern for the Australians may well have been prompted by W. S. Robinson, who visited London in early October as part of his mission to gather industrial secrets. Robinson was also supposedly an unofficial emissary of the Australian Government, communicating 'a slightly more intimate expression of the views of some of the Leaders than is possible through official channels'.[29] Robinson was well placed to be such a conduit and could have presented the Australian case for reinforcements much more effectively than Bruce. But he did not. After two weekends with Churchill, Robinson came away convinced of the British case and anxious to still Australian appeals for assistance.[30]

Like Robinson, Bruce was also anxious to restore calm to Anglo–Australian relations, meeting with Attlee to urge that Churchill reply to Curtin about the 9th Division and take him into his confidence about the counter-attack at Alamein. Although Attlee nodded his agreement, Bruce was 'doubtful if he will do anything about it'. So Bruce sent a cable of his own to Curtin, advising that the withdrawal of the Division was complicated by its involvement in imminent operations, but that Australia's manpower problems were 'fully recognised and the necessity for the return of the 9th Division [was] fully appreciated'. He advised that Whitehall was 'examining the problem with a view to return at the earliest possible date'.[31]

Bruce's message should have satisfied the Australians but the news of the Division's involvement in the forthcoming offensive alarmed Canberra. They needed the Division returned in good shape for the country's defence. If the Division were badly mauled in the Middle East after earlier calls for its return had been ignored, Curtin might face a political challenge. Curtin immediately cabled Bruce, stressing that there would be no reinforcements for the Division and that it was not to be reinforced with its ancillary units. This message, dispatched on the eve of the battle, was designed to prevent the Division's use in the front line, since any casualties it suffered could not be made good. Curtin instructed Bruce to 'emphasise the crucial importance' of these points to Whitehall and ensure that the British Commander in Chief in the Middle East, General Alexander, 'should have regard to this position in his use of the Division'.[32]

Curtin's message was duly relayed to Whitehall and passed to Alexander, but apparently it was not brought to Alexander's notice.[33] Even if it had been, it was too late to alter the disposition of the Australians as they waited on 23 October for the artillery barrage from one thousand British guns to light up the freezing moonlit night and so signal the beginning of the long-awaited offensive. As for Rommel, he was recuperating in Austria when the battle began but

Hitler soon ordered him back into command. There was little he could do. As he wrote to his wife on 29 October, 'I haven't much hope left. At night I lie with eyes wide open, unable to sleep, for the load that is on my shoulders'.[34]

Churchill slept much more soundly than Rommel, knowing the German supply difficulties, and that decoded Axis messages allowed the RAF to sink tankers as they crossed the Mediterranean with the life-blood of Rommel's army. With only three days supply of fuel in reserve, Rommel's freedom to manoeuvre was all but gone and his only hope lay in a timely retreat to a defensible position in Libya.

On 27 October Churchill reported to Curtin on the opening of the 'great battle in Egypt', noting that 'you will have observed with pride and pleasure the distinguished part which the 9th Australian Division are playing in what may be an event of [the] first magnitude'.[35] In his reply, Curtin reminded Churchill of the 'vital importance' to Australia of the 9th Division being returned 'as soon as possible' and 'in good shape and strength'.[36] However, the Chiefs of Staff had still not considered when, how, or even if, this would be done. General Alexander seems not to have been informed of the Australian decision, and only learned of it, and its possible effect on his use of the Division, after the Australian Commander of the Division, General Morshead, visited Alexander's headquarters. Alexander immediately signalled to London that it would be 'quite impossible to lose their magnificent services until present operations are brought to a successful conclusion'.[37] Only then did Churchill, who had seemed committed to the return of the Division, prod the Chiefs of Staff to submit their advice on the matter.[38]

Roosevelt complicated the issue on 29 October by offering to dispatch an American division from Hawaii to the SWPA. The 'common cause' would best be served, Roosevelt claimed, by the retention of the 9th Division in the Middle East. He sent a copy of this cable to Churchill, hoping he would approve.[39] Churchill not only approved, he was, as he immediately informed Roosevelt, 'deeply grateful for your help . . . Later on, if Curtin insists, it will have to go, but I trust your telegram will be decisive'.[40] In the event, it was far from decisive, particularly after MacArthur advised Curtin that the American division had already left Hawaii and was committed to the theatre anyway.[41]

Meanwhile, the Chiefs of Staff reported that, in January 1943, five fast liners which had been scheduled to transport 30 000 Polish refugees from Persia to Mexico, could transport the Australians instead. They recommended that Churchill wait till Canberra had considered Roosevelt's request. Churchill agreed. 'There's no hurry', he noted.[42]

Australia would not have agreed. As Britain's Joint Staff Mission reported from Washington on 27 October, Allied operations in the SWPA had entered a 'critical phase'. The previous day, the Americans had fought another carrier battle with the Japanese as part of their campaign to hold on to Guadalcanal. The aircraft-carrier *Hornet* was sunk and another carrier damaged. The American Pacific Fleet now had only one operational aircraft-carrier to cover that entire ocean. As the British report observed, it made the American Navy's

> ability to command the seas problematical . . . Should Guadalcanal fall, as well it may, and should Espiritu Santo also be taken, which appears less likely, the Americans will be thrown back on New Caledonia and Fiji. Australian fears will be multiplied and difficulties of a renewed American offensive will be greatly increased. The drain of aircraft and shipping to the Pacific will increase.[43]

To cover their temporary weakness in the Pacific, the Americans appealed for British help. The carrier *Victorious* sailed to the rescue, arriving in Pearl Harbor in March 1943 where she remained for two months being refitted. By then the *Hornet* had been on the bottom for more than six months, and the danger was long past. The Royal Navy's official historian, Captain Stephen Roskill, was able to make the spurious claim that 'in spite of its overriding responsibilities in connection with the defeat of Germany, the British Government were anxious to contribute to the Pacific struggle'.[44]

The Pacific War Council, a forum where all the countries involved in the struggle against Japan could discuss progress, met belatedly on 21 October. The official British attitude was succinctly put by the permanent head of the Foreign Office, Sir Alec Cadogan, who wrote in his diary: 'Had to attend Pacific War Council at 6. Review by P.M. (they hadn't met since May!) but no business done'.[45]

Churchill continued to stall on the return of the 9th Division. He had instructed Attlee, as Dominions Secretary, to 'await the effect of the President's telegram and do nothing meanwhile'.[46] Attlee was concerned that London's silence might harm Anglo–Australian relations. In a note to Churchill, he argued that Curtin's cable was 'nothing more than a general, and rather out-of-date, repetition of his wish to have the Division back in Australia as soon as practicable and a request for your personal interest in the matter'. Accordingly, Churchill agreed to send a non-committal cable in which he assured Curtin of his 'personal interest in the question'.[47]

Although the fighting in the Solomons remained on a knife edge, the war of attrition in the Solomons would eventually favour the Americans as their immense productive capacity was progressively geared up for war. During October and November nearly nine thousand Allied combat aircraft were produced, most of them American.[48] In November, American Lightning fighter planes were dispatched to the Pacific, the first aircraft capable of meeting the Japanese Zeros on more than equal terms. Escort carriers were being mass-produced in American shipyards, and would soon sail out to tip the naval balance irrevocably in the Allies' favour.

At the end of October, substantial risks were still being run with Australian security, and the possibility of an invasion could not yet be lightly dismissed. On 2 November the Australians had recaptured the airstrip at Kokoda, and the 9th Division was in the forefront of the fighting at Alamein, helping to ensure Churchill's military victory and his political survival. Australian city dwellers rushed home from work, snatching a glimpse of newspaper headlines that told of these battles and of the tussle for supremacy on Guadalcanal where troops, ships and aircraft were committed to a struggle that continued to be crucial to Australia's survival. As they fought their daily battle for a seat on rush hour trams and trains, the seas and swamps of the nearby Pacific island ran red with the blood of the combatants.

6

Curtin Stands Firm

November 1942

FOLLOWING THE JAPANESE naval victory on 26 October, in which the United States' Pacific Fleet was reduced to just one operational aircraft-carrier, the battle for Guadalcanal was turning into what one Japanese officer termed the 'decisive struggle between America and Japan'.[1] Australia, whose survival might depend on the outcome, could take some comfort from America's willingness to trade blow for blow. So much military and political prestige had been invested in the island that neither side would retreat until the cost became too great. In early November, General Imamura left Tokyo to take command at Japanese headquarters in Rabaul, with orders to finalise the capture of the Solomons and prepare for a major offensive in New Guinea.

Meanwhile, Australian forces pursued the Japanese along the treacherous Kokoda trail. The capture of the airstrip at Kokoda on 2 November assured air supply and fighter support for the troops, but expelling the Japanese from New Guinea was a formidable task. Because of the Allied adherence to the 'Germany first' policy, the Japanese had been allowed to build up a strongly fortified bastion at Rabaul. They controlled the seas north of New Guinea, and still disputed Allied control of the air. However, in both men and machines, the Allies were fast surpassing them. Japan could not make up the huge losses of highly trained airmen and their aircraft. They desperately needed another major victory or the time and the means to secure their empire so that, politically, the Allies could not contemplate its military conquest.

The situation looked even brighter for the Allies at Alamein where the British overwhelmed the German forces with sheer numbers of men, tanks and aircraft. Although the British lost nearly two hundred tanks during the course of one battle compared to a much smaller German loss, the Germans had so few to begin with that the war of attrition in the desert was working in Britain's favour. By 2 November there were twenty British tanks facing each German tank. The following day, Rommel began his retreat, halting momentarily on Hitler's orders, but then continuing in a display of military skill that left the British forces wallowing in his wake.

Rommel's retreat was a political victory for Churchill. In a late night meeting with his army chief, General Brooke, Churchill exclaimed that Alamein 'has the making of the vast victory I have been praying for'. If his forces had met with another failure, he confided, 'I should have had little else to suggest beyond my relief by someone with fresh ideas'. Even as he spoke, a great Anglo–American armada was approaching Gibraltar in the second prong of the Allied plan to capture the whole of North Africa. The invasion of the Vichy French territories of Morocco and Algeria would begin on 8 November. If it succeeded, Rommel would be caught between Montgomery and the Allied forces of General Eisenhower. It would mean, Churchill informed Brooke, that Britain was 'beginning to stop losing this war and [was] working towards winning it'.[2]

For the Australians, there were disturbing signs that Curtin's conditions for the 9th Division would not be met and that its prompt return was still in doubt. On 4 October Curtin cabled to Bruce in London that it was 'very difficult to reconcile' the Division's leading role in the fighting 'with the observance of the steps necessary to give effect to our repeated requests for the return of the Division'.[3] Churchill had instructed General Alexander not to feel hampered by the Australian conditions but to use the troops 'freely', and that 'no further reference to Australia is needed'. Once the battle was over, Churchill observed, 'we will make the best arrangements possible to send them home *if their return is still demanded*'.[4]

Churchill may have been influenced by a message from Richard Casey, the former Australian conservative politician and now British Minister of State in the Middle East. On 3 November Casey had warned Churchill from Cairo that the return of the Australians would put pressure on the New Zealanders to withdraw their division. The South African division was about to be withdrawn for conversion to an armoured force, and these moves, Casey claimed, would be 'a serious matter for this Theatre'. He informed Churchill that the 9th Division had suffered more than two thousand casualties during the past ten days but had four thousand reinforcements in reserve. Unaware of Roosevelt's proposal, Casey suggested that an American division sent to Australia might allow the 9th Division to be retained.[5]

On 8 November the Anglo–American invasion of North Africa was launched, with attacks on Casablanca, Algiers and Oran. The Germans retaliated by moving into unoccupied France and beating the Allies into Tunis to protect the narrow strait to Sicily, which might act as a stepping stone to Europe. Nevertheless, the Allies scored a resounding victory. Within three days, the French defenders surrendered and part of the French Fleet joined the Allies. The remainder of the French Fleet was also lost to Hitler when it was scuttled in Toulon harbour as his troops raced to seize it. On 13 November at a grand banquet in London, Churchill announced that, though the war would be long, it was possibly 'the end of the beginning'.[6]

Churchill urged his military advisers to press on with plans for an invasion of France in 1943 and not to allow the Anglo–American forces to be 'stuck in North Africa'. It was, he thundered, meant to be 'a springboard and not a sofa'.[7] When his Chiefs of Staff disparaged the possibility of a landing in France in 1943, Churchill rounded on them with the accusation that their conclusions would 'probably encourage the "Japan first" elements in America'.[8] Churchill was adamant that Allied resources should not leak away from the struggle in Europe. The Russian armies also needed support in their massive struggle. On 19 November they had broken the siege of Stalingrad and quickly besieged the Germans in their frozen fortifications.

It was at this point that Curtin rejected Roosevelt's division and asked Churchill to return the 9th Division to

Australia.[9] He claimed that Australia's attitude towards the return of the Division had always been 'quite definite and clear'.[10] This was far from the case. Churchill could be excused for doubting the strength of Curtin's convictions and playing on that. Initially, though, Churchill appeared to relent, informing the Chiefs of Staff on 18 November, that 'We shall have to let them go', although 'shipping cannot be unduly deranged'.[11] This was an important rider, although the Chiefs of Staff had previously been able to find the shipping if it were required. Then came confirmation of Casey's warning.

The British High Commissioner in Wellington advised that New Zealand, under pressure from Curtin, was likely to demand the return of its division as well. Churchill was concerned but hoped that the issue could be delayed sufficiently to 'enable us to see more clearly'.[12] His prayers were answered almost immediately when Roosevelt refused to countenance the withdrawal of Australia's division before victory in North Africa and, even then, only after the Combined Chiefs of Staff in Washington had given approval. As Roosevelt confided to Churchill: 'The primary consideration must, of course, be the relationship of the Ninth Division to the Africa campaign and after that there is the secondary consideration of building up the Australian strength for use north of Australia.'[13]

Churchill immediately ordered his Chiefs of Staff to take no action about the withdrawal of the Australians and New Zealanders.[14] However, the Chiefs had already met that morning to discuss the transfer. General Brooke concluded from Curtin's cables that it was 'clear' that the Australians 'must return', but argued that their motor transport and equipment must remain behind and be made up direct from America once the Division reached Australia. Moreover, he called for a last-minute examination of the implications for Britain of the Division's return.[15]

While the British Chiefs were discussing the issue, the Combined Chiefs of Staff in Washington were also meeting. Churchill was heartened to learn that General Marshall, the powerful American army chief, was 'strongly opposed' to the return of the Australian Division and disputed Curtin's point that it was required to defend Australia. General Dill, the British Army representative on the Combined Chiefs committee, pointed out Australia's difficulty in maintaining the Division's strength if it were retained in the Middle East and its value in the SWPA as a 'highly trained nucleus'. Eventually, Marshall relented on condition that the move would not unduly inconvenience personnel shipping. On Dill's report, Churchill highlighted Marshall's initial objections but not his final, conditional approval. Anyway, Marshall's comments still left room for manoeuvre if it could be shown that shipping would be seriously disrupted.[16]

In a message to Roosevelt on 23 November, Churchill asked him to delay his reply to Curtin until the issue had been studied in London, 'especially in its shipping aspect'. Churchill also alerted Roosevelt to the probability of the New Zealanders being withdrawn.[17] Meanwhile, the Chiefs of Staff reconsidered the issue in accordance with Churchill's instructions, observing that 'on military grounds, there are grave objections to the withdrawal of these Divisions from the Middle East theatre. If they were British troops there would be no question of their relief. They realise, nevertheless, that political and Imperial considerations may be overriding.'[18] The Chiefs also approved messages that Churchill had drawn up for Roosevelt and the Dominion leaders.

Even before receiving his cable from Churchill, Curtin briefed journalists on 23 November with the 'news' that 'negotiations for the return of the 9th Division from the Middle East had been successful'. Curtin informed the press that, since Washington would not provide an extra three American divisions, the minimum considered necessary for Australia's defence, 'he must get back as many of our own men as he could'.[19] It is difficult to know on what information Curtin was basing his confident announcement. It may have been a deliberate leak to put Britain under pressure to concede.

The Australian Government was under pressure of another kind at this time, as its troops pushed the Japanese back to the northern coast of New Guinea only to find them dug in for a determined battle. Meanwhile the American forces on Guadalcanal were still locked in bitter fighting for the island. On a world view, these were small-scale battles, but their outcome was of vital importance for Australia.

Australian confidence was not helped by the American troops in New Guinea recoiling at their first taste of battle.[20]

On 23 November Churchill referred to the issue of the 9th Division during a meeting of the War Cabinet and, somewhat to Bruce's surprise, 'took the line of obvious acquiescence in the Australian Government's demand' letting Bruce clearly understand that he was 'very unhappy about it'. Bruce shared Churchill's anger. It was yet another issue on which he was out-of-step with his own government, and he assured Churchill that he had relayed all the British objections to Curtin.[21] Despite Churchill's 'obvious acquiescence', he conspired to prevent the Division's return and to shift the blame to Washington.

On 24 November Churchill informed Roosevelt that Britain could not oppose the Australian demand in principle but that Roosevelt certainly could, and should do so. He told the President that he had 'every right to express an opinion, more especially as American armies are also engaged in North Africa and it is arguable their position might be affected. Moreover there is a great case against the uneconomical use of our limited shipping'.[22] The Chiefs of Staff had not yet presented their report on the shipping implications, but Churchill was obviously going to play his trump card for all it was worth.

Churchill now assured Curtin that the decision was for Australia to make, but threatened that, if the 9th Division was removed, an American division might have to be withdrawn from Australia to make up the deficiency in the Middle East. If the Australians were removed from the Middle East as an isolated shipping movement, this could only be done at the 'expense of our general power to move troops about the world'. This raised questions that could only be settled by the Combined Chiefs of Staff in Washington. As for himself, he pledged not to oppose Curtin's wishes but pointed out that 'it would appear more helpful to the common cause if fresh troops were moved from the United States into the Pacific' rather than remove the Australians from the Middle East.[23] Churchill's opposition was not entirely bloody-minded; the situation in the Mediterranean was not proceeding as smoothly as he had hoped. German troops were reinforcing Tunisia on a large scale and holding up the Anglo–American forces trying to attack Rommel's rear while Montgomery's pursuit of Rommel was bogged down by bad weather.

When Bruce was handed a copy of Churchill's cable to Curtin, he realized the adverse reaction it would have in Canberra and once again rushed in to smooth the Imperial relationship. He informed Curtin that he was surprised by the harsh tone of the cable given Churchill's apparent attitude at the War Cabinet meeting. Bruce concluded that Churchill must have discussed the matter with Roosevelt who, Bruce advised, was very opposed to the Division's withdrawal. In fact, the cable had been drafted before the War Cabinet meeting and Churchill had not been 'fortified by the President's attitude' into expressing objections based 'to a considerable extent upon American views and wishes'. If there was any fortifying going on, it was mainly by Churchill.[24]

The War Office went ahead with planning in case Churchill's efforts failed and the troops had to return. They asked Alexander whether the troops could be released—the Australians in January 1943 and the New Zealanders in March 1943—'without prejudice [to the] completion of operations in Tripolitania'. The question disconcerted Churchill who may have feared that the weight of bureaucratic planning would make the troops' return inevitable, for he immediately informed the War Office and General Alexander that he was 'still steadily resisting' such a move. While he might lose the Australians, he hoped to 'save the New Zealanders'.[25] Churchill's message arrived too late. Alexander had already admitted to the War Office that the divisions could be removed without harming his military operations.[26]

Alexander's message was received at the War Office around midnight on 24 November. When Churchill learned of it the following day, he realized that his own officials were undercutting his efforts, albeit unknowingly. He immediately sent a minute to General Brooke outlining his opposition to the break-up of the Middle Eastern army. Without the Dominion troops, he asked querulously, 'what is going to be left? It seems to me that we have got to think of the whole picture in relation to the next six months.

Please report. I am disquieted'.[27] Part of Churchill's disquiet may have been caused by the need to maintain the relative strength of British forces as American troops poured into a region over which America previously had little influence. Similarly, when the time came to move back into Europe, the more divisions that could be counted as British, the more influence Britain would have on the European peace settlement.

Brooke was not as 'disquieted' as Churchill. Even without the Dominion divisions, the Middle East Command would have sufficient troops for its 'minimum defensive requirements' and could draw on those British troops presently fighting alongside the Americans in their bid to capture Tunisia. Churchill would have none of this. He instructed Brooke that 'We must resist strenuously. Let nothing slip off without my knowing. Let no measures be taken to facilitate dispersion. Watch it and warn me in good time'.[28] The Chiefs of Staff had actually considered the issue that morning, before Brooke had received Churchill's peremptory instructions. They discussed a paper by Brooke in which he opposed the withdrawal from a 'purely military point of view' but recognized that 'political pressure from the Dominions will be difficult to resist and may therefore be an overriding consideration'. As such, he recommended that the Australians be withdrawn in January 1943 when five fast passenger ships were available to make a speedy and economical move. The troops' equipment, though, could not be shipped.[29]

The irresistible political pressure that Brooke was anticipating came in a cable on 30 November, in which Curtin took a strong and clear line that allowed no further prevarication by Britain. He informed Churchill that he 'had hoped . . . this matter was finally settled' and that Churchill would have actively supported Australia's case in Washington. He disputed the importance of the shipping factor since the victories in North Africa and the opening of the Mediterranean route were expected to save some two million tons of Allied shipping. Although the withdrawal would affect Allied operations elsewhere, it would not, Curtin claimed, be crucial. Moreover, 'it may in certain circumstances mean everything to us'. Although Churchill had threatened to withdraw an American division from Australia, Curtin persisted, threatening to support and stimulate the 'Pacific first' elements in the United States if sufficiently provoked. Lastly, Curtin demanded that the Division's equipment also be returned to Australia.[30]

Curtin's cable clinched the matter, but also further embittered the British attitude towards Australia. Churchill drafted a 'very stiff' reply calculated to damage Anglo–Australian relations although Attlee had it modified. Churchill stoutly resisted the dispatch of the Division's equipment, again enlisting Roosevelt's help to bring the upstart Dominion to heel. Meanwhile, the New Zealanders submitted to Churchill's plea and allowed their division to remain in the Middle East. Its eventual return would still be required since 'it would be neither wise nor proper' to permit the Pacific war 'to be conducted entirely by the Americans without substantial British collaboration'.[31] Churchill was overjoyed although he had suspected that the New Zealanders would prove more malleable than the Australians. He told Fraser of his feelings of 'admiration for New Zealand and all that she stands for'. Roosevelt said simply that he was 'delighted' that New Zealand had 'done the right thing' which was 'altogether generous'.[32] Both leaders left no doubt that their attitude towards Australia was plumbing new depths.

Meanwhile, P. G. Taylor was looking for support for his alternative Pacific air route. Lord Gowrie and Evatt had been encouraging, but Curtin had bowed to the opposition of MacArthur, who had opposed any British route that might compete with the existing American one. Taylor was undeterred and left for London hoping to persuade British officials of the sense of his plan.[33]

Taylor arrived in London in late October and made a strong appeal to the Air Ministry, claiming the support of Evatt and advising that the Australian Government was concerned about the safety of the existing Pacific route. Taylor warned that the American Air Transport Command, with its strong commercial representation, was 'determined to dominate the world'. Although Australians were grateful to the United States for defence assistance, Taylor claimed that 'they had no intention of becoming Americanised. Australia was, as a whole, pro-British, and there was every intention that it should remain in the British Empire'. Taylor's plan received a sympathetic hearing in Britain; officials were enthusiastic but regretted the practical difficulties which prevented immediate implementation.[34]

Taylor was hoping that the Under Secretary of State at the Air Ministry, Harold Balfour, who had been Taylor's flying instructor in the First World War, might still approve the plan. However, Balfour feared the route would divert scarce resources away from the war effort.[35] Moreover, Bruce did not help matters when he pressed for a quick decision, confiding to Balfour that if the Air Ministry 'wish to throw cold water on it—that is all right by him, provided we decide quickly'.[36] Bruce was confirmed in his opposition to the plan when Curtin informed him that Taylor had no government backing for his mission.[37] With no support from Australia House, and the Air Ministry opposing it as a war measure, Taylor's mission crashed to the ground. Balfour accepted the advice of his planning staff that, since the Pacific was an American sphere of strategic interest, 'it is up to them to develop it'.[38]

Certainly there was some force in this argument, although it had not prevented them from enlisting American and Chinese assistance in the defence of India and Burma, a sphere of British strategic interest.[39] On 5 November, the day after Taylor's interview with Balfour, the Admiralty advised Churchill that American naval inferiority in the Pacific had reached such an alarming level that 'the security of the trans-Pacific air and sea routes may be endangered'.[40] As we will see, Taylor's plan was eventually accepted in 1944 when it was felt to be in Britain's post-war interest to develop such an Imperial route across the Pacific.

Taylor was unsuccessful in 1942 largely because of Britain's 'Germany first' strategy, which was also the basis of Churchill's opposition to the withdrawal of Dominion troops from the Middle East. Churchill was also convinced that Australia did not face the imminent prospect of a Japanese invasion. Nevertheless, he was sensitive to criticism about Britain abandoning parts of her Empire or not playing a proper part in the Pacific war. When the Americans had requested the transfer of a British aircraft-carrier to the Pacific to cover their temporary deficiency, Churchill responded more positively than he would have if the request had been made by MacArthur and Curtin.

Churchill had offered to send not one, but two aircraft-carriers to the Pacific provided that an American aircraft-carrier be allowed to cover British convoys in the Atlantic. Such moves, Churchill claimed to Curtin on 2 December, would 'provide an additional and important reinsurance for the safety of Australia'.[41] This had not been the rationale behind the move but it was handy to claim it. In fact, it was planned to take the aircraft-carrier from the Eastern Fleet as one of the transferred ships. So what Australia gained in one area, it would lose in another. As it happened, the transfer never went ahead after the Americans declined to let Britain use an American carrier in the Atlantic.[42]

Churchill was still concerned to fix American attention on Europe during 1943, despite the impossibility of mounting an invasion of France. He feared that when the Americans realized the invasion was off for that year, they would turn to the Pacific for a decisive victory. To the annoyance of General Brooke, Churchill even joined the Americans in November to press for a cross-Channel invasion in 1943, but abandoned it in December when German resistance in Tunisia promised a longer campaign in North Africa than he had anticipated.[43]

Australians at this time hoped that the close Anglo–American relationship would continue into the post-war period but would leave Australia firmly within the British, rather than the American orbit.[44] Thus, for all his conflicts with Churchill, Curtin wanted the judgement of history to record that his leadership had been one of 'successful struggle to assert Australian sovereignty as a Self-Governing Dominion'.[45] It was not a struggle to assert independence but a struggle for recognition within the Empire. Shedden later recorded with regret the effect of 'Churchill's dictatorial attitude which implied that Australia was still a colony, and [which] eroded rather than strengthened sentiment in the British Commonwealth'.[46]

Even in retrospect, Shedden failed to grasp the inevitability of this conflict. With his Victorian conceptions of Empire, Churchill was personally predisposed to trample Australian sensibilities, but Curtin and his colleagues found little sympathy for Australia even among anti-Imperial counterparts in the British Labour Party. The truth of the matter was that, particularly after the outbreak of the Pacific war, British and Australian interests were fundamentally and diametrically opposed as Britain pursued her national interests and concentrated her resources in the European theatre to the detriment of her distant Empire. Australia was reluctant to question the Imperial propaganda that equated British interests and Imperial interests. Whenever a clash developed between London and Canberra, it was usually ascribed, not to a clash of fundamental interests between the two countries, but to a flaw in the personalities or the decision-making process in London that prevented Britain pursuing her 'real' interests.

Although Australia had not seen a ship of the Eastern Fleet since the beginning of the Pacific war, she proceeded with work to accommodate such ships in Sydney should they ever arrive. The British Admiralty had requested the construction of berthing facilities at Australian ports shortly after the attack on Pearl Harbor and at a time when Churchill was absent in Washington. Although they had been approved immediately in principle, it had taken ten months of discussions between Canberra, London, and Washington before the plans finally re-surfaced in mid-November 1942, for the approval of the Australian War Cabinet.[47]

The fact that the expenditure was approved at a time when there was little sign British ships would use the facilities, indicates the Dominion's responsiveness to the tugs of Empire. As Evatt promised in a secret Christmas message to Churchill in which he sent 'deep and affectionate greetings', when Australia had 'finished with the Japs with his assistance we shall put our full force into Europe'.[48] First, though, Australia would have to finish with the Japanese, and at the end of 1943, Allied victory in the Pacific seemed to be receding more often than it was advancing.

7

Return of the 9th

January to February 1943

TWELVE MONTHS AFTER Pearl Harbor, the Japanese empire in the Pacific had not been seriously challenged. The superior Japanese navy inflicted heavy losses on the Americans in successive naval battles, but, unfortunately for the Japanese, the Americans could afford to lose warships. Their massive ship-building programme would soon be replacing them faster than they were being sunk. Until that time, the battle for Guadalcanal continued to absorb much of the human life, both American and Japanese, that was poured into that backwater of British colonialism. By December 1942 the cost to the Japanese became too great; their outnumbered troops clung to their positions hampered by constant American attacks on their naval lifeline. On New Year's eve 1942 Emperor Hirohito finally consented to evacuate his troops.

The Allied forces in New Guinea had pursued the Japanese back to Gona and Buna on the north-east coast where they had retreated behind heavily defended fortifications in some of the most difficult country imaginable, a mixture of 'thick jungle, kunai grass, sago swamps and mangrove swamps'.[1] Australian supplies were flown in over the awesome barrier of the Owen Stanley Range, so heavy weapons were scarce and the battle was mainly won in face-to-face struggles, in which the Japanese put up a determined resistance. Battle fatigue and malarial mosquitoes took their toll of the Australian troops, although they could look forward to some relief now that Curtin and Churchill had settled the return of the 9th Division.

Allied efforts in the Pacific were hampered by the continuing struggle for resources. In early December Churchill pressed for a cross-Channel invasion of France in 1943. He proposed that he meet Stalin and Roosevelt somewhere in North Africa to settle plans for the year.[2] Stalin begged off; he intended to continue the Russian offensive right through the winter. Roosevelt proposed that they meet in Casablanca, formerly the Vichy French capital of Morocco and lately made famous to American cinema audiences.

Churchill intended to confirm the commitment to an invasion of Europe in August 1943, but his hopes gradually faded as the German and Italian resistance in Tunisia toughened, with German reinforcements from as far away as Russia pouring across the narrow strait from Sicily and the 'toe' of Italy. Churchill's easy victory slipped away; his forward troops retreated in the face of overwhelming odds until both armies bogged down in the winter mud. On 16 December, at a meeting with his Chiefs of Staff, Churchill was finally made to see sense. A cross-Channel invasion of Europe in 1943 would be impossible.[3] Now he would have to ensure that a sufficiently large operation could be mounted in the Mediterranean to prevent resources leaking to the Pacific war.

Churchill met Roosevelt in Casablanca in January. By then, British intentions in the Pacific were so suspect that Churchill had to pledge British assistance for America's war against Japan to guarantee continued American assistance for Britain's war against Germany, and particularly for his strategy of fighting that war in the Mediterranean. The Americans had foreseen that the 'Germany first' strategy would almost certainly mean a two-stage ending to the war. Once Germany was defeated, the Allies would have to gather their strength to defeat Japan at a time when governments would be under pressure to devote resources to domestic reconstruction. German bombing and the interruption to her peacetime trade meant that Britain faced more domestic reconstruction and might be less inclined to see the war against Japan through to the end. Moreover, any industrialized country that could quickly take advantage of the expected post-war boom in world trade would have enormous advantages over its allies. Both London and Washington suspected each other's economic motives and imperial ambitions.[4]

In November 1942 Britain's War Secretary, Sir James Grigg, submitted a report to the War Cabinet warning of a

> definite American Imperialistic policy which aims at the building up of American power and prestige in various parts of the world . . . under the guise of the war effort, this is already being attempted or achieved, at Britain's expense, in the Argentine, New Zealand, Persia, Australia, India, Egypt, Turkey, to mention only a few places.[5]

Grigg's report confirmed British suspicions and was buttressed the following month by a similar report from the Production Minister, Oliver Lyttelton. During his trip to Washington, he had been struck by the attention American industry was devoting to post-war trade. He claimed that the Ford Motor Company had new models ready for production as soon as the war ended and that 'the aircraft industry, and in particular Pan American Airways, are largely engaged in preparing for post-war Civil Aviation'.[6]

Likewise, American suspicions of Britain were very widely expressed and were probably rooted in the country's colonial past as much as anything. Eisenhower's Chief of Staff, General Bedell Smith, confided to Churchill's military adviser, General Ismay, that the American military opposed heavy involvement in the Mediterranean, which some Americans saw as serving purely British interests. General Dill confirmed this when he warned from Washington of the many leading Americans 'who feel that we have led them down the Mediterranean garden path and although they are enjoying the walk are fearful of what they may find at the end of it'.[7]

To allay such suspicions, Churchill assured the Americans at Casablanca that Britain would not leave them to fight Japan alone. He claimed that both our 'interest and our honour were alike engaged' and that Britain would 'devote their whole resources to the defeat of Japan after Germany had been brought to her knees'. Roosevelt rejected as unnecessary, Churchill's offer to formalize the pledge in a treaty between the two countries. Churchill often repeated the pledge over the succeeding years until it began to haunt him as the time approached to put it into effect. Britain also promised to mount a major operation in Burma. The two leaders then announced that they would continue their struggle until they had forced the unconditional surrender of Germany, Italy and Japan.[8]

Although Australian interests were intimately involved in the Casablanca discussions, Curtin was neither informed nor consulted about them. When he heard rumours of the meeting he tried to influence the discussions, but such was the secrecy surrounding the meeting that Curtin wrongly believed it was being held in America. He cabled Washington, appealing this time for two thousand additional aircraft. Curtin informed a group of Australian journalists about the meeting and confided that he was 'taking a hand in the conversations'.[9] By the time Curtin's message reached the Allied leaders, they had reaffirmed the 'Germany first' strategy.

In Australia, the number of army units on paper exceeded the ability of the population to sustain them. By March 1943 the army was under strength by 79 000 and required a monthly injection of 12 500 men which was simply not available.[10] Men were increasingly reluctant to enlist for the AIF rather than the militia, so in mid-November Curtin announced that militia troops, both conscripted and voluntary enlisted men, would be made liable for service in the region north of Australia.

This was a momentous decision for Curtin, given the strong anti-conscriptionist tradition of the Labor Party and his own involvement in the anti-conscription movement of 1916 and 1917. Before the Great War, Labor had introduced conscription for the militia, but it was opposed to conscription for overseas service, believing that Australians should not be forced to fight the battles of other nations. In 1942, though, conscripted troops fighting the Japanese in the islands near New Guinea were arguably fighting for the defence of Australia. It would have been nonsensical if militia troops fighting in the Australian colony of Papua were debarred from crossing the unmarked border to Dutch New Guinea in pursuit of the Japanese. Opposition MPs were harrying Curtin over conscription, and he faced a federal election at some point during 1943. By moving early and decisively, Curtin made the issue a problem for the Opposition, which soon split apart on their reaction to this system of limited conscription.

There was at this time a real fear that Australia would not bother to continue fighting Japan once the continent was secure.[11] Despite the most strenuous fighting close to Australian shores, the society was slipping back into pre-war apathy. With free-spending American troops thronging the pavements of Australian cities, many felt it was time to turn a few 'quid' into a small fortune. Horse-racing continued to captivate so many Australians that Curtin complained of not being able to make a phone call on a Saturday because the lines were tied up with punters placing their bets with illegal bookmakers.[12]

On 30 December Curtin spoke bitterly to journalists of the

> buggers in Australia who won't work. Coal mines are idle, and everyone is thinking about holidays just at a time when a few extra tons in our war effort would have a crucial effect. We are like people who have just got contagion out of the house, and just over the back fence. Apparently we are not worrying how dirty the yard is.

It was no good, Curtin said, looking to a great power for the country's salvation. In fact, Australia's plight earlier that year had been, so Curtin claimed, 'the proper fate of any country which did not build its own defences . . . [and] was also the proper fate for a country that thought it could fight anybody's war before it made its own position safe'.[13] Curtin was reaffirming the Labor Party's defence policy of the 1930s which called for the defence of Australia to be secured before contributions were made to Imperial defence.

While Curtin railed against the complacency of his fellow citizens, Evatt and Robinson supported a push for greater regional power for Australia. To this end the Australian Government fought to change the relationship between European powers and their colonial territories, with regional committees of interested nations supervising their administration and helping to propel them towards independence. The Australian and British governments were prodded along by Bruce who feared that China would otherwise come to dominate South East Asia, 'probably the richest prize of the world'. From Canberra, and to the considerable consternation of Whitehall, Evatt backed freedom for India and sought greater Australian influence within the Netherlands East Indies (now Indonesia) and Malaya.[14]

As Evatt worked to expand Australia's regional role, Menzies sought to restore the Empire in all its tarnished glory. From his position on the Advisory War Council, he argued unsuccessfully for Australia to align itself militarily with Britain's fight in Burma, rather than with MacArthur's struggle to the north. He loudly predicted a gloomy future of stringency and belt-tightening after the war, and sought his own political salvation in London. In February 1943 he asked in vain for Churchill to sanction his permanent return to London, this time as 'an out-and-out supporter of Winston and as a Britisher devoted to Australia and the Empire'.[15]

Nearly all Australians expected Britain to continue to play a large part in Australian life after the war, differing only on the degree of British involvement. Curtin, as Menzies had done before him, blamed Churchill rather than Britain for Australia's shabby treatment. Following his trip to London in mid-1941, Menzies had concluded that Australia was too far away for Churchill to give it his proper attention. Curtin took a similar view; in late 1942 he called Australia Churchill's 'forgotten land'.[16] Britain's failure to fulfil its defence guarantees appeared to be at least partly a result of misguided leadership rather than the inevitable result of national self-interest. Hence the various attempts by Menzies, Bruce and Evatt to change the composition of the British War Cabinet and to curb Churchill's largely untrammelled power or cause his complete downfall.[17]

At the end of the year, Curtin told journalists he was 'fed up with the way [Churchill and Roosevelt] played ball with one another, quite regardless of the world at large'. Events over the past year had convinced him, he said, that both Allied leaders had 'made their minds up that if the British

Empire in the Far East had to go then it had to go'.[18] Despite Curtin's warm personal relations with MacArthur, America would not necessarily treat Australia's requests for assistance more generously than Britain had till now. Curtin was given a good lesson in *realpolitik* when he tried to oppose Roosevelt's plan to replace a respected career diplomat with a political crony as American Minister in Canberra. On 30 December 1942 Curtin had told journalists off the record that 'I'm not having him'. Two days later and 'very downcast', Curtin was forced to back down after he learnt that Roosevelt would be 'deeply offended' if Australia refused to accept his nominee.[19]

In the aftermath of this minor crisis in Australian-American relations Curtin tried to woo Churchill, or perhaps one of his ministers, to visit Australia and help restore amicable relations between the two countries.[20] A visit might also awaken British sympathy for Australia. But the suggestion went no further than the British High Commissioner, Sir Ronald Cross, himself a former British minister. He petulantly pronounced that the suggestion betrayed the Australian Government's lack of confidence in him. In fact, both the British and Australian Governments regarded Cross as an embarrassment that had to be borne stoically.

In February 1943 Curtin asked unsuccessfully for 'some demonstration by the United Kingdom of its interest in the Pacific by the despatch to Australia of a token force, e.g., a Brigade or a naval unit'. He watched enviously as Britain sent hundreds of bombers in almost nightly raids over Germany but, as he complained at the end of 1942, 'by Christ you can't get any here'. All Britain had on offer were superannuated Lancaster bombers which it planned to convert into civil aircraft after the war and provide free to the Dominions to compete with the superior American machines that would be available. The Government asked if the Lancasters compared with 'American designed passenger aircraft embodying comfort and speed',[21] an indication of Australia's changing attitude to Empire.

Australia did not want to forego superior American aircraft in order to buy British, so she sought to hedge her bets. At the same time, the Australian Air Minister proposed that Australia produce one hundred American C47 transport aircraft, the military version of the versatile DC3 aircraft, intended to service Australia's internal air transport needs after the war. If Australia were able to manufacture such large aircraft she would be less dependent on either America or Britain in future.[22]

Australia demonstrated her political immaturity when she was faced with the problem of recommending a new Governor-General to replace Lord Gowrie. The Scullin Labor government had set a valuable precedent in 1930 when it insisted that its nominee, Sir Isaac Isaacs, should be appointed, but more than a decade later, Curtin was anxious to avoid taking a similar stand. The appointment of the Duke of Kent to the position in 1939 had been postponed for the duration of the war, while Lord Gowrie soldiered on. When the Duke was killed in a plane crash in August 1942, it not only exacerbated Curtin's fear of flying but also forced him to find a replacement for the dead Duke.

Labor policy required that the Governor-General be Australian, and Curtin was no doubt reluctant to provoke Britain over a matter of symbolism when matters of survival were at stake. So Curtin did nothing, hoping that the elderly Gowrie would soldier on indefinitely. As Gowrie confided to his counterpart in India, the Duke's death had 'made it more difficult for me to retire—as labour policy is that an Australian should be G.G., and as there are several candidates but nobody suitable, Curtin does not want to be faced with the problem'.[23]

At the same time as it was seeking to restore harmony to its links with London, the Government was ensuring that the wartime links with America would be continued into the postwar period. In January 1943 Curtin successfully argued that the wartime communication links between Australia and the United States should be retained, thereby altering the pre-war imperial system.[24] This was a policy change of considerable symbolic and practical importance and one that had been resolutely resisted by the Menzies Government.

Britain stoutly opposed any change in the pre-war colonial status quo. In November 1942 Churchill said he had not become Prime Minister to preside over the break-up of the British Empire, a remark designed to kill speculation about a new post-war imperial order. The British Labour Party, outsiders such as Evatt and Bruce and, most importantly, Roosevelt, all sought to prise Britain loose from an empire she no longer had the will or power to defend.[25]

Churchill rejected any change to the colonial status of Britain's territories not only because of his romantic Victorian notions of Empire, but because Britain needed the Empire if it were to join America as an equal partner in any post-war relationship. The Foreign Office was not so committed to preserving the Empire in its pre-war form, but it realized the value of the Empire in its dealings with America. Good relations between London and Washington continued to be marred by the issue of Empire.

The old imperialist order, led by Leo Amery, Secretary of State for India, fought a vigorous delaying action against the forces of change. Amery's view was clear: Britain should keep clear of any post-war alliance, whether with America or the countries of Europe, 'so long as our main interests, our primary obligations and our instinctive loyalties are concerned with our partners in the British Commonwealth'.[26] Amery was more forthright in a private message to the Governor of Burma, now cooling his heels in Delhi while the Japanese settled into their occupation of Rangoon. Amery assured him that he was

> at least as Colonel Blimpish as you are, and 'By Gad! Sir' am not at all prepared that anyone, Yank or Chink, should poke either projecting or flat noses into the problem of the reconstitution of Burma. In that we shall certainly have Winston behind us and I think increasingly the Cabinet as a whole.[27]

Other British leaders may have doubted that the Empire would be Britain's salvation, but they were nonetheless concerned to preserve Britain's overseas interests. Gerald Wilkinson, the British security agent, had strong business links throughout the Pacific and was anxious that they be resurrected after the war. He was also concerned that Britain's neglect of the Pacific would lead to her losing economic influence after the war. For this reason, Wilkinson felt it to be vitally important that he tag along with MacArthur and that Britain contribute forces to MacArthur's command.

In February 1943 Wilkinson left Australia for consultations in London. He passed through Washington on the way, and impressed upon the British Ambassador, Lord Halifax, the danger that, once Germany was defeated and American forces were switched to the Pacific, MacArthur would be tempted to make his expedition against Japan a 'Pan-American' one,

> in which case British status in the eye of the Native peoples of the areas under MacArthur's command, would not increase with the defeat of the enemy, as would American status . . . we might find ourselves not only in an inferior position to take advantage of the immediate opportunities of post-war reconstruction, but of the greater reaching opportunities . . . [when those areas] become due for a great economic and social up-surge.[28]

In late 1942 Wilkinson had floated a plan for Britain to support MacArthur's promotion to the command of all Pacific land forces while ensuring that MacArthur was made 'unofficially aware' of the British support 'so that we might hope for his friendship and understanding in future Anglo-American matters such as Pacific and Asiatic post-war settlements'. Although MacArthur at first gave guarded approval, the paranoic general later backed down, warning Wilkinson on 1 November 1942 that 'Roosevelt has his spies right down to the kitchen sink' and that the time was not ripe for such a move. MacArthur had decided that Wilkinson was no longer useful and that this shadowy figure with his top-level contacts in Whitehall should not be attached to the American headquarters. Despite strenuous efforts by Churchill and Wilkinson, MacArthur refused to let him return.[29]

Meanwhile, Japan continued to be hamstrung by its commitment to the war in China and in Manchuria by the need to guard against a Russian move into the Pacific war. It was more difficult to regain the offensive in the Pacific or take the bold steps that might have staved off defeat. Just as the Russian struggle absorbed the German strength, so too was the war in the Pacific won to a great extent by the fighting in China.

As for Australia, its immediate defence was secure, but

its future security depended on the Japanese empire being reduced to its original size. Curtin had fought hard for the return of the 9th Division, but the victory almost became pyrrhic, when the troop-ships set off from Cairo with the lightest of naval escorts to cross the dangerous expanses of the Indian Ocean. It had been axiomatic that troop-ships should have what the British naval historian, Captain Stephen Roskill, called 'special protection'.[30] As far as Australia was concerned, this meant that troop convoys should be escorted by a capital ship (a battleship). In March 1941 the Government of the acting Prime Minister, Artie Fadden, rejected a suggestion that it was sufficient for troop convoys in the Indian Ocean to have capital ship 'cover' (a capital ship force in the general area of the convoy, providing protection from a distance). Fadden insisted that 'capital ship escort should be provided for the larger troop convoys, in accordance with the principle agreed to at the 1940 Singapore Defence Conference'.[31]

Five fast liners were allocated for the return of the Division. Because of their speed, they often sailed unescorted since any accompanying destroyers would not be able to maintain the pace. The speed of the ships was their security, allowing them to outpace submarines and reducing the risk of being torpedoed. This was all right for the Atlantic and the western part of the Indian Ocean, where only isolated enemy raiders might be encountered. However, the convoy would pass within reach of the Japanese naval and air forces stationed at the former British bastion of Singapore. The troops' departure from Egypt could hardly be kept a secret and their destination could be easily guessed. There was a very real danger that a force of enemy cruisers, or even a battleship, might intercept and annihilate the convoy with its thirty thousand troops.

On 29 October 1942 when the British Chiefs of Staff were contemplating the practical details of the Division's withdrawal, they had informed Churchill that the Australian Government would have to agree 'to the running of the ships without escorts'.[32] Three weeks later, the naval chief, Sir Dudley Pound, observed at a meeting of the Chiefs of Staff that 'it could be assumed that the Australian Government would not consent to the return of their ships unescorted'.[33]

So, in reporting their revised conclusions to Churchill, the chiefs allowed for the convoy to have naval escorts, although without specifying the strength of the escorts.[34]

In mid-January, the troops of the 9th Division lined the rails of their camouflaged ships as they sailed from Suez. Joining them for the voyage were two escorts—the cruiser *Devonshire* and an armed merchant ship—and destroyers for their passage through the narrow waters of the Red Sea. There was no battleship to protect them from raiding cruisers or enemy submarines. Roskill wrote in his official history, 'in the event, this large movement took place without any untoward incidents'.[35] This was more by good luck than good management.

A German armed merchant raider was crossing the lonely wastes of the Indian Ocean at the very time as the troop convoy, intercepting, then sinking or seizing any merchant ships it encountered. Heavily armed and complete with spotter planes, the raider arrived in Batavia on 10 February without sighting the convoy. If it had, it may itself have been sunk by the *Devonshire*, which had sunk a similar German raider fourteen months earlier.[36] But there was no certainty in naval warfare, as the unfortunate sailors aboard the Australian cruiser *Sydney* learned to their horror when their ship was destroyed by just such a raider in 1941.

Japanese submarines were also active across the Indian Ocean and a German U-boat 'wolf pack' operating off the west coast of Africa sank twenty-three ships in early 1943 as it tried to cut the British supply line to the Middle East. In February 1944 a lone Japanese submarine chanced upon a similarly escorted troop-ship convoy, albeit a slower one, and succeeded in evading the accompanying escort. It fired two torpedoes that struck a troop-ship, sending it to the bottom with almost all of its complement of 1500 servicemen and women. The subsequent inquiry into this incident judged that the number of escort vessels available in the Indian Ocean was 'totally inadequate for the protection of valuable convoys on widely separated routes'.[37]

At the beginning of 1943 the inadequacy of the Eastern Fleet became even more marked when the carrier *Illustrious* was transferred to the Pacific in response to the American request for assistance.[38] Although Admiral Somerville protested at the depletion of his strength, it was to no avail. One of his officers commented that even with the aircraft-carrier the fleet did not amount to much and if the Japanese 'really go for us they'll put in forces which would eat us'.[39] Somerville shared this concern, observing that, 'We have formidable antagonists and if you start taking liberties . . . you are bound to come a cropper sooner or later'.[40] He was particularly worried that he would 'come a cropper' over the Australian troop convoy.

When Somerville received orders in mid-January to 'cover' the passage of the Australian convoy, he was also ordered to dispatch two of his destroyers to South Africa to help deal with the German submarines. 'It doesn't make sense', commented Captain Edwards, one of his senior officers. On 3 February the much-reduced Eastern Fleet raised anchor in the Kenyan port of Kilindini and set course for the Seychelles islands, where they were due to rendezvous with the troop convoy. The force of three battleships, one cruiser and six destroyers was calculated to impress the returning troops who would not have realized that two of the battleships were of First World War vintage and had been described by Churchill as 'floating coffins' unfit for use against the Japanese.[41]

By 6 February the ships were in the Seychelles waiting for the overdue Australians. Edwards vented his anxiety in the pages of his diary, writing that 'I don't like this convoy and I'm thankful at all events that the Jap'n major forces appear to be fully engaged to the Eastward . . . It's one horrible gamble . . .'[42] Just how horrible a gamble was made plain the following day when a report was received of a Japanese submarine in the area; the fleet had few ships capable of dealing with it.

Somerville planned to fool the Japanese into thinking that a substantial British fleet was operating in the central Indian Ocean. He hoped that the Japanese would then retain their surface fleet in port, at least until the British intention became clear.[43] The other part of his plan was to impress the passing Australian troops with a false sense of British strength. He had hoped, he told Admiral Pound, 'to steam close past them and give them a good chuck up'. However, just over an hour before the convoy sailed into sight on 10

February, another submarine was detected. The convoy was immediately ordered to change course while keeping to its zigzag pattern. Somerville's ships could not venture any closer than a few miles of the convoy as the troops strained to catch sight of the distant specks before they sank away below the horizon and the darkness of the tropical night descended to cover their inadequacies.[44] Apart from the light escort, the thirty thousand troops were now on their own for the four thousand miles of ocean that separated them from their homeland.

Fortunately, Somerville's subterfuge seems to have worked; the convoy arrived in Western Australia unscathed, on 18 February. The single railway line across the continent would take months to transport them in a slow shuffle of steam trains, and Australia could not afford the delay. They were sent back to sea, but this time with all the escorts, both sea and air, that the country could muster. Curtin confessed to journalists that he had not slept well for three weeks while awaiting the arrival of these troops.[45] Somerville was similarly relieved at their safe return, confiding to Admiral Pound that

> I still feel we took a bit of a chance over this since it is unlikely the Japanese were unaware of the movement of the Australian Division from Egypt, and its probable destination must have been obvious. Had they sent out a cruiser force with an auxiliary carrier, they would have had no great difficulty in locating the convoy, and would have made hay of it if they had found it.[46]

So, by a stroke of considerable luck, Australia had most of its forces returned to its control, or at least that degree of control MacArthur allowed them.

While the 9th Division had been zigzagging its way across the Indian Ocean, the fighting on Guadalcanal had finally stopped, in a completely unexpected way. The remaining Japanese troops, exhausted and half-starving, left the island in a daring night-time operation reminiscent of the brilliant Australian evacuation from Gallipoli in the First World War. When the American trap clanged shut on the Japanese positions, there was not a live Japanese in sight. The enemy had left. More than twenty thousand men had died fighting for control of an unknown British island. The Americans had suffered seven thousand casualties and lost twenty-four warships to achieve a victory that mattered little in the final defeat of Japan.

The core of Japanese strength in the south-west Pacific remained untouched and unthreatened at Rabaul because of the concentration of Allied bomber strength in Europe. While the victory on Guadalcanal was not decisive enough to allow MacArthur to attack Rabaul, it did, along with the Allied victory in Papua, mark the last defensive struggle against the Japanese. The holding war had stemmed the Japanese tide. It would now be up to the Allies to take the offensive and gradually chip away at the briefest empire in history.

8

Battle of the Bismarck Sea

March to April 1943

FOLLOWING THE SURPRISE Japanese withdrawal from Guadalcanal in February 1943, Australia's immediate security problems were resolved. Japan had a huge empire to defend, and could not seriously contemplate extending it. Although the threat of invasion was lifted, the threat of attack remained. The Japanese were deeply entrenched at Rabaul, within striking distance of northern Australia and of the Allied forces in New Guinea. Tokyo intended to make the war too expensive in men and machines for the Allies to prosecute. At Casablanca, the Allies had demanded unconditional surrender, but the Japanese hoped to reach a compromise that might allow them to retain part of their territorial booty, which could then, at some later date, again pose a threat to Australian security. Australia was determined that the Japanese empire be destroyed.

From mid-1942, MacArthur had intended to mount an offensive to dislodge the Japanese from Rabaul. It could have worked had he had the wherewithal to do it, but long-range bombers were almost exclusively restricted to the so-called strategic bombing offensive against Germany, spreading death and destruction and terrifying both civilians and soldiers. This blunt instrument, wielded by Allies and Axis alike, bludgeoned entire populations senseless and sent the stench of death wafting across Europe. Unlike the gas ovens of Treblinka and Auschwitz, it was death delivered from a distance, but it was no less 'scientific' and cold-blooded. Entire cities were laid waste by bombs calculated to maximize civilian casualties, but despite the slaughter, it did not produce the victory that its eager advocates predicted. However, it did prevent an early and strong blow being directed against the Japanese.

Following the Casablanca conference, Curtin realized for the first time that the Allied strategy would necessarily involve a two-stage ending to the war, that, as he put it to a group of reporters on 2 February, 'we will be at war in the Pacific when we have ceased to be at war anywhere else'. This was a daunting prospect for the former anti-war activist. In a burst of prescience, Curtin realized that the war was likely to outlast him, that his long and successful campaign to unify the Labor Party under his leadership had culminated in it presiding over a military struggle for the nation's life rather than the social revolution that he had been seeking.[1]

The two-stage ending to the war also threatened the reconstruction effort that the Government had begun planning as soon as the Battle of Midway had tipped the balance in the Allies' favour. Australia would have to fight Japan and provide supplies for the Allies long after Europe had returned to peacetime pursuits. Although Curtin could no longer claim that Australia needed additional forces to ensure her defence, he was frustrated that Allied strategy had pushed Australia into a military backwater. He complained privately of Russia 'getting more planes in a week than we got in almost a year', although he had to concede that 'they were of course doing a useful job' and that Russia and China 'probably had stronger claims than Australia'. Despite the apparent logic of Allied strategy, Curtin complained that the country deserved better—that Australians were 'the only white race in the southern hemisphere and almost the only Empire country doing anything'. His fantasy was to have half a million pounds available, free from public scrutiny, to scatter around the United States to reinforce Australia's arguments.[2]

It was at this stage of the war that Curtin understood Australia's relative weakness. In the 1930s the Labor Party had criticized Australia's dependence on Britain and argued for a strong defence force capable of resisting invasion and perhaps extending its power over the adjacent Pacific region. The experience of the war, with the nation fully mobilized and a substantial American force stationed in the country as well, revealed the limits of Australia's power to influence events in the region by force of arms. According to Curtin, the country could never raise more than five divisions for service outside Australia, and could not tackle a great power by itself.[3] Once the war was won, Australia would have to return to the protective umbrella of a great power.

Australia accepted the Casablanca decisions but interpreted them in a way that would favour Australia, with Curtin quickly requesting additional forces for the holding strategy.[4] MacArthur and Curtin combined to ask Churchill to visit Australia in the confident, but mistaken, expectation that Churchill 'could not fail to be impressed if he saw our difficulties'.[5] Australia naively failed to recognize that her relative insecurity was caused by the Anglo–American pursuit of national interests. Churchill was well aware of Australia's position; he discounted the danger and felt, along with other British officials, that a 'taste' of bombing would do Australia good.

Although Churchill declined to visit Australia, he extended an invitation for Curtin to visit Britain. He had succeeded in transforming the normally irascible Dr Evatt into a simpering faun, and was keen to exert his influence on Curtin. But the Australian Prime Minister would not be wooed, claiming that parliamentary business prevented his leaving Australia. Curtin knew the Labor Party was facing elections with only a tenuous grip on parliamentary power. The arduous trip by plane would not help Curtin's failing health, particularly in view of his morbid dread of flying. He could only watch in amazement Churchill's 'energy' and 'enterprise' as he criss-crossed the globe. In fact, travel also took its toll on Churchill who was just then struggling with a serious bout of pneumonia, the 'old man's friend'.[6]

Churchill had recuperated by 15 March when he lunched with the British intelligence agent, Gerald Wilkinson, who was in London to confer with his superiors at MI6. Churchill asked him to 'dilate' on the 'question of Mr. Curtin'. Wilkinson described Curtin as 'a small and parochial man and obviously not equipped by experience and environment to take naturally to a global view of strategy'. He urged that the British Government 'should forgive him any waverings that might have occurred in connection with Australian overtures to Washington last year'. As Wilkinson argued, Britain 'might manage [Curtin] more easily than we might manage some of his less scrupulous associates'. But Churchill was unrepentant. He merely 'chuckled rather grimly', saying that the Dominion had paid the price for its disloyalty when Washington rebuffed its advances in early 1942.[7]

Meanwhile, MacArthur dispatched a mission to Washington headed by his Chief of Staff, General Sutherland, to seek sufficient forces for an assault on Rabaul, including amongst a request for 1800 additional aircraft, a number of long-range bombers. When Washington was slow to respond, MacArthur took up Curtin's offer of help, warning on 17 March that any supporting appeal should be made in such a way that 'it does not lay itself open to the suspicion of definite collusion'. He suggested that Curtin try the circular route of approaching Roosevelt via Churchill, claiming that it was 'astonishing' that Churchill had not yet replied to Curtin's appeal of 19 January.[8]

There was another Australian security scare when forty-four Japanese aircraft raided the north Australian port of Darwin on 15 March. In a phone call to Curtin, MacArthur claimed that the raid was part of a general Japanese buildup in the region to the north-west of Australia, perhaps in preparation for a serious attack on northern Australia in a few months. The Japanese were developing sixty-seven airfields within reach of Australia, from which they could operate up to two thousand aircraft, although MacArthur admitted that there was no sign that such numbers of aircraft were about to be deployed. In fact, the Darwin raid had been easily repulsed, and although it was part of a Japanese buildup, it was a buildup of a defensive rather than an offensive nature. MacArthur deliberately exaggerated the danger to Australia and claimed that he was asking Washington for only enough forces to defend what he already held, rather than to mount an offensive against Rabaul. He told Curtin that the south-west Pacific was no longer threatened, but the north-west approach to Australia was.[9]

The following day, 18 March, Curtin sent cables to both Churchill and Roosevelt reminding that his request for additional aircraft was still unanswered after two months. On 2 March, in the Battle of the Bismarck Sea, a force of low-flying American bombers had intercepted and destroyed ships carrying several thousand Japanese reinforcements to New Guinea.[10] Curtin claimed that the attacks on Darwin indicated the 'paramount importance' of maintaining air superiority in the north and west of Australia. 'Of particular importance', he wrote, was 'the vital base of Fremantle [Curtin's constituency] where owing to the depletion of the Eastern Fleet a heavy attack of the tip and run variety might be carried out by naval bombardment and carrier-borne aircraft'.[11]

The chances of the Curtin–MacArthur appeals succeeding were extremely slim. Churchill had already rejected one without bothering to reply to Curtin. In Washington, American service chiefs rejected MacArthur's proposed assault on Rabaul in deference to the Casablanca call for an all-out bombing offensive against Germany.[12] The assault on Rabaul was never mounted, but Curtin's repeated requests now demanded an answer, particularly as he intended to dispatch the dreaded Evatt on another mission to the Allied capitals in support of the Australian demands.

Curtin's cable to Churchill on 19 January 1943 had been sent to Washington where Curtin thought the two leaders were meeting. The message was then passed through American sources to Roosevelt in Casablanca who handed it to Churchill. Churchill promptly tossed it aside. When Bruce came to follow up the requests with General Ismay on 20 March, he found that Ismay knew nothing of the request; Churchill had not passed it to the Chiefs of Staff for consideration. Bruce moved to head off yet another dispute between Australia and Britain, urging Ismay to dissuade Churchill from sending any intemperate reply to Curtin until the Chiefs of Staff had considered the issue.

In the meantime, Bruce put the Australian case to Ismay to gauge the likely British reaction. He then agreed with Ismay that he should 'telegraph to Australia and raise all the points that were discussed between us, and see what reaction I got'. By unofficially advising Curtin of Britain's objections, Bruce might thereby avoid the necessity of Britain blankly refusing the Australian request. At the least, it would soften the blow when it came.[13]

And come it certainly did. On 27 March Churchill informed Curtin that his request was right out of court, that Australia was sufficiently protected by the aircraft already allotted to the region. This was an arguable point; the British counted aircraft in the adjacent South Pacific Command, carrier-borne aircraft throughout the Pacific, and those under MacArthur's command, with little idea of the great distances over which these aircraft had to operate. Counting aircraft in New Zealand and Fiji was akin to counting aircraft in London as part of the force available for the defence of Moscow.

If the arithmetic was questionable, Churchill brandished Allied strategy, reminding Curtin that the war in Europe had 'first charge on the forces of the United Nations' but that, once this was won, 'every man who can be carried and every suitable ship and aircraft will be concentrated on Japan'. Attlee, probably on the advice of Bruce, had warned that Australia would not be satisfied by a blank refusal, and had suggested that Churchill undertake to discuss the whole question with Evatt once the External Affairs Minister arrived in London.[14]

In an illuminating off-the-cuff remark in a cabinet meeting Churchill referred to the 'troublesome attitude of the Colonies' when he was referring to the Dominions; he had not really absorbed the gradual moves towards independence by the various Dominions.[15] Churchill continued to treat the Dominions like colonies, as he had done as Colonial Secretary before the First World War. He treated Canada and South Africa with a little more consideration, but New Zealand's deference to Whitehall only encouraged him in his Victorian ways.[16]

Contrary to the claim of Bruce's biographer, the High Commissioner was very much side-lined in Whitehall because of Churchill's hostility.[17] As Bruce's naval adviser and former chief of the Australian navy, Admiral Colvin, confided to Gowrie, 'Winston has no use for people who got on well with Chamberlain'. Although Bruce's admission to the War Cabinet 'promised better things . . . it all came to nothing and he [Bruce] only attends the weekly War Cabinet meeting which decides nothing of importance'.[18] British officialdom, when faced with Australian views, 'simply shrug their shoulders, say "what queer people" and pass on to the next business'. Colvin suggested that Curtin visit London to correct the poor impression Evatt created and also to imbue Curtin with 'a better conception of the magnitude and scale of the system of Government and of running the war'.[19] In other words, Curtin could be given the 'world' view of the war, which usually distilled into the view from Whitehall once the imperial rhetoric had evaporated.

Britain's negligence over the defence of its Pacific interests had worrying repercussions in other areas. Whitehall found that the Pacific Dominions, both now governed by labour parties, held disturbing views about the future of European colonies. Australia was salaciously eyeing the Dutch East Indies, the New Hebrides and the Solomon Islands, but worse still, the Dominions wanted to dissolve the European empires altogether.[20] This was dangerous talk when Roosevelt was continuing his personal pressure in the same direction.

Lord Cranborne, Britain's former Colonial Secretary and soon to succeed Attlee as Dominions Secretary, thought that Britain must somehow involve the Dominions in the administration of the colonies so that the Dominions would cease to regard the colonies as Britain's private property. Cranborne was adamant that Britain make clear that the Empire is 'an inter-dependent whole'. 'The Australians', he wrote, 'are particularly bad about this. They want all the advantages of the Imperial connection without any responsibilities'.[21]

The problem with Cranborne's scenario for a new co-operative imperial nirvana was that it ignored the relative decline of Britain's military and naval might. Historically, Britain's strength had largely guaranteed Australia's continuing adherence to the Empire. Now that this strength had waned in the Pacific, there was no telling how much longer Australia would stick close to Britain. As the head of Britain's military mission in Australia, Major General

Dewing, informed Whitehall:

> The future relationship of Australia and England will be greatly influenced by the extent to which the latter can within the next year or so, demonstrate, in a manner which will get across to all Australians, her ability and desire to give practical help to Australia in the war against Japan.[22]

Certainly the relationship was not helped by the negative messages from London.

Curtin refused to be diverted by vague promises and platitudes from London. Australia had not, he said, acquiesced in the 'Germany first' strategy, the Dominion had been 'confronted with a *fait accompli* and we had no alternative but to accept the decisions, much as we disliked them'. Curtin disputed the Allies' confident assessment of the situation in the SWPA, pointing out that there were only 1450 modern aircraft in the region, less than half of which were in a serviceable condition. Delivery dates for aircraft from the United States were not being met and planes were being put out of service faster than they could be repaired or replaced. It was not enough, Curtin argued, for Australia to have a few more planes than the Japanese. What was needed was the 'provision of such air power as will enable the forces in the South West Pacific Area to prevent the consolidation of the Japanese in their positions to the north of Australia and so render reasonably feasible the task of ultimately defeating them when the war in Europe ends'.[23]

Curtin hoped Evatt's trip to Washington and London would help, but Evatt regarded the trip at least partly as an opportunity to display his prime ministerial qualities. Rumours were already circulating about his professed willingness to change sides and lead a 'national' administration as Billy Hughes had done during the First World War. Evatt had made such a proposal to Menzies in 1941 without success and there is no reason to suppose that he had dropped the idea. There were even suggestions that Evatt, with the business backing of W. S. Robinson, might switch sides and lead the conservative opposition. People were as disillusioned with the political process as they had been during the 1930s when an economic crisis had also tested the measure of the Dominion's political leaders and found it wanting.[24]

Evatt moved quickly to claim political credit when the Spitfires finally arrived. In a radio broadcast from Sydney, he extolled the virtues of these now elderly aircraft, claiming that they represented the 'greatest qualities of the British race—their engineering genius, their refusal to be satisfied with anything less than the best, and above all their dauntless courage in the face of adversity'. Their temporary diversion to the Middle East was, he claimed, an 'excellent' move that had 'played a vital part in the battle for Egypt'. Now that the Spitfires were operational and Australia 'gradually emerging from the shadow of death and disaster', they would be able to 'share in that process of transformation and salvation'.[25]

With Curtin husbanding his health, Evatt was the obvious, though by no means ideal, choice to press Australia's case for more aircraft. His mission was announced by Curtin on 24 February, at a time when Australia was keen to expand the RAAF but was encountering opposition from Washington over its size and future role. Evatt was meant to overcome American opposition, although his ability to do so was seriously questioned before he left by Bruce's military liaison officer, Colonel Wardell, who was recalled to Canberra at Evatt's insistence. Wardell denounced the robustious minister and, in a talk with Shedden on 17 March, claimed correctly that Evatt's previous visit had been a disaster, that Evatt had 'created a very poor impression . . . due to his bad manners in discussion with people' and that his return to London 'would not be received with very great enthusiasm in the circles which had had contact with Dr. Evatt during his previous visit'.[26]

Meanwhile, Britain was preparing for some effort in the Far East. The Casablanca conference in January had approved a British offensive in Burma, code-named Anakim, but resources for its success were not being made available. From afar, Churchill had dispatched querulous cables to Wavell, questioning why such 'very great forces' were standing idle in India.[27]

Wavell pointed limply to the 'natural and climatic conditions especially poverty of communications' that were making his task difficult, claiming that his forces were not excessive for the defence of a country two-thirds the size of

Europe.[28] Although the War Office soon confirmed Wavell's protests, Churchill refused to accept the report.[29] However, his Chiefs of Staff, and even Roosevelt, realized that Anakim could only be mounted at the expense of the planned invasion of Sicily. Eventually, on 2 April, Churchill called Wavell and Somerville to London for talks on options in the Far East.[30]

Churchill was very conscious that America would hold Britain responsible if China fell because insufficient Allied support reached her via Burma. As Churchill informed his War Cabinet at the end of April, the recapture of Burma had been agreed by him at Casablanca 'largely as a concession to United States opinion'.[31] In Washington's view, keeping China in the Pacific war was almost as important as keeping Russia in the European war. Both these countries had vast fighting forces capable of grappling with the enemy until the Americans could deliver the *coup de grâce*. If the Chinese were defeated, or accommodated the Japanese, this strategy would be in tatters. The Japanese would be able to strengthen their Pacific forces, thereby drawing American forces away from Europe; they might even attack the Russians with forces previously tied down in China. It was for these reasons that Churchill was also anxious to secure the lifeline to China. The implications in Europe of the defeat of China were simply too terrible to contemplate.

Roosevelt had agreed with Eden, during the Foreign Secretary's visit to Washington, that Anakim could be dropped and China sustained by aircraft flying over the mountainous 'hump' from India to Chungking, but Churchill refused to let the planned offensive fizzle out. In a withering message to his Chiefs of Staff in early April, Churchill blasted Wavell's campaign in India, describing it as going 'from bad to worse' with British forces being 'completely outfought and out-manoeuvred by the Japanese'. Churchill foresaw political problems if public opinion ever focused upon 'this lamentable scene'. Rather than tackling him head-on, Churchill's military advisers recommended that Anakim be postponed for practical reasons rather than cancelled altogether.[32]

On 7 April Montgomery's Eighth Army had finally

joined forces with Eisenhower to close the ring on the Axis army in Tunisia. The next step was to leap-frog across to Sicily and thence perhaps to Italy itself. If the Allies were to exploit any victory in Sicily, shipping and landing craft would have to be on hand in the Mediterranean. This not only precluded an amphibious offensive in Burma but also killed off the Allied plan to capture a bridgehead on the northern French coastline in the autumn of 1943. It remains a matter of dispute whether the British preoccupation with the Mediterranean delayed the D-Day landings in Normandy, thereby unnecessarily lengthening the war.

Although the Allies were clearly winning the battle for Tunisia, the fighting was more prolonged and bitter than they had anticipated. Hitler reinforced his shattered *Afrika Korps* with fresh divisions and armour. With their backs to the Mediterranean, these troops resisted so strongly that Eisenhower questioned whether the invasion of Sicily was feasible for 1943. In Washington, General Marshall scotched Eisenhower's fears and ordered plans for an invasion of Sicily even if the Germans still held on in Tunisia. This was 'splendid', Churchill cabled, and on 17 April invited Marshall to London to set their joint stamp upon the decision. But Marshall begged off, confiding that first he had to travel to the Pacific theatre to 'orient the various people concerned, who [were] reacting to public clamour for intensification of [the] war effort in that theatre.[33]

With the invasion of Sicily still in some doubt for 1943, Churchill no longer championed the Anakim operation in Burma. When Wavell met Churchill and the Chiefs of Staff in London on 22 April, it was Churchill who put the case *against* Anakim and proposed an operation against north Sumatra as a stepping stone to an eventual assault on Singapore. Wavell's carefully laid plans were nullified. Anakim, he persisted, 'seemed the only possible plan particularly in view of the political necessity of gaining touch with China'. Churchill agreed that Japan would only be defeated by bombing from bases in China and Russia, but claimed that this could not be done until after the fall of Germany. Until then, he argued, the Allies should sustain China in her struggle against Japan with Roosevelt's proposed aerial lifeline. Burma could be bypassed and the danger of embroiling large numbers of British troops in jungle warfare could be avoided. Churchill urged the use of sea power to tackle Japan elsewhere; the effect would be more dramatic and the cost would be much less.[34]

So far so good. Then the American Chiefs of Staff invited Wavell and his fellow Far Eastern commanders to Washington to discuss Allied plans for the war against Japan. The last thing that Churchill or his military advisers wanted was the 'Pacific first' elements within the American Services and his own Far Eastern commanders joining in a push for a greater effort against Japan. On 28 April Churchill's Chiefs of Staff warned that accepting the invitation 'might well give the Americans the impression that we were weakening on the "Germany first" policy. In any case the American Chiefs of Staff would be certain to put considerable pressure on the Commanders-in-Chief to endorse a more forward policy in the Pacific.'[35] Churchill needed no urging. He was feeling politically vulnerable over the inactivity of the sizeable British forces in India and had already prevented Wavell visiting Australia on his way to London at least partly out of fear that he might support Curtin's call for more Allied resources.

Churchill was caught in a cleft stick. As he informed the War Cabinet on 29 April, the American commanders in India and China, Generals Stillman and Chennault, had accompanied Wavell's team to London and were going on to Washington. It would be 'awkward', Churchill said, if the British officers refused to go with them. On the other hand, any such visit carried the 'grave danger that the main strategy of the war would be altered and greater emphasis put on the Pacific'. The only solution, Churchill argued, was for himself, his Chiefs of Staff and the Minister of War Transport to accompany them to Washington.[36]

When news reached the United States that Japan had executed captured American aircrew, Roosevelt's influential adviser, Harry Hopkins, alerted Eden on 23 April that the 'shooting of our fliers in Tokyo has started all the isolationist papers promoting the war in the Far East as against Germany'. Hopkins thought that the public still supported Roosevelt, and that even the 'Australians or Mme Chiang [the persuasive wife of the Chinese leader]' could not change the strategy, but the danger could not be dismissed.[37] The combination of MacArthur, Ernie King, the 'isolationist' press and the governments of Australia and China was potentially a formidable one.

Evatt was already in Washington trying to increase the allocation of forces for the SWPA. MacArthur had suggested that Evatt extract a commitment for sufficient aircraft to more than double the size of the RAAF. Even as Evatt, his wife and an American nurse were flying across the Pacific in a converted Liberator bomber (assured by MacArthur to be the 'best and safest' aircraft available), Roosevelt signalled that Evatt's trip would be in vain. Although additional forces would be provided, they would not be sufficient to mount an assault on Rabaul but would, Roosevelt argued, be sufficient 'to preclude any serious attack on the continent of Australia'.[38]

Curtin was acutely aware that Washington believed Australia was only concerned with the unlikely possibility of invasion; he tried to rebut it in his reply to Roosevelt. Protecting Australia from attack was not his primary objective, Curtin maintained. Sufficient aircraft should be made available to keep the Japanese off-balance until the Allies were ready to deal them the final blows.[39] With Evatt, Churchill and Roosevelt all together in the American capital, Australia had the chance of having its case properly heard for the first time. The signs were, however, that Evatt's advocacy would be in vain.

9

Evatt Takes Off

May to June 1943

ON 3 MAY 1943 IN London, the British intelligence agent, Gerald Wilkinson, met Admiral Somerville, commander of Britain's much-depleted Eastern Fleet. Somerville was in the uncomfortable position of having to do too much with too little. Apart from defensively covering the Middle Eastern reinforcement route with his elderly battleships, he was also under pressure to go out and attack the Japanese. He was most definite that no part of his fleet should be transferred to Western Australia at present. All Japan could hope to do against Australia was a hit-and-run raid from the air or sea and this did not justify any transfer of British forces. In any event, Somerville confessed, his strength was simply not up to the task. His fleet still relied for its success on deceiving the Japanese, Somerville citing with relish 'the fake photographs of absent British battleships which we are making in India and selling to the enemy'.[1]

Meanwhile, Churchill was moving to head off the expected demands from Evatt for increased airpower in the south-west Pacific. On 19 April he had informed his War Cabinet colleagues that he had ordered a 'searching examination into the situation in the Pacific, with a view to establishing the extent to which the Australian demand for reinforcements is justified from the military point of view'.[2] Two weeks later, while crossing the Atlantic in the blacked-out *Queen Mary*, Churchill and his Chiefs of Staff discussed a paper prepared by his Joint Planning Staff in which the Australian argument for further supplies of aircraft was firmly rebutted. It acknowledged that Japan's strategy would be 'aimed at making her existing Sphere impregnable to Allied assault'. This is just what Australia wished to counter, in the hope that a sustained and timely hammering at the Japanese defences would cause them to collapse. Britain assumed that Australia still feared an invasion and the British argument was designed to counter this rather than deal with the question of keeping the Japanese off-balance.[3]

The danger posed by Evatt, in Churchill's view, centred upon his ability to mobilize those pro-Pacific elements within the American services, and it was to neutralize such forces that Churchill was travelling to Washington with his high-powered entourage. As he informed Stalin in a cable from the *Queen Mary* on 8 May, his talks with Roosevelt would concentrate upon the Allied plans for Europe and would also 'discourage undue bias towards the Pacific'.[4] Churchill was particularly concerned that the scarce supplies of landing craft and amphibious shipping would threaten operations planned for the European theatre. The operations in Tunisia and the planned invasion of Sicily had drawn off so much shipping from Britain that a cross-Channel operation was now impossible for 1943. This meant that the American troops accumulating in Britain would have to cool their heels until at least the spring of 1944. As Churchill realized, this would increase the pressure in Washington to divert even more American forces to the Pacific where they could at least grapple with the enemy.

Meanwhile, the British Air Ministry was trying to persuade Australia to accept superannuated British bombers for conversion to civil aircraft once the war was won. The proposal arose during the visit to London of an Australian mission investigating new types of aircraft suitable for production in Australia. This mission, under the leadership of Daniel McVey, head of the Department of Aircraft Production, returned with recommendations for the production of Mustang high altitude fighters and Lancaster heavy bombers, both British designed planes. The choice of British aircraft was at least partly due, as Bruce observed, to McVey's personal commitment to 'maintaining the link with the United Kingdom'.[5] But there was a deeper significance for Australia in the report of the mission.

The production of Lancaster bombers in particular reflected Australia's view of her position in the post-war world. The bombers would not be ready until at least 1946, by which time the war would probably be over. However, the Lancasters would extend Australian power over the adjacent Pacific region and increase the independence of her defence industry. The bombers were also a partial repudiation of the country's pre-war reliance on the Royal Navy. During a press conference, Curtin said that Australia had to be mindful that the war might end soon and therefore she must choose an aircraft for production 'the machinery and dies for which could without much trouble be turned to the production of civil aircraft'.[6]

In 1941 Menzies had promoted aircraft production in Australia, intending it to be the precurser of a post-war automobile industry controlled by British capital. Such an industry, Menzies believed, would power the Dominion's economic development.[7] The Labor Government's plan was similar but more grandiose: an aircraft industry established in wartime would persist in peacetime and lay the basis for the country's post-war power and prosperity. The Dominion might also avoid in future having to go cap in hand to Washington or London, as Evatt was then doing, to secure sufficient aircraft for its defence forces.

British authorities were keen to support the Australian programme and were obsessed with the idea that the Americans were stealing a march on them in civil aviation. Much time was spent debating the issues of civil aviation in the widespread belief that it would comprise a key sector of the post-war economy. It was assumed that the rapid progress in aircraft development would continue until aircraft could challenge ships as carriers of people and goods. In January 1943 Hugh Dalton, President of the Board of Trade, pointed in alarm at the 'long start' the Americans had and the danger to Britain 'of being put by them at a great disadvantage in export trade, both of aircraft and of other goods'. He argued for an immediate diversion of some war resources to civil production.[8] The following month the War Cabinet agreed to new civil aircraft being designed immediately so that Britain would be able, after the war, to produce aircraft 'on a scale and quality in keeping with our world position'.[9]

In all the discussions of post-war civil aviation, the concept of internationalization loomed large. This basically meant that civil aviation would be removed from competition between nations, that aircraft production and operation, as well as the airfields themselves, would all be controlled by an international authority. Some proponents believed that an international air force would supersede national air forces. There was considerable support in Britain for the concept, both for idealistic reasons and because, although Britain had a virtual stranglehold on the world's airfields, she could not compete with the Americans in aircraft production. As Anthony Eden's secretary observed, the alternative to internationalization was 'cut throat competition between national lines, heavy subsidies for British lines and final defeat by the richer, more efficient Pan American Airways. Yet there are fools here in the Government who cannot see it and are for starting up British national lines again'.[10]

Australia also supported internationalization, for idealistic and practical reasons, but despite official support, particularly from Bruce and Evatt, there were important groups eager for Australia to go it alone, either as part of the British Empire or as an independent initiative. The Japanese thrust in 1942 had cut Australia's commercial air links with Britain, but the Qantas Empire Airways chief, Hudson Fysh, was soon seeking to restore his airline's pre-war hold on these routes.

When Britain proposed establishing a route from Ceylon to Perth using RAF planes, Fysh objected strongly, pointing to the 'danger to post war Australian Empire aviation should the Commonwealth lose its interest and equity in the operation of the Indian Ocean section'.[11] With the backing of the British High Commissioner in Canberra, Fysh flew to London to put the case for Qantas operating the route. To avoid arousing American suspicions, he was 'requested to travel as unostentatiously as possible'.[12] Fysh spent several months campaigning successfully in London, where he also backed moves for the immediate development of civil aircraft if Britain was to avoid being beaten by the Americans. Fysh warned of American encroachment in Australia and called for a 'strong and virile lead from the United Kingdom'.[13]

Fysh won back the right for Qantas to operate the Indian Ocean route, but there were few aircraft to operate the routes. Since the war began, Australia's stock of civil aircraft had dwindled to such an extent that the American air transport command provided the limited civil air service within Australia and maintained her links with the outside world. The Civil Aviation Minister, A. S. Drakeford, observed in April 1943 that Australia faced the awful prospect of the 'total loss of Civil Aviation in this country to American airline companies'.[14]

For a decade before the war, Australia had refused landing rights to any American airlines, but now, to her dismay, 'every day large United States airliners arrive in Australia operated on behalf of the U.S. Air Transport Command by personnel of American Companies such as Pan American and United Airline Companies'. To prevent these companies dominating Australia's post-war civil aviation, Drakeford urged that Australia either secure her own transport aircraft from overseas sources or else build them herself. He suggested that twenty-four such aircraft might be built from wood, plywood and plastic, but the War Cabinet thought better of it, opting instead to approach MacArthur for an allocation of American aircraft.[15]

Shedden informed MacArthur of the forthcoming arrival of the Lancaster bomber and its British sales mission. He was being careful to avoid any dispute with MacArthur, and also believed that MacArthur sympathized with Australia's position, and that the American general had 'gone native' during his Australian tenure.[16] MacArthur responded sympathetically, informing Shedden that Australia's civil aviation should be left at the end of the war 'in at least the same condition, insofar as the number of up to date aircraft was concerned, as it was at the beginning'. At the same time, he warned that 'powerful influences were at work in regard to the post-war civil aviation position and these may have exercised some influence on the refusal of the request for transport aircraft from America'. In fact, MacArthur opposed the production of such aircraft within Australia in case it used resources that might otherwise be better used for his command.[17]

MacArthur's word was almost holy writ in Canberra. He and Curtin had a cosy, symbiotic relationship. Some of MacArthur's prestige as the 'saviour of Australia' rubbed off on Curtin to his political advantage, while MacArthur liked a docile Dominion over which he could exercise his military reign and where he could construct a reputation that might eventually stand him in good stead in Washington.

In trying to boost British aircraft interests in Australia, London had to be wary of American sensitivities. If Australia decided in favour of Lancasters, the bomber that had carried the sales team would be presented as a gift. Because of the difficulties of a flight across the disputed reaches of the Indian Ocean, Britain would have to seek American permission for the plane to land at Hawaii for refuelling. In order to disguise the purpose of the mission from the Americans, Australia was asked to request the aircraft 'to assist in the raising of subscriptions for the next War Loan'.[18]

Bruce suggested that Evatt make the request to the American Government when he was in Washington. The plan almost came unstuck when Evatt asked suspiciously 'under what circumstances [the] loan of the bomber has been obtained from [the] British Government and how the idea originated'. He observed that, if it were to be used for publicizing a war loans campaign, it 'seems to me to be far too early as there will be no loan appeal for several months'. Bruce immediately alerted Evatt that the real reason for the Lancaster's flight was Britain's 'desire to show the flag in Australia in relation to transport aircraft'.[19]

Evatt needed no further prodding. W. S. Robinson had already bombarded Evatt with advice about the importance of air power underpinning Australia's place in the post-war world. In a paper submitted to Evatt in early June, Robinson argued that Australia's future survival

> depends more on strength in the air than on any other factor. A thousand million coloured people are her nearest neighbours—

> but a few hours flight from our shores. Not only for our defence but for the maintenance of our economic activities we *must* have efficient aircraft in sufficient supply—we *must* be able to build them and we must have trained men to fly them.[20]

Robinson also believed that Australia must retain for her own use the 'vast oil reserves in New Guinea' to fuel the large air force and the extensive civil air services that he envisaged would link various parts of the continent to each other and to the outside world. 'It is not too much to say', wrote Robinson, 'that the areas of Australia's geographical markets after the war which it will be necessary to cover by air will be twice as large as the Continent of Australia is today'. Defending this area would require a greater degree of Australian self-reliance. According to Robinson, the most important lesson from the war was that 'we cannot rely on others to make and sell us aircraft when they are in danger themselves'.[21]

Of course subsidized post-war aircraft production would benefit Robinson's metal interests, but even so, his criticism of the British industry was well made. Even in 1943 the British aircraft production companies only grudgingly accepted the Australian industry and urged the British Government to keep all aircraft research and development work in British hands.[22]

Ironically, it was Churchill who had the clearest idea about post-war civil aviation despite his well-known aversion to considering peacetime matters in the heat of battle. He argued with some accuracy in June 1943 that 'there is no industry which will undergo a more intense and severe contraction than the aircraft industry'. Although he conceded that, 'for strategic and political reasons' Britain must develop 'large, efficient air lines binding the British Commonwealth and Empire together', he was careful to point out that these routes would be uneconomic. British aviation might make a killing in Europe, where Churchill urged that they 'should make a strong British effort to excel in the profitable European traffic'.[23]

While Britain was trying to give away one Lancaster bomber to induce Australia to build them in peacetime, Churchill was ensuring that Evatt would not get any such bombers for wartime use. On 12 May, the day after Churchill arrived in Washington, Evatt sent him a note setting out his plan to obtain sufficient aircraft for the RAAF, which was to be built up to forty-five squadrons by the end of 1943. The Americans were most averse to using it as a striking force and generally relegated it to the defence of base areas and the servicing of American squadrons. Nevertheless, MacArthur wanted to maximize the forces under his command and had suggested that Evatt secure American produced aircraft on allotment to Britain. If successful, this would mean that the Australian aircraft in his command would be increased without detracting from his American air forces.[24]

So desperate was Evatt to achieve such a commitment that he pitched his appeal in terms that were difficult to refuse. 'All that is asked for', he informed Churchill in a note on 12 May, 'is that the United States and the United Kingdom agree to equip the additional 27 squadrons over such a period and at such a rate as is regarded as reasonable'. Evatt followed up this note with a long talk with Churchill at the White House after which he confided to Curtin that he was 'hopeful of securing Churchill's support at any rate for a substantial part of the programme'.[25]

Evatt was to offer Churchill something concrete in return for his support. At that time, the vital relations between London and Moscow were strained after Stalin broke off contact with the Polish government-in-exile, based in London, because the Poles publicly, and correctly, accused the Russians of the notorious Katyn Forest massacre of Polish officers. In order to raise its own international profile, as well as to help Britain out of a sticky situation, Australia offered to represent Polish interests in the Soviet Union. Evatt put the proposal to Churchill in a meeting at the White House, adroitly linking it to his proposal for additional aircraft. He was pleased to find that Churchill was 'greatly impressed' with the Australian approach to the Polish problem.[26]

Evatt also requested that the Allied leaders consult Australia in making their war plans. Australian leaders had repeatedly learnt of war plans from the newspapers. Evatt readily admitted that the Australian tail could not expect to wag the British dog, but pleaded that the dog at least keep the tail informed as to its future movements. Citing the decisions at Casablanca, in which Australia had no part but which vitally concerned her, Evatt reminded Churchill that

> Good heart and high morale can best be maintained if the people feel that their government is fully consulted in advance of major decisions vitally affecting them. From the Empire point of view the public appearance of consultation is almost as important as the fact of prior consultation.[27]

In other words, token consultation would do so long as the public could be beguiled into believing that Australia was playing a full part in the councils of war.

Evatt's entreaty marked a switch in Australian tactics away from hectoring demands and fearful appeals. MacArthur had suggested the switch, claiming that it would be 'inadvisable' to repeat previous Australian demands 'unless specifically called for' by Roosevelt and Churchill. Accordingly, Curtin told Evatt to 'angle for an opportunity to represent our case rather than to awaken disfavour by a forcible intrusion'.[28] The silky voice of moderation worked. On reading Evatt's appeal for aircraft, Churchill instructed that 'We must do something for him. Observe that he has not asked for any timetable. Pray make me proposals'.[29] In other words, Churchill saw an opportunity to make a more or less meaningless promise to supply American aircraft on an extended timetable at the Allies' convenience.

Just two days after his conciliatory note to Churchill, Evatt called upon Admiral Somerville to discuss plans for the Eastern Fleet. As an impolitic afterthought to the discussion, Somerville said he was sorry the Australians had not remained in the Middle East long enough to participate in the final victory that was then being celebrated in Tunisia. With some astonishment, Somerville observed Evatt's response. In a report sent to Churchill, the Admiral claimed that

> Dr. Evatt immediately flared up and said this was the 'dirtiest crack that anyone could make'. I asked him to explain what he meant; then he embarked on a long rigmarole that neither I nor anyone else had the right to question Mr. Curtin's decision [to

> recall the 9th Division] and added once again that I had made a very dirty crack.[30]

Somerville's stance of injured innocence would have been at least partly contrived, but his remark was perhaps a further sign of the widespread sense of superiority among British officials.[31] Neither deference nor diplomacy (at least in the conventional sense) were Evatt's strong suit.

In vain Somerville tried to repair the damage by limiting their talk to the Eastern Fleet, but found that even this provoked a storm from the tempestuous Evatt, who now claimed that Britain had broken an agreement to maintain the strength of the Eastern Fleet at no less than five battleships and three aircraft-carriers and that, as a consequence, Western Australia had been exposed to Japanese attack. Certainly Western Australia was potentially vulnerable but no attack had been made and there was no agreement in the sense that Evatt claimed.

In his report on the incident, Somerville warned that Evatt was 'singularly ill informed of the general strategic situation, that he was anxious to pick a quarrel and that he took no trouble to avoid being offensive'.[32] According to Somerville, Churchill later expressed his approval for the manner in which the admiral had stood up to Evatt, but then sought to smooth over the dispute during a social function at the White House where he 'brought Evatt over to Somerville and proceeded to put their two hands together with his own over the top and said that they had all got to be friends in Washington'.[33]

Although Churchill wanted to placate Evatt and satisfy his vague appeal with an equally vague promise, he found that the Air Ministry head, Air Chief Marshal Sir Charles Portal, had practical objections to such a course. Portal had little inkling of Evatt's post-war vision of a virtual 'Greater Australian Co-Prosperity Sphere' enforced with air power. The Air Ministry could only examine the facts and these did not support Evatt's case, at least as far as Britain was concerned. Allied air power in the Pacific far outweighed that of Japan and was more than sufficient for the agreed holding strategy. As for the Cinderella treatment of the RAAF, Portal saw this as a matter for the Australians to take up with

Washington rather than London. He suggested to Churchill that Evatt should ask MacArthur to allow the RAAF to man a greater proportion of the aircraft allocated to the SWPA.[34]

On 18 May, the same day that Portal was urging Churchill to reject Evatt's request, Evatt claimed that the proposed expansion of the RAAF had been 'committed to my charge by the Australian Government and I specially ask that you enable my mission to be fulfilled before your present conferences terminate'.[35]

Churchill was certainly anxious to prevent Evatt allying himself with the 'Pacific first' elements within Washington. At a meeting of the Pacific War Council on 20 May, and with Evatt and Roosevelt in attendance, Churchill expressed his support for the RAAF's expansion. This was what Evatt had wanted to hear, but his legal background as well as his political sense demanded that it should be committed to writing if it were to have any effect. So he 'respectfully' called on Churchill to do just that and for the two Allied leaders then to make 'a suitable public announcement' that would, he claimed, 'create an excellent impression from an Empire point of view.[36]

For once, Churchill and the Australian Government were as one, mainly because Australia was demanding something that would not affect the 'Germany first' strategy. As Churchill saw it, any increase in the strength of the RAAF would subtract from the aircraft allocated to the American forces in the SWPA and there would be no increase in the overall aircraft strength of the Allied forces in the Pacific. In fact, it might release American fliers for the fight in Europe. Following the Pacific War Council meeting, Churchill informed Portal that he had,

> with the President's permission, strongly urged the supply of this small number of aircraft from the vast American production in order that the Australian fliers might take a larger part in the defence of their own country. I consider this is most desirable. The President promised to consider it sympathetically. Dr. Evatt was extremely pleased.

Churchill urged Portal to 'get this matter moved forward on a lower level, as I wish to clinch it with the President before I leave'.[37]

Evatt was elated. On 19 May he had watched from the gallery of the American Congress as Churchill addressed the assembled politicians and made his familiar promise to fight the Pacific war to a finish with all the resources of the Empire once Germany had been defeated. Churchill was at the top of his form and easily cast a spell over Evatt despite his failure to acknowledge in his speech Evatt's presence or the effort of Australia while nevertheless making 'a pointed reference to the presence of Mr. Mackenzie King [the Canadian Prime Minister]'.[38]

Evatt was also encouraged by a meeting on 20 May when Churchill assembled various Commonwealth representatives at the White House for a garden party followed by a discussion on the war. As the music of the military band died away, the counsellor at the Australian Legation, Alan Watt, watched with approval as Evatt reined in his impetuous personality to address the meeting on the issues facing Australia. According to Watt, Evatt 'spoke firmly but politely, and I thought that his remarks were well received by Mr. Churchill and had some effect'. For this part, Churchill redressed his omission of the previous day and, in the privacy of this Commonwealth conclave, spoke warmly of 'our beloved Australia and New Zealand' and of the splendid work of the 9th Division.[39] Evatt seems to have been impressed, informing Curtin two days later that the British Prime Minister was 'far keener on Pacific activities than during last year' and was 'really anxious to join with Roosevelt and push against Japan whenever it is practicable to do so'.[40]

In fact, there was no change of heart; Churchill's course was still set firmly against Germany. As he told a group of British correspondents on 23 May, the Japanese Air Force was weakening so that it had become 'possible to hold Japan aggressively until the time is ripe for all out attack'. Again he promised that, 'when the European war is over, Britain will send every available man against Japan to assist the Americans'.[41] It was this repeated promise that helped to captivate Evatt and that allowed the Allied effort against Germany to continue by offering a future effort against Japan. As we shall see, Britain never fulfilled her promise in the terms that Churchill repeatedly set out.[42]

Churchill also remained deeply antagonistic towards Australia after the bitter disputes of the previous two years. Churchill was careful to conceal his feelings from Evatt but revealed them to a colleague when the question arose of using the New Zealand division in the capture of Sicily. Using the New Zealanders, Churchill claimed, would provide

> a great opportunity for them to win honour, and the fact that Australia has failed us makes it all the more necessary. . . Of course, if they refuse there is nothing more to be said. They will then place themselves on the same level as Australia. But I do not think they will refuse.[43]

Churchill was right. The New Zealanders left their troops in the European theatre despite Curtin's severe displeasure.[44]

Evatt remained on guard lest Australia's needs be overlooked at the Washington conference. Accordingly, the day after his cautiously optimistic cable to Curtin, Evatt again approached Churchill asking that he intervene with Roosevelt to have the RAAF expansion programme formally approved. Writing from Baltimore where his wife was recovering from an operation, Evatt claimed that 'it would be heart-breaking to have the matter shelved further'.[45] Evatt again blamed British service officials for dragging their feet over the proposal as he had in London in 1942. Churchill responded favourably, informing Portal that he attached 'great importance to getting this settled now' and asking imperiously, 'What has happened?'.[46]

Evatt also requested a conference with Churchill and Roosevelt to settle the issue. Under his plan, the aircraft would be delivered over the next twelve months, with nearly two-thirds not due until the first half of 1944. So the RAAF would reach its maximum strength at a time when the Japanese would probably be located far from Australian shores; but Evatt had more than wartime considerations in view. To the consternation of the Dutch government, he had already spoken publicly in New York of Australia's desire to be involved in a 'partnership' with the Dutch in the post-war development of the Netherlands East Indies and to take control of Portuguese Timor from Lisbon.[47]

Although Churchill threw his weight behind Evatt's plan, the Air Ministry opposed any programme that would increase the level of aircraft in Australia. As Portal advised Churchill, MacArthur and the Australians seemed set on using the argument for an expanded RAAF as a cover 'to obtain a further overall increase of air forces in the Pacific for which neither the American nor the British Chiefs of Staff see any justification'. If Evatt was prepared to restrict himself to expanding the RAAF within the limit of the agreed aircraft deliveries, Portal was prepared to dispatch by the end of 1944 sufficient aircraft to satisfy Australia's request.[48]

Evatt had hoped to tie up the agreement before Churchill left Washington, realizing that it might otherwise be lost in the labyrinth of American bureaucracy. On the eve of Churchill's departure, Evatt asked W. S. Robinson to make a personal plea to Churchill. From his temporary office in the Australian legation, Robinson informed Churchill that

> Bert [Evatt] has just returned and told me of his most encouraging talk with you regarding aircraft. . . Knowing what this means to the Empire and the justifiable anxiety of Bert to have it completed immediately, I trust that you and the President will be able to complete it before you leave.[49]

Although Churchill wanted to satisfy Evatt's request, it was a matter on which the Americans had the final say. So, before boarding his flying boat for the journey to Newfoundland, Churchill told Evatt that he should direct his enquiries to Harry Hopkins, Roosevelt's close adviser, claiming that Roosevelt was 'very favourably disposed to your request'.[50]

Churchill had more pressing matters to attend to. While in Washington he and Roosevelt had agreed to share research into the atomic bomb, and he had diverted the Americans from switching the focus of the Allied war effort to the Pacific. He had failed, however, to convince the Americans that, once Sicily was captured, the Allied armies should descend upon the Italian mainland and knock Italy out of the war altogether. To do so, Churchill left Washington via Newfoundland and Gibraltar for Algiers, for talks with the American Supreme Commander in the Mediterranean, General Eisenhower. Churchill prevailed upon Roosevelt to send the American army chief, General

Marshall, to accompany the British group to secure a final decision on the issue. Churchill knew that Marshall had been about to visit the south-west Pacific, where he may well have become convinced of MacArthur's case for reinforcements. By getting Marshall to visit North Africa first, Churchill ensured that the general did not waver from his attachment to the 'Germany first' strategy. Although no final agreement was reached at this eight-day conference in Algiers, Churchill left satisfied that his view about eliminating Italy would prevail as soon as Sicily had been conquered.[51]

Meanwhile, in Washington, Evatt was frustrated. The formal agreement to provide the requisite numbers of aircraft had not been finalized despite the timetable for deliveries slipping back twelve months to the end of 1944. To put the Australian request for less than five hundred aircraft in some perspective, the British and Americans supplied Russia, which already had a huge aircraft production industry, with more than eighteen thousand aircraft during the course of the war.[52]

On 4 June, as Churchill was travelling back to London from Algiers, Evatt again complained that service officials were not accepting the wishes of Churchill and Roosevelt for supply of the aircraft. It would now depend, he said, upon the support of the American Army Air Force chief, General Arnold, who was off duty until the following week. In order to await Arnold's decision, Evatt postponed his departure for London, while imploring Churchill to instruct British service officials in Washington to support the Australian case more actively and so 'clinch the matter'.[53]

Churchill arrived back in London on the morning of 5 June after narrowly escaping death at the hands of the German air force, which shot down a flying boat taking the same route as Churchill from Gibraltar to Britain and killing, among others, the British actor Leslie Howard. Churchill, though, made a last-minute switch, abandoning his flying boat for a converted bomber that avoided the German interceptors. That same day in Tokyo, a grand official funeral laid to rest the ashes of Japan's naval hero, Admiral Yamamoto, the architect of the attack on Pearl Harbor, who had been caught in a similar assassination attempt by the Americans but had not been lucky enough to survive it. American intelligence had learned of Yamamoto's planned flight to Bougainville on 18 April, and ordered an ambush. His escort of Zero fighters was overwhelmed by eighteen American fighters which then turned upon Yamamoto's bomber, sending it crashing into the island's jungle-clad terrain.

Before reading Evatt's message, Churchill presided over a lunchtime meeting of his War Cabinet where he reported upon his prolonged trip. He informed his colleagues of his meeting with Evatt and of his hope that he 'had been of some help to Dr. Evatt in obtaining the aircraft which the Australian Government were anxious to obtain from the United States'.[54] The bulk of Churchill's report centred upon his success in keeping official American attention fixed upon the European war. Claiming that the visit had been 'amply justified', Churchill reminded his ministers that American public opinion 'was much more concerned about the war against Japan—it was almost true to say that the American public would be more disturbed if China fell out of the war than if Russia did so'. Churchill added that his idea of forcing Marshall to accompany them to North Africa rather than Australia had been a triumph. 'It seemed likely', Churchill crowed, that Marshall would 'return to the United States a convinced supporter of the projected operations in this theatre'.[55]

When Churchill finally did read Evatt's renewed appeal for help, he instructed General Dill in Washington to use his influence to produce an agreement on the RAAF expansion programme and thereby prevent any further delay of Evatt's departure for Britain. Dill immediately promised to do all he could to help Evatt since, he said, 'we are equally anxious to speed Dr. Evatt on his way'. The problem remained, however, that 'none of us can justify his claims on strategical grounds'.[56]

Since it was clear that Evatt would not leave without a piece of paper to wave in Canberra, Roosevelt finally relented and promised to supply 475 additional aircraft to Australia by the end of 1944 but with no commitment to specific types of aircraft or by which dates they would be delivered. Evatt accepted Roosevelt's promise gratefully, realizing that this rather vague commitment would have to suffice. When the Combined Chiefs of Staff allocated future Allied aircraft production between the various theatres, Evatt hoped Australia would receive at least some of the heavy bombers she had requested. Without such aircraft, as Evatt informed Roosevelt, the long-range striking power of the RAAF was almost negligible.[57] This meant that Australia might be prevented from partaking in the final blows against Japan and would be denied adequate power to hold sway over the south-west Pacific in the post-war period. Evatt's bombers would hold together the Australian Pacific empire that he intended to construct from the ruins of the European empires left by the Japanese.

On 12 June Evatt trumpeted his success, informing Curtin that 'I cannot tell you how relieved and proud I am . . . that I have discharged the sole mission entrusted to me by you in relation to aircraft'. He claimed that the initial American refusal of his request was only overcome when Churchill arrived in Washington and lent his support, despite the opposition of Portal who, Evatt confided, pointed out 'truly enough that my application was an indirect way of increasing the overall allocation to the theatre'. According to Evatt, it was only as he was about to leave for London that Roosevelt finally agreed to supply the aircraft and make them additional to those already committed to the theatre. Lastly, in order to maximize his political benefit, Evatt wrote that he was 'most anxious' that Curtin 'make a public statement in Australia that I have fully carried out the mission entrusted to me'. W. S. Robinson sent a similar signal to Curtin claiming a 'major Australian victory' for Evatt.[58]

At the same time, Evatt cabled to Colonel Hodgson, the former intelligence officer heading the External Affairs department, reiterating his call for Curtin to 'issue an appropriate[ly] worded statement marking the conclusion of my mission here and its complete success in relation to the outstanding object of obtaining equipment for the expansion of the RAAF'. Moreover, he suggested that Hodgson should 'discreetly' tell various key Australian political figures, as well as the editor of the influential *Sydney Morning Herald*, 'something of the fight I have had and of its ultimate success. I am happy to know that—whatever befalls us—something big has been done for Australia and the Empire'.[59]

This final comment was presumably a last political testament in the event of his flight across the Atlantic ending in disaster, as a disturbing number of such flights did. To Evatt's chagrin, political kudos was denied him; Curtin sent his congratulations but refused to make any public statement. Curtin had Evatt where he wanted him—far from political conspiracies in Canberra and entrusted with a mission that was, or so he would have thought, impossible to achieve.[60]

Evatt also claimed as a success Churchill's pronouncement at a Washington press conference that the war against Japan would be prosecuted with the same vigour as the war against Germany. This message was interpreted in Australia as meaning that resources would be equally provided to both wars. Even the normally sceptical MacArthur, while witholding final judgement on the statement's significance, argued that it 'cannot be ignored by those who made it'. In a week-long series of meetings at his headquarters in Brisbane with Shedden, MacArthur suggested that the political pressure on Roosevelt, whom he lambasted as 'utterly ruthless and unscrupulous', might reach the point where he would pull American forces out of Europe altogether in order 'to save his own position'. The presidential election of 1944, according to MacArthur's cynical assessment, would concentrate Roosevelt's energies upon gaining a military victory somewhere and avoiding too many casualties.[61] This might be good news for the war in the Pacific.

Following Shedden's return from Brisbane, and with Evatt's advice that he express cautious optimism, Curtin announced to the press that there had been a change in the Allied attitude towards Australia. The only problem now, said Curtin, was the 'physical problem of giving this theatre equality in strength'. As a consequence, Curtin claimed that it was now 'possible to see daylight in the ultimate result . . .'.[62]

Four days later in Sydney, Curtin met MacArthur who talked of attacking the Japanese stronghold at Rabaul and forcing the Japanese back to their strategic naval base on the island of Truk in the central Pacific. Although MacArthur assured Curtin that he wished Australian forces to accompany him as his campaign worked its way north towards Tokyo, he forecast that, with the fall of Rabaul, Australia would be able to

> resume a more reasonable and rational basis of national activity, more closely harmonised with its normal structure and post-war aims. It was important that the National effort must not be so devoted to war as to result in the Commonwealth being left behind in the post-war situation.

MacArthur told Curtin that Australia could alleviate its manpower shortage by reducing its land forces south of Brisbane.[63] As a result of this optimistic picture, and perhaps as a means of holding the Allied leaders to their supposed commitment, Curtin told journalists that the first stage of the Pacific war was over. Australia was safe from invasion and available as a base for future action. Australians should 'sit tight' as they waited for the 'second phase which involves certain limited offensive action'.[64]

On 12 June Evatt informed Churchill of his apparent victory in Washington and of his likely arrival in London three days later, when he would be 'anxious to have the honour of conferring with yourself'.[65] Britain's Air Ministry representative in Washington also warned of Evatt's imminent arrival in London and confirmed that the Australian had extracted from Roosevelt a commitment of additional aircraft for the Dominion. Although Evatt refused to show the British official the terms of Roosevelt's letter, Australia's Air Force representative did reveal the contents, which were promptly passed on to London. As the British representative confided, there was still much opposition within the American Services to Roosevelt's commitment.

With American military opposition and the vague wording of the commitment, Roosevelt's paper promise might not prove to be convertible into American planes. However, the British official warned Portal that Evatt had telephoned him after receiving Roosevelt's letter and had

> said that in the absence of any firm assignment of American heavy bombers he would be asking you [Portal] for Lancasters. On second thoughts he withdrew this and said he hoped I would do my best to get American heavy bombers for Australia. I made a non-committal reply. Nevertheless I think it possible that Dr Evatt will ask for British aircraft and in that case you will I am sure bear in mind that any such allotment especially of heavy bombers would of course prejudice our chances of securing additional aircraft for RAF.[66]

Of course, Portal needed no reminding on this point and Evatt would soon find that the political corridors of London and Washington were like halls of mirrors in which illusion often merged with reality to create a false sense of security. Once again, his lawyer's training overwhelmed his political sense as he sat in his aircraft rigid with apprehension about the flight but confident that Roosevelt's letter would compensate for the danger that he was forcing himself to endure.

10

Evatt in London

June to August 1943

BY MID-JUNE 1943, as Evatt was flying across the Atlantic to London, Australian and American troops were continuing to prepare for an eventual assault against Rabaul. MacArthur's plan was to fight a series of small battles to seize the whole north-eastern coast of New Guinea before encircling the Japanese fortress. Enjoying air superiority and moving with increasing boldness at sea, the Allied forces had the upper hand over the Japanese. However, such was the difficulty of the terrain in New Guinea, and the courage of the defenders, that each minor battle became a major struggle.

In Burma, the Japanese launched a pre-emptive strike against the British forces assembling for their own offensive and to establish defensive positions across the border in India. From Washington, Churchill watched in horror and embarrassment as his troops retreated in a sorry end to a campaign that he described privately to his army chief, General Brooke, as 'one of the most disappointing and indeed discreditable which has occurred in this war'.[1]

In the European theatre the crucial battle of the Atlantic, on which Britain's war effort depended, was won by mid-1943, when new Allied tactics led to the withdrawal of German submarines from the North Atlantic. As in the Great War, the submarine offensive had pushed Britain close to defeat. In July the Allies were building ships faster than the submarines were able to sink them. If that continued, the final victory against Germany was almost assured.

In Russia, the Germans were pushed out of the Caucasus region and faced a determined Russian offensive along much of the eastern front. Russian numerical superiority was taking its toll of German technical superiority. On 25 July, fifteen days after the huge Allied invasion force landed on Sicilian beaches, Mussolini was overthrown by an Italian government anxious to make peace with the Allies. Meanwhile, the build-up continued in Britain for the cross-Channel invasion of France planned for May 1944.

On 16 June, the day after Evatt's arrival in London, Churchill cabled to Roosevelt thanking him for his 'kindness in meeting my wishes by giving additional help to the Australian Air Force'.[2] At the same time, he agreed with Portal that Evatt must not be given any heavy bombers from Britain. Evatt had quickly pointed out at a British cabinet meeting that the bombing of Germany was possible because of Australian aircrew. There was much truth in this. Some 21 500 Australian airmen served in Britain during the war, of whom about a quarter were killed. The flow of such men was accelerating as Evatt spoke; more than 10 000 arrived in Britain between July 1943 and June 1944.[3]

Portal warned Churchill that Evatt was likely to use this as the basis for a 'moral claim for the supply of British heavy bombers'. Portal marshalled his arguments to resist the Australian claim: it would harm the effort against Germany, be a blow to the morale of Bomber Command and make it more difficult for Britain to obtain heavy bombers from America. Portal resented stories put about by Evatt after his previous visit, that he had scored a victory over the Air Ministry with the commitment of Spitfire squadrons which Churchill approved after rejecting opposition from Portal. 'I am sure', wrote Portal, 'that you would not wish the Australians to think that there is any difference between us about the allocation of British Heavy Bombers'. He requested that Churchill approve the answer he was proposing to give Evatt which was 'to say that we cannot possibly give him any'.[4] Churchill agreed, but suggested that the Air Ministry at least help Evatt get his American aircraft.[5]

Meanwhile, as Portal had foretold, Evatt suggested

during a stay at Chequers that Churchill commit British squadrons to the Australian theatre. He linked his request to the great increase in Australian aircrew for the war in Europe. Following this casually contrived suggestion with a letter on 2 July, Evatt proposed that Britain dispatch three additional squadrons to Australia—one Lancaster bomber, one Sunderland flying boat and one additional Spitfire squadron. Two of them could be Australian squadrons serving in Britain and one a British squadron. Evatt made clear that any such commitment would be subtracted from the aircraft promised from American sources, leaving Australia's net position unaffected. Evatt's request was 'intended to be a personal unofficial suggestion to the Empire's leader, upon which you [Churchill] are left free to take the initiative if you feel you can make a favourable response'.[6]

Portal remained adamant that no such commitment could be made. It would mean, he claimed, that there would be 'nothing left for Turkey, Portugal or Russia'. Of course, neither Portugal nor Turkey were at war, although Churchill hoped to draw them into it. Portal suggested that, if necessary, they include a Sunderland and a Spitfire squadron among the American allocation to Australia, provided that America sent equivalent aircraft to Britain. Portal knew that the Americans were retreating already from Roosevelt's commitment. General Arnold had already refused to provide the heavy bomber aircraft for the RAAF expansion programme.[7]

On 8 July Evatt told Curtin that the purpose of his London mission had been 'completely achieved' and that Australia's stocks stood high in the British capital.[8] He left London for Washington confident that his relationship with Churchill would result in the Air Ministry objections being overruled once again. When he heard nothing, Evatt advised Churchill that he would shortly be leaving North America for Australia, and hoped that 'it will be possible for you to tell me soon what I can put into my kit for Australia'.[9]

Evatt's air adviser in Washington, Air Marshal Williams, had informed him that the promised American aircraft were those rejected by the RAF and USAAF, including five more squadrons of dive bombers than Australia had requested. The United States was discontinuing such units, and the

RAF wanted no more of them. Williams pointed to the irony of Britain being allocated American bombers that she was only able to accept because there were Australian air-crew to operate them, while Australia received no bombers at all.[10] Evatt's great victory had been transformed into a farce. Australia would achieve a numerically strong force but one so ill-balanced that it would hardly be able to project its power over its own continent, let alone over the adjacent Pacific region.

Evatt received Williams's messages when he arrived in Washington, and his apparent success faded into relative insignificance. As Williams warned, it would be 'madness and a waste of valuable manpower' for Australia to accept such aircraft; they would provide an Australian force that would be 'incapable of little other than attending merely to the passive defence of its own possessions'. He urged Evatt to take the matter up with Roosevelt, and perhaps even with Churchill.[11]

Churchill was certainly keen to do something for Evatt. Although he had refused to safeguard Australia in the critical days of early 1942, Churchill now wished to restore the links of Empire that had been worn so thin by neglect. On 12 July he overruled the objections of his Air Ministry, citing the 'high importance for the future of the British Commonwealth and Empire that we should be represented in the defence of Australia and the war in the Pacific'.[12] Although it was too late to defend Australia, which now stood secure behind a solid barrier of American air and naval power, Churchill recognized a perfect opportunity to increase the British presence in the Pacific at minimal cost to Britain, while securing Australian aircrew essential for the bombing of Germany.

As Churchill pointed out to Portal, the token RAF Spitfire squadron sent out to Australia in late 1942 had 'played a part out of all proportion to the size of the unit'. He now suggested that another three RAF squadrons be dispatched to Australia, using forty or fifty of the 945 fighter pilots in Britain for whom serviceable aircraft were not available. The pilots would not be missed in Britain, and neither would the aircraft, since Churchill proposed to ask the Americans to subtract an equivalent number of fighter aircraft from their allocation to Australia and send them instead to Britain. If Churchill's plan worked, Britain would establish a significant presence in the Pacific with eight Spitfire squadrons, four of them with British pilots, and all without diminishing the war against Germany. 'It is my duty', intoned Churchill, 'to preserve goodwill between the Mother Country and this vast continent of Australia, inhabited by six million of our race and tongue'.[13] He soon found, though, that it took more than rhetoric to shift the Air Ministry.

The Air Minister, Sir Archibald Sinclair, questioned whether more RAF squadrons would have a similar political benefit in Australia, and claimed that from the 'military point of view their despatch would be quite indefensible'.[14] On 21 July Churchill's staff suggested a compromise to satisfy both Evatt and the Air Ministry while 'taking into account the political importance of carrying Australia along with us'.[15]

Meanwhile, Evatt was on his way, having collected his recuperating wife from Washington. He had taken up Williams's suggestion and appealed to Roosevelt to ensure that his promise of aircraft was put into effect in the spirit in which it was made.[16] As he waited to leave San Francisco on 23 July, Evatt confided to Churchill that he was 'anxiously hoping to hear from you'.[17] With Evatt's appeal in hand, Churchill once more assailed his Air Ministry for at least one or two Spitfire squadrons in addition to the extra aircraft proposed by the Ministry. As Churchill pointed out, his proposal was 'going less than halfway towards meeting the Australian requests' and would not include any of the Lancasters or Sunderlands that Evatt had requested.[18] Still the Air Ministry resisted, with Sinclair begging Churchill not to help Australia 'at the expense of Royal Air Force squadrons now closely engaged in the Mediterranean with a more formidable enemy than the Japanese'.[19]

Lord Cherwell, Churchill's scientific adviser, dissected Sinclair's argument and agreed that it was 'a pity to deprive ourselves of the most modern types of Spitfires when almost any fighter could cope with the Japanese Zeros', but was conscious that Britain 'must try to make a good showing in Australia and it is probably important for Evatt to have some trophies to announce before the election'.[20] Churchill, however, was apparently convinced that the Labor Government would be tossed out of office and suddenly saw a way of helping to ensure it by withholding Evatt's vote-catching election trophy.[21]

On 1 August, after Evatt arrived back in Australia and had again implored Churchill to 'please let me know whether you can help me along the lines discussed',[22] Churchill suggested to Portal and Sinclair that he offer Evatt two Spitfire squadrons on condition that Australia provide their groundcrew. 'Far better sentimental results will be produced in this', Churchill argued, 'than by giving even a larger quantity of Spitfires and pilots without the Squadron personality'. Churchill also agreed not to make any offer to Evatt until after the Australian election was over. Given the personal hostility between Portal and Evatt, it was unlikely that the Air Ministry would agree to anything that might allow Evatt to retain power in Canberra. Meanwhile, Churchill assured Evatt that he had been doing his best and hoped 'shortly to have something to report to you'.[23]

Evatt was probably more relieved to touch Australian terra firma after his dangerous trip than he was to receive Churchill's encouraging news. In fact, he had already exaggerated the success of his mission. In a national radio broadcast on 2 August, just nineteen days before the federal election, Evatt announced that he had convinced Churchill and Roosevelt to increase assistance for Australia. Without giving any details, Evatt assured his audience that they could 'rely absolutely upon Mr. Churchill's statement that unremitting pressure of an offensive character is being and will be employed against Japan'.[24]

Evatt also described Australia's secure and prosperous future, based upon the development of a local aviation industry that would 'provide industrial leadership in the South-west Pacific, where there was already a population of 130 millions'. Evatt suggested that Japan had done Australia a favour by the war. Because of the conflict, the 'lazy, peaceful islands in the Pacific' of pre-war times 'have become acquainted with the latest aircraft, with the means of mechanised warfare, and the products of industrial production. Demands have been created, markets have been made possible, and Australia has the machinery and resources to supply the demand.'[25]

Despite Evatt's vague assurances, the Labor campaign benefited from his timely return and his optimistic speeches, and the Government won a massive victory over its disunited and discredited opposition.[26] The Labor Government had majorities in both houses of parliament for the first time since 1916. The elderly opposition leader, Billy Hughes, resigned from the leadership. Menzies, who had been denounced by Hughes as 'the great self-seeker, the man behind the scenes in every intrigue, the fountain head of every whispering campaign, the destroyer of unity',[27] took over as leader and began the laborious task of constructing a credible opposition.

Although Sinclair had agreed to Churchill's proposal on 13 August, Churchill maintained his silence. The day after the election, Evatt, who was recovering from a bout of pneumonia, informed Churchill that 'you will greatly help me by letting me know the results of your plans in relation to aircraft, as to which naturally I am still most anxious'. It was only after the election results were clear that Churchill, with the agreement of the Air Ministry, finally informed Evatt of the two squadrons.[28]

Britain had achieved maximum impact at minimum cost to herself. Instead of the three complete squadrons requested by Evatt, one Lancaster, one Sunderland and one Spitfire, Britain was sending just two Spitfire squadrons without their groundcrew. Evatt's request for an increased flow of Spitfires was refused, which meant that an additional Spitfire squadron that Australia had formed for operations in New Guinea had to be disbanded.[29] Despite this, Evatt informed Curtin that, 'from the highest political and imperial angle, the result is eminently satisfactory'.[30]

In London some were promoting the idea of Empire as the core of Britain's post-war political life, but their voices were drowned out by those calling for Britain to concentrate on interests she had fought for, in particular the Middle East, which London wished to regard as a purely British sphere of interest. As for Churchill, he took a longer view of Britain's place in the world than many of his contemporaries. As a young man he had fought personally to seize and hold various part of the Empire, but the experience did not blind him to the fact that Britain had been great before the Empire and could continue to be great with its strategic position off the coast of Europe and its links with the United States. The Empire would buttress Britain's position in the world rather than provide its foundation. The war had certainly shown that it could not be held in its entirety for ever.[31]

As for the Empire in the Far East, Britain could not defend it and those who might otherwise seek commercial outlets in such distant parts had lost interest. As Wilkinson observed in March 1943, 'the British apathy . . . towards commercial opportunities of the Far East [was] rather in contrast to a keen interest by Wall Street Imperialists'.[32] Churchill was content to leave the Pacific to the Americans. Despite serious Anglo–American disputes before the war over the sovereignty of various Pacific islands, Churchill now maintained that it 'should be the basis of our policy that we would not object to any American proposals to acquire further bases in this area'.[33]

It was in civil aviation that the Empire bulked large since it provided the landing fields that might link the world in a web of British air routes and offered a market that might underpin the commercial success of the British aviation industry. In June 1943 a Minister at the Dominions Office reported that a parliamentary debate on civil aviation had made it 'quite clear that the House in general, and the Conservative Party in particular, are taking a real and lively interest in the Empire'.[34]

The following month, W. S. Robinson circulated a memorandum on post-war aviation among British officials which he claimed was in line with Evatt's views and which extolled the importance of air domination. Australia's future survival, wrote Robinson, 'depends more on strength in the air than on any other factor'. Australia was mainly a fabricator rather than a manufacturer of aircraft, and would have to do much more, Robinson argued, to avoid 'quick death and destruction' in the future. He pointed ominously to the fact that there were

> 1,100,000,000 of the coloured races lying above Australia . . . Today the ability of our enemies to produce and utilise the Light Metals is fortunately limited. But the powers of the coloured

> races in this direction are capable of almost unlimited expansion within a single generation.[35]

According to Robinson, Australia would have to establish a defence perimeter some 1500 miles from northern Australia covering Singapore, the Netherlands East Indies, New Guinea and the adjacent island chains. Canberra would also have to come to an arrangement with Wellington to cover a similar arc to the north of New Zealand. 'These are the areas', wrote Robinson,

> in which we cannot afford to permit those opposed to our ideals of life and to our ways of living to establish himself [sic] in the air. These are also the areas to which Australia and New Zealand must have free economic access if not economic direction and in the Government and control of which they must have a definite voice.[36]

Robinson's agenda for Australia reappeared in the Australia–New Zealand Agreement of January 1944, in which Evatt tried unilaterally to extend Australia's boundaries to encompass the adjacent resource-rich territories.

Like Robinson, Curtin also wanted Britain, America and Australia to combine as the 'policemen of the Pacific', taking over the defence of those islands that the controlling European powers had been unable to defend against Japan. He was even prepared to consider a post-war American military role in Australia's New Guinea territory.[37] Australia's leaders might wish to extend her connections and encourage a high American profile in the post-war Pacific, but isolationism remained strong within the Australian population, boosted, according to the head of the British military mission, General Dewing, by 'Anti-British Irish Catholics and [the] worst elements in the Trades Union organisation'.[38] This isolationist tendency had little influence on the formulation of Australian foreign policy, which sought to prevent a repetition of Pearl Harbor and its aftermath.

Robinson addressed the problem when he claimed that aircraft would allow Australia to dominate potentially threatening territories to her north and would also allow their resources to be exploited on a vast scale. Robinson pictured undeveloped regions abundant with 'tropical products, plus oil, timber, water power, minerals and metals' that would 'provide opportunities for the enterprising Australian which he has not had for more than a generation'. If Britain, America and Australia combined to exploit these resources, they could produce prosperity 'great enough to attract the millions of white people we so urgently require'. Latching on to Churchill's vision of an English-speaking alliance, Robinson called on the 'White Races' to put aside their national rivalry and combine their aeronautical resources to ensure the 'future safety of the world'.[39]

Australia would prefer to buy British aircraft, Robinson wrote, but only if they fitted Australia's needs and were competitive with American types.[40] Qantas Empire Airways chief, Hudson Fysh, who claimed in an interview with Beaverbrook at London's RAF club that, although the Australian people were keen to develop air links with the Empire, the Government took a broader view. Qantas certainly wanted to resume the route to London, Fysh said, but also wanted to fly to China, the Soviet Union, and across the Pacific to North America.[41] Australian needs were appearing in ways that would have been unthinkable even two years before.

Churchill recognized that, as a declining Imperial power, Britain should hive off those defence burdens capable of being borne by friendly powers such as America. Australia would also have to shore up neighbouring parts of the battered Empire, but Britain looked to India in particular to support the eastern Empire and conceal the relative decline in her world-wide power. Thus Ernest Bevin, Churchill's Labour Minister, proposed that India be built up to protect the region from Burma to the Persian Gulf. It would also provide a base from which to threaten the Russian underbelly, for Bevin already saw Russia as a post-war threat to Britain.[42]

However, if she was to have any say in the post-war structure of the Pacific, Britain would have to play a larger role in the defeat of Japan. British ministers recalled with trepidation the unrest within the British army following the Great War and the refusal of the troops to consider Churchill's plan to suppress the Bolsheviks in Russia. Similar problems might arise once Germany was defeated again, and if the troops refused to fight, the Empire might well be lost for ever.

This was the fear of the India Secretary, Leo Amery, who together with the War Secretary, Sir James Grigg, urged the War Cabinet to prepare the British people for the need to fight Japan to a finish. He also proposed that servicemen fighting Japan once Germany was defeated should be rewarded with better wages and conditions. Since many of them would be Indian, the morale of the Indian Army also needed attention.[43] Grigg wanted the British people to realize that their country's 'destiny as a great Imperial power in the future as in the past, demands that we British should recover, stabilize and develop our Imperial possessions in the East'.[44]

As War Secretary in 1919, Churchill had had the difficult task of keeping the lid on unrest within army ranks and the memory bulked large when considering the problem of demobilization. Churchill was also worried about Britain's discreditable war record in the Far East. With the Japanese pressing hard on India's eastern flank, Churchill considered dispatching Eden to India as Viceroy in June 1943, lamenting to his Foreign Secretary 'how bad it would be if after doing so well in the West we lost India and boggled the war against Japan'.[45]

However, Churchill had not been converted to the 'Pacific first' strategy. In mid-July, when the invasion of Sicily went off much better than anticipated, he moved quickly to ensure that the Allies invade the Italian mainland rather than divert resources to the war against Japan. The Allies had been unable to agree on the next step in the Mediterranean campaign. Now a decision was urgently needed and Churchill proposed that he meet with Roosevelt in Quebec in early August. As Oliver Harvey observed, Churchill was 'anxious to pin the Americans down before their well-known dislike of European operations except across Channel gets the better of them again, and they pull out their landing craft and send off their ships to the Pacific'.[46]

Churchill had announced that there would be partial demobilization in Britain after the war against Germany, allowing part of the armed forces and the munitions industry to return to peacetime activities. But Britain could not plan for this until she knew exactly what her contribution would be to the war in the Far East. Already Churchill was hoping that he would not be held to his promise to send a great army against Japan. As he informed MacArthur on 20 July, there simply would not be the shipping 'to carry more than a portion of our total military forces to the Far East'.[47] So Churchill began to scale down his grand promise to devote Britain's 'whole resources to the defeat of Japan'.[48] He hoped Russia would declare war against Japan and allow her territory to be used for a great air onslaught against Japanese towns and cities. Perhaps also at the back of his mind was the thought of the atomic bomb, the American development of which he had so recently approved.

Churchill confided to his Chiefs of Staff on 19 July that the Americans would be 'gratified at the interest we are taking in the Japanese war and at our making earnest preparations to undertake it, and the introduction of this topic as a major issue at "QUADRANT" [codename for the Quebec conference] will perhaps make other nearer decisions more easy'.[49] He had been criticized by the press during his previous trip to America for not taking the Pacific war seriously.[50] Now, by raising the Pacific war himself, he would dodge this criticism and keep the Americans on-side over Europe.

Churchill did not intend to translate the talk at Quebec into greater resources for the Pacific until Germany was defeated. In fact, by capitalizing on the Sicilian success and invading the Italian mainland, any major operation in Burma would have to be postponed. The amphibious shipping and tank-landing ships would be tied to the Mediterranean and, given the cross-Channel operations planned for May 1944, might perhaps never leave the European theatre. Churchill wanted more action from the forces already committed to India rather than commit more forces to the region.

Wavell, who had been shifted to India after being denied sufficient forces in the Middle East to beat Rommel, was now facing a similar rather ignominious move. This time, Churchill wanted to appoint him Governor-General of Australia, where he might oversee MacArthur from the comfort of Yarralumla.[51] When he suggested this to his Dominions Secretary, Clement Attlee, and that he planned to 'enlist Evatt, who is in the most friendly mood, behind the proposition' to appoint Wavell, Attlee quite properly reminded Churchill that Dominion Prime Ministers 'nowadays advise on appointment to [the] Post of Governor General' rather than having them imposed from London. But Churchill was adamant; Australian sensitivities would not thwart his plan to remove Wavell. He petulantly informed Attlee that there was 'no harm in my asking Curtin what he thinks and I will do so to-day'.[52]

Churchill's hopes were dashed when Curtin refused to accept Wavell, partly for fear of upsetting MacArthur who was determined to keep British military representatives, even superannuated ones, at arm's length.[53] As well as this Curtin wanted a 'Royal' to replace the dead Duke of Kent who had been appointed Governor General in 1939 but had never taken up the post.[54] Labor Party policy stipulated that the Governor-General should be Australian, but Curtin knew a royal appointment would convince the public that his government was not hostile to Britain and repair some of the damage to the Anglo–Australian relationship.

The choice of Governor-General was heavy with symbolism: an Australian would emphasize how much the Dominion's political life had matured during the war, and would have helped to re-establish her relations with the great powers on a more equal footing. With a British official in Yarralumla, it was no wonder that American leaders were confused about Australia's constitutional status, thinking her a colony in principle and practice.

When Curtin rejected Wavell, Churchill considered appointing a forceful and vigorous Supreme Commander to the Far East. The appointment would also establish British authority over the American and Chinese forces using India as a base for operations.[55] Churchill informed his Chiefs of Staff on 26 July that he wanted a 'young, competent soldier, well trained in war, to become Supreme Commander and to re-examine the whole problem of the war on this front so as to infuse vigour and audacity into the operations'. Despite the limited resources in the theatre, pessimistic planning must be avoided, he continued, for it would 'rightly excite the deepest suspicions in the United States that we are only playing and dawdling with the war in this theatre'.[56]

Churchill also wished to plan a bold stroke against Japan, not in Burma, but using Britain's advantage in sea and air power to strike at a distant point such as the northern tip of Sumatra. During a meeting of the Defence Committee on 28 July, his military advisers pointed out that such a stroke would have to be followed through in Malaya to capture the Japanese airfields. This would, in turn, make necessary the re-capture of Singapore, and resources were simply not sufficient to do all this until Germany had been defeated.[57] Game and set to the Chiefs of Staff.

Churchill persisted. A week later, at a meeting of the moribund Pacific War Council, he was still speaking of launching 'a stroke more adventurous and further flung' than an offensive in lower Burma.[58] Roosevelt had finally succumbed to Churchill's pressure for an August meeting. On 5 August Churchill sailed down the Clyde in the *Queen Mary*, heading for Canada determined to keep America in Europe and to settle the nature and extent of the British contribution to the defeat of Japan.

11

Quebec

August to October 1943

ON 6 AUGUST 1943, as the *Queen Mary* ploughed through the heavy seas of the north Atlantic towards Canada, Churchill suggested appointing Lord Louis Mountbatten to the Supreme Command of Britain's war effort in South-East Asia. Although Mountbatten was not Churchill's first choice for the position, he was perfect for his purpose. As the King's cousin, he would be a dashing commander, projecting an image of daring onto what Churchill angrily described as 'this decayed Indian scene'.[1] A Supreme Command under British control would also debar the Americans from independent action within India and the adjacent British colonies.[2] However, as Mountbatten would find to his dismay, he would have no more resources than Wavell for pursuing his allotted task. His war experience up until then gave little indication that he was even capable of fulfilling it.

Mountbatten had been a destroyer captain before being elevated to the head of combined operations where he had supervized the disastrous commando raid on Dieppe in August 1942. To make this royal figurehead militarily effective would obviously be a problem, as General Sir Alan Brooke realized when he observed that Mountbatten 'will require a very efficient Chief of Staff to pull him through'.[3] The most important consideration, though, was not

Mountbatten's lack of qualifications but, as Churchill wrote, 'whether the Americans liked the idea'.[4]

The second strand of Churchill's plan for the Far East featured the rather eccentric figure of Brigadier Orde Wingate who had led a commando group through occupied Burma, attacking the Japanese lines of communication. Wingate had arrived in London on the day of Churchill's departure for Quebec and was peremptorily ordered to join the *Queen Mary*. Churchill hoped to obtain American approval for Wingate's dramatic but low-cost operations and thereby cancel the major offensive that the Americans wished the British to mount in Burma. Churchill also planned to raise his north Sumatra scheme, disputing his military advisers' claim that it would necessarily have to lead to an invasion of Malaya.[5]

Although the Americans did not want the British to take a part in the Pacific, they had expressed concern about the lacklustre campaign against the Japanese in Burma which they believed was threatening the Chinese war effort and allowing Japan to concentrate its forces against the Americans in the Pacific. Washington suspected, not without cause, that Britain sought the recovery of her colonial territories rather than the final defeat of Japan. If this suspicion were not allayed, the Americans would divert even more resources away from Europe to the Pacific. Churchill instructed his Chiefs of Staff to reach agreement with him on proposed operations against the Japanese so they could present a united front at Quebec and convince the Americans 'that we mean business'.[6] But the differences between Churchill and his advisers were too deep to be bridged.

Britain had sizeable forces in India but communication problems prevented them going into action against the Japanese on a large scale. Churchill wanted a bold stroke against Japan but could not provide the means to launch it. A major land offensive in Burma, as favoured by the Americans and the Chinese, would take all the forces in India and possibly forces from Europe. Hence Churchill's idea of a limited but dramatic attack on Sumatra using a minimal number of troops, but to his chagrin, his advisers would only countenance such an operation if it were mounted with sufficient forces and within the context of a strategic plan that made military sense.[7] For nearly two years this fruitless debate went on, prompting General Ismay to declaim in May 1944, that the 'waffling' over Britain's Far Eastern strategy 'will be one of the black spots in the record of the British Higher Direction of War'.[8]

Although the British were divided on operations in the Far East, they did catch the Americans somewhat off guard in Quebec by their willingness to discuss Pacific operations and to support the American build-up in Britain for a cross-Channel invasion, now timed for 1944. At an earlier meeting with Roosevelt in Washington, Churchill ceded control of this invasion to the Americans, in return for regaining supreme command of Allied forces in the Mediterranean and having Mountbatten's appointment approved.

Meanwhile, the Australian Government accepted Churchill's declaration about the Pacific war being fought with equal vigour, little realizing that it would result in largely cosmetic change at the top. Australian forces were struggling to regain control of the north-east coast of New Guinea, before launching an Allied invasion against the Japanese bastion at Rabaul where some 100 000 troops protected this major air and naval base. Other Allied forces, including some New Zealand troops, hacked away at the Japanese defences in the Solomon Islands on the south-east approaches to Rabaul. All this activity was being overshadowed in the central Pacific, where Admiral King's ships converged on successive Japanese islands, unleashing a lethal bombardment before the marines waded ashore to confirm their capture with bullets and bayonets. Rabaul had been a base too tough to crack without the enveloping movements in New Guinea and the Solomons, but it was becoming a base not worth cracking. By March 1944 it was neutralized by intensive bombing and isolated by the Allied capture of the surrounding islands.

After his overwhelming election victory, Curtin confidently turned his attention to the shape of the post-war world and Australia's place within it. Ever since 1939, the Dominion had been pursuing three different strands in defence and foreign affairs. One strand stressed greater Australian self-reliance, albeit under the protection of one or other great power; another strand encouraged stronger links between Canberra and Washington; while the final strand was aimed at preserving and strengthening the ties with Britain. All three strands operated concurrently, whether under a conservative or labor government. Only the emphasis changed according to circumstances and the political colour of the government. Throughout 1943 and early 1944 Australia's capacity to exercise independent power within her immediate region was emphasized. The Australian people supported this mild surge of nationalism at the federal election despite conservative criticism that Curtin had turned Australia's back on Britain.[9]

However, greater self-reliance could not be achieved by a population of seven million people occupying a continent the size of Europe. Accordingly, a post-war immigration scheme was planned so that Australia would have a 'population requisite to its safety';[10] and this population would be white, in line with Australian 'principles and traditions'. Curtin had been prepared to import temporarily thousands of Asian labourers in early 1942 to repair Japanese bomb damage in Darwin and Port Moresby,[11] but that emergency had passed and the strict colour bar was reinstated at Australian ports. After all, that is what the war was all about as far as many Australians were concerned. When Curtin appealed for help from the United States, he had consciously pitched his appeal in racial terms with his supposedly poignant picture of Australia as 'the only white man's territory south of the equator'.[12]

In late July 1943 Macarthur had advised that the American Army was employing several hundred Chinese labourers in Brisbane and sought Curtin's approval to import a 'substantial number' of labourers from India and China for use in the islands and perhaps in Cairns and Townsville, where Australian labour was difficult to recruit and Australian troops had to be deployed in the canefields to ensure a sufficient supply of sugar to make ice-cream for the American troops.[13] This request cleverly placed the Labor Government in a quandry; it had been steadily reducing the manpower to support American forces and could hardly refuse MacArthur's request to import Asian labourers without undermining their own pleas about the Australian manpower situation.[14]

After deliberating for five weeks, the Government gave its approval 'subject to the proviso that Asiatic labour should not be employed on the mainland of Australia south of the 20th parallel, nor permanently anywhere on the mainland'. Moreover, the Government turned the request to their advantage by handing over to MacArthur those Chinese phosphate miners rescued in early 1942 from Nauru and Ocean island and then banished to the hostile heart of Australia to mine tungsten. For obvious reasons, these Allied citizens had not been grateful for their treatment and proved troublesome as workers. So they were placed in American custody for use outside Australia, the Government requesting that at the end of the war the Americans assume responsibility for getting rid of them.[15]

Curtin was also at this time exploring ways of strengthening the frayed links to the Empire, but the conception of Empire had changed. Curtin used the annual dinner of the United Commercial Travellers' Association as a platform to call for an Empire Council. No longer was it sufficient, argued Curtin, for Britain to 'manage the affairs of Empire on the basis of a Government sitting in London'. Instead, he said, the Empire must establish a 'standing consultative body with all the facilities for communication and meeting'. He envisaged an Empire of equal members consulting together to develop a common policy and meeting not only in London but doing the rounds of the Dominion capitals.[16]

Curtin's proposal was based on the Australian view that the Empire belonged to the British race, rather than to the Government in Westminster; therefore it was not inconsistent for the Australian Government to criticize the British Government while upholding the Empire. It saw the British government of the day as a transitory political entity, while the Empire was a racial brotherhood that would endure. Westminster, however, believed that the Empire was something that Britain possessed. In fact, the word 'possessions' was used interchangeably with 'colonies' to describe the territories under British control.

That the leadership of the Empire might pass to one of the Dominions, an idea not uncommon in Australia, was totally alien in London. Curtin's proposal for an Empire Council also failed to take account of the opposition such proposals always met from South Africa and Canada, neither of which was anxious to be seen supporting closer integration of the Empire. Curtin recognized that, during the war, Australia's industrial and military strength had grown while Britain's relative strength had declined. He therefore thought it logical that Britain should turn more towards the Empire for support, while Australia sheltered under the shadow of Imperial defence arrangements and constructed its 'sub-empire of the south seas'.[17]

Curtin was partially correct: Britain did seek greater support from the Dominions. But Curtin was wrong in assuming that Britain might climb down a step or two from her pedestal, nor was she prepared to have the Dominions climb up alongside her. This was emphasized symbolically by British leaders, particularly Churchill, who referred to the 'Empire', stressing possession, whereas Australians preferred to use 'Commonwealth', stressing fraternity.

In Curtin's view, the British shield was needed as much against America as against a resurgent Japan. In a confidential briefing to journalists on 6 September Curtin advised that he intended to build up the External Affairs Department as it 'would be necessary to appoint men to go into the islands after the Japanese had been driven out to watch Australia's economic and commercial interests'. He warned that Australia had to 'watch her interests very carefully' since there was already evidence that 'the Americans would not give up economic claims to some of the Pacific islands'.[18]

Curtin predicted that there would be a fierce economic war after the present military war was won and that Australia could not 'allow her economic position to be not known or misunderstood with a Pacific studded by bases occupied by half a dozen nations shut out behind tariff walls'. It is easy to disparage Curtin's fears in hindsight but many people expected that the pre-war economic stagnation would return. Had Curtin realized that Australia would enter the longest boom period in its history, his attitude would have been different, but he could only foresee a return to markets closed behind high tariff walls with Britain, Australia's biggest customer, tasting military victory but unable to afford the champagne to celebrate.[19]

In such a situation, Australia would have to seek new

markets closer to home in the resource-rich Pacific territories. Obviously, the United States would control many of these territories and to counter possible American expansionism Curtin called for Australia to enunciate her own Pacific claims. He felt that Australia would have to stand on the faltering shoulders of Britain after the war so that she could look Washington squarely in the eye in the event of territorial or trade disputes.[20]

As part of Australia's continuing attachment to the Empire, Curtin made it clear that Bruce would remain in London as High Commissioner, despite Churchill's antipathy, something that Bruce managed largely to conceal from Canberra. During October and much of November 1943, Bruce and Churchill were at loggerheads over Churchill's refusal to consult properly with the High Commissioner. When Bruce threatened to resign, Churchill successfully called his bluff, refused to meet him and forced him to make a humiliating climb-down. Only afterwards did Bruce inform Curtin of the conflict and then he glossed over its bitterness and the extent of his defeat.[21]

Relations between Evatt and Bruce had also been very poor, with Bruce convinced that Evatt and Robinson were conspiring to have him removed. During Evatt's visit to London in 1943 he and Bruce had engaged in a shouting match over who was the ruder, and in 1942 Evatt had accused him of being disloyal. There was a certain amount of truth in this. Although Bruce had developed a good range of sources for information in London, he privately admitted that he did not pass on to Canberra all the intelligence that he gleaned. Curtin was unaware of this, exclaiming that Bruce 'had been a marvellous man in the job and Australia had been well served by him'. Even Evatt now professed support for keeping Bruce in London, but by retaining this former conservative Prime Minister in London and pressing for a royal appointment as Governor-General, Curtin was closing off important options for change in Australia's external relations.[22]

The essential contradictions in Curtin's scheme for a 'practical fraternity' of the British Commonwealth were made plain within days of his announcement. Churchill did not consult Canberra about the Allied discussions at Quebec despite their having direct relevance to Australia and despite Evatt's plea that Churchill at least maintain the appearance of consultation with Canberra.[23] When Australia suggested that the British Commonwealth, rather than just Britain, be treated as one of the four Great Powers, the British War Cabinet rejected it as 'impracticable'.[24] In Britain's realistic view, the practical fraternity would be neither practical nor particularly fraternal.

Britain did not want to discourage closer Commonwealth co-operation, but she was not ready to recognize a relative diminution in her own power. But Curtin persisted, conscious that Australia needed the Empire for protection and as a source of population if her sphere of influence was to be carved out of the Pacific; he also believed that the Anglo–Australian relationship would become increasingly symbiotic.

However, Australia's ambitions exceeded her ability to achieve them. Evatt wanted to create a post-war shipbuilding industry; Robinson wanted the production of transport aircraft to be made a top priority; the RAAF wanted the manpower for an expanded air force while Blamey wondered how he could keep even three divisions in the field; Britain and America demanded foodstuffs from Australia; and MacArthur expected Australians to service his forces. There was also the daunting prospect of a British Pacific Fleet being based in Australia and placing intolerable burdens on her manpower.[25]

Despite the meeting at Quebec in August 1943, Britain's effort against Japan remained a matter of conjecture. British military planners, for want of something better, continued to favour an offensive in Burma. Churchill, for want of something easier, continued to favour the capture of northern Sumatra while insisting that it need not be the first stage in the recapture of Singapore. Far from it, he claimed, warning that

> Singapore would be regarded by the Americans as a purely British target and it is certainly divergent from the main line of advance into China. It is more probable that Singapore will be recovered at the peace table than during the war. For the present it should be ruled out altogether . . .[26]

Churchill's disclaimer does not ring true. He was obsessed with the disgrace of Singapore's surrender and determined that it be forcefully redressed. If he could win support for his Sumatran operation, the case for an attack on Malaya and Singapore could be argued afresh and with much more strength.

In fact, the British planners had already proposed to the Americans the recapture of Singapore instead of Rangoon, claiming that it would 'electrify the Eastern world and have an immense psychological effect on the Japanese'. The Americans, who were intent on recapturing their Philippines territory, suspected that Britain wanted to recover her Empire rather than defeat Japan. They pressed for an Allied commitment to defeat Japan within twelve months of VE Day, otherwise, the Americans warned, the Allies would face the 'serious hazard that the war against Japan will not, in fact, be won by the United Nations'. Both Allies sought to use Dominion forces in their favoured campaigns.[27]

Churchill warned his advisers not to make any agreement with the Americans at Quebec. If they did, he would refuse to take any responsibility for it and would refer it to the War Cabinet for decision upon their return to London.[28] This threat worked. Churchill was able to report to London that 'all differences have been smoothed away except that the question of the exact form of our amphibious activities in the Bay of Bengal has been left over for further study'. Exhausted after the conference, Churchill took a few days rest before accepting an honorary degree at Harvard University.[29]

Churchill had achieved at least some of his goals. Mountbatten's appointment had been approved, as had Britain playing a 'full and fair' part in the war against Japan once Germany was defeated. Britain's role was not spelt out, and Churchill continued to plan for a substantial demobilization of British forces once the European war was won. Churchill feared that the cross-Channel invasion might degenerate into the static slaughter of the Great War, but nevertheless the invasion was agreed upon for May 1944. The development of atomic bombs was also made a joint Anglo–American project, with secrets shared between the two partners. Only the Sumatra operation remained unsettled; Roosevelt sided with the British Chiefs of Staff and pointed out to Churchill, without success, that an attack on Sumatra 'would be heading away from the main direction of our advance to Japan'.[30]

Despite division over this issue, Churchill had succeeded once again in committing America to Europe while convincing Roosevelt of his sincerity in the Far East. It was not because of America's wasteful use of shipping resources in the Pacific that Britain postponed the Burma operations, as the official British naval historian, Stephen Roskill, alleged. Churchill would have been opposed to his army fighting its way back through the jungles of Burma, and the lack of shipping provided a convenient excuse. The Americans were little the wiser. As Admiral Pound privately observed during the talks, 'I think that the U.S. were very gratified by the interest which we were taking in Far Eastern operations'.[31]

While still in North America, Churchill had received yet another appeal from Australia, stressing the political benefits for Britain if she dispatched British troops to the Dominion. This time the appeal came from Churchill's High Commissioner in Canberra, Sir Ronald Cross.[32] It was, perhaps, a measure of the British disregard for Australia that Cross was kept in Canberra throughout the war despite the most bitter denunciations of him. One highly placed British official returned from a visit to Australia in early 1943, and 'drew the most appalling picture of Cross' ineptitude and quite frankly suggested that he had got to be removed from Australia'. The official claimed that Cross was 'a complete snob, disliked Australia and the Australians, and the only associations he had were with the so-called "best people" in Melbourne and Sydney'.[33]

Cross's suggestion was backed up by Sir Owen Dixon, Australia's Minister in Washington, during a meeting with Churchill on 10 September. Churchill made it clear to Dixon that no British troops would be available before the end of the European war, by which time he predicted that they would be used 'far to the north and west of Australia'.[34] Instead of troops in the Pacific, Churchill suggested to his War Cabinet that, with the imminent defeat of Italy, and the consequent reduction in Britain's naval commitments in Europe, she should build up the skeletal Eastern Fleet. Churchill wished to send these reinforcements to the Indian Ocean but pass them through the Panama Canal into the

Pacific Ocean where the Fleet could 'put in at least four months of useful fighting in the Pacific before taking up its Indian Ocean station'. It would magnify the British war effort against Japan and help satisfy Britain's 'obligations to Australia and New Zealand', and 'it would surely give satisfaction in the United States as being a proof positive of British resolve to take an active and vigorous part to the end in the war against Japan'.[35] Over the following months, Churchill's suggestion was gradually whittled down before being finally abandoned.

To Churchill's horror, the announcement of Mountbatten's appointment had raised such wild public expectations about future British operations against Japan that Churchill was forced to dampen them down. On 30 September the Information Minister, Brendan Bracken, met with press representatives to quell talk of an offensive in Burma. He asked Churchill whether he should have Mountbatten beside him during the briefing. 'Better not', replied Churchill who wanted to kill the speculation without dulling Mountbatten's keenness for action. At the same time, Churchill instructed that Mountbatten's 'Order of the Day', an uplifting statement for the forces in his command, must not be generally published because it would only 'draw more Japanese to this theatre'.[36] In the House of Commons Churchill stressed the long period of training required before an offensive could be mounted. 'I cannot too strongly emphasise', he warned, 'the importance of damping down all publicity about this theatre for at least three months'.[37] Mountbatten was undaunted. Ignoring warnings that his command would not have a high priority he exulted to a fellow officer: 'It's grand. They have promised me everything'.[38]

Mounbatten was to be disappointed, but there were good reasons for optimism in the Allied camp. On 3 September 1943 the Allied armies had crossed from Sicily to land on the continent of Europe for the first time since the expulsion of British Empire forces from Greece two and a half years before. Within a month, American troops had entered Naples. The end of the war looked closer than ever. Deceptively so, as Churchill reminded people, warning them of the many battles before Germany would accept the Allied demand of unconditional surrender. Conscious that Britain would be relying on the Dominions for greater support after the war,[39] the War Cabinet decided to call a meeting of Dominion Prime Ministers, the first since 1937.

In July 1943 Attlee had proposed that Britain promote the role of the Dominions in the proposed United Nations organization. 'Australia', he wrote, 'might play a leading part among the countries of the Pacific Region (after the World Powers) . . . if it were known that she would be backed by the full weight of the British Commonwealth, if the need arose'. Attlee also suggested that Britain dispense with the term 'Far East' as it was 'often taken in Australia and New Zealand to imply that that area is far from our thoughts and our plans'.[40]

As Dominions Secretary, Attlee was eager for the Prime Ministers' meeting to proceed since so many post-war issues, such as trade policy and civil aviation, demanded discussion. Alarmingly, the divisions between Britain and the Dominions seemed greater than those between Britain and the United States. Somehow, the Commonwealth had to develop a united view on these issues, although, as Attlee would discover, there was much to prevent the Dominions aligning with each other and with Britain on such issues as civil aviation.[41] As well as the future of the British Commonwealth, 'There is also', Attlee argued, 'the important question of the prestige which a meeting between the Heads of the Governments of the Commonwealth at this stage would have in the world generally'. In other words, it would emphasize to Russia and America, as their forces dominated the fighting and they jockeyed for post-war supremacy, that Britain was more than just a small island off the coast of France.[42]

'Imperialists' such as Amery proposed a strengthened system of Imperial preference to defend Britain against American economic domination.[43] When the Americans dragged their feet over the Italian armistice terms, the War Secretary, Sir James Grigg, wondered whether 'the Yanks are merely suspicious mutts or whether (what is more likely) they are busily laying the foundations for post-war commercial penetration. Sometimes I think that Winston gives in to them too much, and we are certainly far too polite to the Russians.'[44]

Churchill had never relinquished his free-trade beliefs and wanted trade barriers reduced after the war, and an expansionist economic policy. Australia agreed; the 1930s had demonstrated that the Empire alone could not provide sufficient markets for Australian exports. Although Churchill and Roosevelt had agreed on a reduction in trade barriers as part of the 1941 Atlantic Charter, it was not until July 1943 that Britain and the Dominions were able to agree that an expansionist commercial policy was preferable to a return to high trade barriers.[45]

Curtin, however, would not travel to London before April 1944, and was even reluctant about going to Brisbane to see MacArthur, which he insisted on doing by train. A plane trip to London was more than he cared to think about. He justified his reluctance by citing manpower problems and the 'important operations' proceeding in New Guinea. He then propped up this weak justification with a reason that Whitehall could have read as blackmail. 'I do not feel satisfied', Curtin cabled, 'in leaving Australia until the results of the decision to carry on the war in the Pacific with the same vigour as the war in Europe are evident'.[46] Despite this, the British War Cabinet pressed ahead with their proposal for a conference by January 1944.[47]

The war had been raging for four years and Allied victory seemed likely. Many predicted peace in Europe during 1944, although Churchill was unconvinced, worrying that the long-awaited invasion of France might miscarry. He feared a re-creation of the awful trenches of the Great War sucking the life out of another Allied army, and perhaps robbing him of victory against Hitler. Then, as Hitler's power waned, Stalin loomed over Europe, threatening to rob any Allied victory against Germany of its meaning. Meanwhile, the competition between Britain and the United States intensified as they sought to mould the post-war world in their own image.

Australia was more concerned with her own survival. The war had emphasized her isolation from Europe and the problems of seven million people trying to defend a continent. As the tide of Japanese invasion ebbed from her shores, Australia fervently wanted to prevent a repetition of 1942. A policy of greater self-reliance was called for, both industrially and militarily, although the protection of a great power would still be necessary. Britain, the likely protector, was needed to provide the security within which an Australasian co-prosperity sphere in the south-west Pacific could be developed. Australia would discover that Washington and London had their own plans for the Pacific.

12

Keeping America at Bay

October to December 1943

BY THE BEGINNING of October 1943 the Allies were tightening their grip upon both Germany and Japan. Naples had been taken by the American Fifth Army under General Mark Clark while Montgomery inched his Eighth Army up the 'toe' of Italy in support. The excessive caution of the Allied effort in Italy allowed scanty German forces to hold back superior Allied armies and even to dispatch several of their stronger divisions to Russia, where a series of Russian offensives had drained the strength of the German army and brought the Russian army across the Dnieper River to threaten Kiev. In retrospect, it is clear that Hitler was living on borrowed time, but it was not so clear then, as his forces gradually retreated, increasing in savagery as they went and gaining strength from shortened lines of communication.

In the Pacific, Hirohito's forces were out-manoeuvred as America avoided Japanese strong points to leap-frog from island to island. Allied sea and air superiority counted for everything. Churchill had tried to deter Japan from war by simple arithmetic, comparing Allied steel production with Japanese; now it was beginning to add up. In mid-September the Japanese high command had contracted its forces behind a line that ran from Malaya to western New Guinea and thence to the Caroline and Mariana Islands. They hoped that a smaller nut would prove harder to crack. In her factories and shipyards, America was assembling a sledge-hammer.

As October dawned, the Australian War Cabinet finally resolved to reassess the country's war effort and reconcile the impossible demands being placed upon the nation's manpower. The army and the aircraft and munitions industry were ordered to release forty thousand men by June 1944, with fifteen thousand of them being transferred to rural industry. The men were being sent back to the farms. The plan to build up the RAAF to seventy-three squadrons, on which Evatt had spent so much time in Washington and London, was abandoned. Instead, Australia would make only that military effort sufficient to assure her 'an effective voice in the peace settlement'. It was an impossible balancing act that managed to upset everyone. The Australian public resented increased control when the security threat was receding, although Curtin did relax restrictions on horse-racing, but then introduced meat rationing. MacArthur objected to Australia limiting support for his forces, and both Britain and America wanted Australian food but resented men being transferred from the army to harvest and process the food. Curtin could only afford to contemplate such an impossible situation now he was secure in Parliament.[1]

The war effort was re-balanced to adjust to the change from a defensive to an offensive war. Australia no longer needed such strong garrison forces, and the success of Admiral King's island-hoping campaign across the central Pacific reduced Australia's role as the main base for the invasion of the Philippines and the final assault on Japan. Curtin also stressed the need to prepare the Australian economy for the post-war world.

Shedden made clear in a memorandum that Australia would respond to those Allied demands that best promoted the civilian economy. Thus the dairy industry, the products of which sustained American troops and Britain, could solve its manpower problem by taking men from the armed forces or from other war projects; 'the nearer the national economy can be swung back towards its peace characteristics of our specialised industries, the more simple will be the transition from a war to a peace footing'.[2] In other words, as far as Australia was concerned, it was time to start putting butter before guns. Britain had similar plans. She was stock-piling raw materials for her export industries and proposing large-scale demobilization once Germany was defeated. In order to avoid American criticism, such demobilization was referred to obliquely as 'releases'.[3]

When Curtin assembled his re-elected Cabinet on 22 September for discussions on proposed budget allocations, Australian priorities were clear when the Treasurer, Ben Chifley, savagely cut expenditure for the service departments. Despite resistance from Evatt and the service ministers, Chifley stood his ground, pointing to the economic dangers for Australia of a two-stage ending of the war.[4]

Britain was also worried about a two-stage ending to the war; it would be difficult to encourage her forces to fight if many of their compatriots had been released from the services. Moreover, British society would be reluctant to weather more hardships for the sake of a distant war in which many of them felt they had no stake. The Royal Navy, which would have to shoulder most of the burden of the Pacific war, was 'appalled' at the failure to plan for the switch of resources to the Pacific once Germany had been defeated, and warned the Government of the political repercussions. Churchill was increasingly unable to cope with the myriad demands of his office as illness and alcohol took their toll on his constitution.[5]

MacArthur's initial reaction to the Australian plight had been sympathetic. He observed to Shedden that the country had been 'bled white as the price of the assistance that was being rendered to them' and promised to 'nurse the A.I.F. Corps as much as possible'. But his attitude stiffened when Australia sought to reduce not only her garrison troops, but those servicing the American forces. MacArthur pointed out that, if the American forces had to maintain themselves, fewer American combat units could be shipped to the south-west Pacific. Curtin could not ignore this, and he was most averse to resisting the wishes of the Supreme Commander.[6]

As the Australian army was gradually disbanded to cater for the demands of the Americans, Blamey tried to stiffen Curtin's resolve by claiming that Australia would 'approach such a state of weakness of its armed forces that it may find itself very much in the same position as an occupied country'. It was up to the Australian Government rather than MacArthur, Blamey claimed, 'to determine the strength of the forces it is to maintain, and the extent to which its nationals may be employed by an Allied Force Command in Australia'. He urged that, in the face of MacArthur's intransigence, Curtin should approach Washington directly to have the matter resolved.[7]

Relations between MacArthur and Blamey were far from good. MacArthur accused Blamey of harbouring political ambitions and sought to confine his authority to the Australian mainland. When this failed, he created special commands under his own control to which he allocated all the American troops, thereby removing them from Blamey's control as Allied Land Force Commander.[8] Although Curtin claimed a warm relationship with MacArthur, his concern about post-war American imperialist encroachments in the Pacific must have affected his view of MacArthur. Certainly, from about this time Curtin began to be more forthright towards MacArthur. He still sought MacArthur's opinion but no longer deferred automatically to the Supreme Commander.[9]

On 8 October Curtin informed Churchill of Australia's problems, admitting that they had 'overreached' themselves by taking on more commitments than the available manpower could handle. Although Curtin was not aware of it, many countries experienced the same problem in both wars. The organizational structures drawn up at the beginning of the war were almost invariably over-optimistic and had to be scaled down when the flow of recruits proved unable to sustain them.[10] The German army kept diminished divisions on the Russian front rather than amalgamating them to keep them up to strength, and as a result their commanders overestimated the strength of their forces.[11]

Australia had to meet the various demands placed upon civilian and military manpower, and still maintain a military effort in the Pacific sufficient to guarantee her a place at the peace table. If this core of armed strength were inviolable, other commitments would have to be reined back. Curtin suggested to Churchill that Australians serving in Europe should return home to alleviate the manpower shortage. After all, he argued, the Australian effort was not just for

Australia's sake but for the Empire as a whole. As usual, Curtin did not demand the return of Australian forces but simply set out his conclusions and asked for Churchill's 'observations'.[12]

The tone of the cable ensured that the observations would be slow in coming. Three and a half months after receiving Curtin's cable, the British Chiefs of Staff submitted a reply for Churchill to approve. It held out little hope, calling on Australia to keep its army up to strength and refusing to transfer the three thousand Australian sailors attached to the Royal Navy, or the twelve thousand Australian air crew operating from Britain. As always, everything depended upon the war against Germany and, even after that, only those Australians immediately required for the war against Japan would be transferred to the Pacific. Not only would Australia's manpower shortage not be relieved by the return of its citizens, but the Chiefs of Staff intended to advise Australia that British units arriving in Australia would add to her manpower crisis. The Chiefs urged that the cable be sent in conjunction with advice about Britain's plans for the Pacific war, but this remained a matter of dispute between Churchill and his advisers,[13] so the cable was never sent.

Churchill was willing to accede to Curtin's request when it would no longer affect the war against Germany, and asked his service departments to advise on the consequences of withdrawing the Australians from Europe. There was no certainty that Germany would collapse in 1944, he warned, and therefore it was 'unwise' to lead the Australians to expect the arrival of British troops in the Dominion. Moreover, Curtin's plan to have just three divisions in action against the Japanese at any one time displeased Churchill. 'There is really no excuse for this', he wrote, 'and I should think the United States would complain'.[14] Churchill failed to understand the difficulties of jungle warfare, where disease took more casualties than the fighting. In fact, maintaining three divisions in the field would require six divisions behind the lines. The British Chiefs of Staff were more understanding, having experienced the problems of Burma, and described the Australian military effort against Japan as 'a very remarkable achievement'.[15] Churchill's continuing hostility to Australia deafened him to this praise, and

Curtin's attempt to solve the Dominion's manpower shortage continued to exacerbate the tension between Canberra and London.

Australia's plan to build bomber aircraft as the base of her post-war power was supported by Britain on condition that Australia reassume her dependency on the British aircraft industry, but America was opposed to Australia building anything larger than a small two-engined aircraft. There was sense in their objection since it would waste resources in wartime, but the peacetime benefit to America of restricting Australia's aircraft industry was doubtless not forgotten. Britain had largely concentrated on smaller aircraft while America had concentrated on larger types, such as the four-engined 'flying fortress' bombers which could be built more efficiently in North America, away from German bombing, and then flown across the Atlantic. It made sense in wartime but it also gave America a virtual monopoly on the production of large civil aircraft in peacetime.

When MacArthur was asked to agree to the manufacture of Lancaster bombers in Australia, he refused on the grounds of efficiency.[16] In 1942 this opinion would have ended the matter. By late 1943 MacArthur's word was no longer law. Daniel McVey, secretary of the aircraft production department and leader of the mission to Britain which had recommended the Lancaster, called upon Curtin not to relent on this issue. He claimed that

> great importance is attached in many quarters in Great Britain to the production of Lancasters in Australia. As an Australian, I am also convinced that it is of the greatest importance that we should preserve the most intimate possible relationships—technical and commercial, as well as social and political—with Great Britain.[17]

In fact, Curtin agreed with McVey: he fully supported the development of an Australian aircraft industry, believing that Britain and America were winning the aviation race and fearing for Australia if she were not independent in the air. So when the Lancaster proposal was discussed by the War Cabinet on 11 November, Curtin drew upon McVey's arguments to convince his colleagues to approve the production of fifty Lancasters.[18]

Curtin also sought approval from the Advisory War Council, whose members were drawn from all political parties, emphasizing the importance of the Lancaster programme to the post-war aircraft industry and stressing the 'potential use of these resources for the production of civil transport aircraft'. Not surprisingly, Menzies supported this move towards the Empire, particularly if it would allow the RAAF to enjoy operational autonomy and not be merely an adjunct of the USAAF. 'From the aspect of post war defence', he said, 'it was important that the R.A.A.F. should have an opportunity of functioning as a complete air force during this war'.[19] Four days later, in a radio broadcast on civil aviation, Menzies called for a 'strong and well-knit British Empire' protecting its aviation interests from American encroachments. The two Pacific Dominions and Britain had 'their own proper and vital flying interests in the South-west Pacific and in the Pacific generally', and, he added, 'shall not lightly abandon those interests'.[20]

In mid-October, when Evatt reported to Parliament on his visit to London and Washington, an Australian sphere of influence was explicitly laid out for examination by MPs. It was no good France being responsible for the defence of New Caledonia and Holland for the East Indies if they collapsed at the lightest touch by Japan. Instead, said Evatt, the region adjacent to Australia had to be regarded as 'a great zone of mutual interest which had to be grouped in the same defence area after the war, and be the subject of special efforts for economic betterment and aviation development'. 'Australia's predominant interests', he proclaimed, 'lay in the Pacific, where Australia had a leading part to play'. Pointing to the island territories to Australia's north, Evatt challenged their nominal European owners: 'From the point of defence of trade and of transport most of them could fairly be described as coming within an Australian zone'.[21] If Australia's interest in this zone were to be recognized, Britain would have to be used as a foil to American ambitions in the Pacific. British observers noted with satisfaction Australia's renewed interest in the Empire, without appreciating the underlying motivation.[22]

Curtin emphasized this interest by arranging the appointment of the Duke of Gloucester as Governor-General. His cabinet colleagues deplored the rejection of Labor Party policy and the lack of consultation with the Cabinet. Evatt supported Curtin, who claimed that the King's brother would act as 'a symbol in Australia of the whole Empire'.[23] The decision had the desired effect. It was widely acclaimed in both Britain and Australia, with one British newspaper suggesting that it 'effectively disposes of fears that Australia might cast in her lot with the United States'.[24] In another conciliatory move, Curtin roused himself from his sick bed to rebuke Arthur Calwell, the Information Minister, who had requested that cinemas play *Advance Australia fair* as the national anthem. Curtin said that he did 'not know of any anthem other than the National Anthem, God Save the King'.[25] Calwell's move was thirty years too early. It was not until Gough Whitlam's Labor Government in the early 1970s that the anthem was changed.

Of more practical importance to Britain was Curtin's ready agreement to the appointment of a British liaison officer, this time to MacArthur's headquarters, who could act as a personal and direct link between Churchill and the American general. Curtin had attempted to resist such a suggestion in 1942 when the British agent, Gerald Wilkinson, had been put forward for the post. This officer would be in addition to the British military mission in Canberra. His opposition then had been partly out of deference to MacArthur who opposed the idea, but also because he suspected Churchill's intentions and feared that Australia's connection, via MacArthur, to the decision-making councils of Washington and London would be diminished. Now he submitted meekly in the interests of the Anglo–Australian relationship and despite the opposition of Blamey, who considered such liaison officers would erode the Imperial relationship between the British and Australian armies.[26]

At the end of November 1943 Curtin and Shedden travelled by train to meet MacArthur in Brisbane, the sleepy, subtropical capital of Queensland. They informed him that the Government planned to go ahead with the Lancaster proposal, despite MacArthur's opposition, and despite the crash in Australia of the Lancaster on loan from Britain, and belated advice from the Dominion's air representative in Washington that the Lancasters were inferior to American bombers and not so easily converted for civil use. Curtin justified this uncharacteristic defiance to the Supreme Commander by admitting openly that Australia intended to develop a local aircraft industry for the post-war period. It was essential, Curtin said, 'that the largest type [of aircraft] should be produced, even if it might not be manufactured in time to be of much use in this war. It was the price which had to be paid for the development of the industry'.[27] That same day McVey warned Curtin to conceal from MacArthur the civil dimension of the Australian plan. Surprisingly, MacArthur agreed with Curtin and promised to do what he could to support the long-term project, while also promising to secure smaller American DC3 transport aircraft for immediate use within Australia.[28]

MacArthur was rather like Roosevelt in his approach to provincial politicians. He would express sympathy and appear to approve their requests but leave himself enough room to crawl out from under any commitment. His headquarters had, among its Air Force staff, various officers with interests in commercial aviation eager to claim the south Pacific for American aircraft manufacturers. So, while the DC3 aircraft would suit this purpose, building British aircraft would not.

Aviation was a particularly touchy topic between Washington and London as they fought for post-war pre-eminence. Five powerful American senators had toured the world, including Australia, to report on national defence, returning to Washington in September 1943 with strong recommendations about combatting economic competition from Britain. Claiming that GIs wanted jobs after the war, not Anglo–American collaboration, the senators alleged that Britain was using the war to promote its economic interests while the United States played the role of a 'global sucker', winning battles for Britain but not collecting the prizes. They cited the two million British soldiers in India confronting just sixty thousand Japanese in Burma and the virtual British monopoly of the world's airfields. Senator Owen Brewster observed bitterly that 'America has invested hundreds of millions of dollars in Britain, Africa, Australia and India, yet there is not a single spot in these countries where our planes can land after the war'.[29]

This was dangerous talk, exacerbating rivalry between the two Allied powers and threatening the unity needed for victory. Harry Hopkins, Roosevelt's close adviser, tried to explain to Richard Law, a junior Minister at the British Foreign Office, that the

> real root of the trouble was the Pacific. He did not know what the returned senators might say of their visit to the Pacific area and e.g. of their talks with Curtin. But it was the old doubts about British actions and intentions in the Pacific that were the real danger point.

Law backed up Hopkins' comments, reminding Churchill of Sir Owen Dixon's appeal for British forces made when Churchill was last in Washington. Once again, Churchill refused. His forces were hard pressed in the Mediterranean and there was the forthcoming invasion of France. He dismissed the arguments as 'politics, and bad politics at that'. Anyhow, how could Britain, Churchill asked, divert forces to the Pacific when the Americans were insisting that all their forces be concentrated for the invasion of France?[30]

By mid-1943 it was clear that Australia's preferred option in civil aviation—internationalization of aviation facilities—was impossible.[31] The competition between Britain and the United States intensified as Britain sought to retain her pre-war territorial advantage, warning Washington that America would retain no residual rights to airfields constructed by the Americans in British colonies.[32] This was what the American senators had found so objectionable. Britain was also preparing herself for the expected post-war boom in aviation by developing an infrastructure that could quickly be converted into a web of British-controlled peacetime air routes, serviced initially by converted British bombers and later by civil airplanes already appearing on the drawing boards.

From Canberra, Sir Ronald Cross tried to push this process along when he advised Churchill in August 1943 to use the transfer of British forces to the Pacific to justify establishing a British trans-Pacific air service 'on a scale if possible equal to those now operated by the United States. You will be aware that employees of the big United States Air Lines have been put into naval and military uniforms . . . and are in effect establishing their companies upon this route'.[33] Churchill, however, was not prepared to allow Allied competition over civil aviation to complicate the establishment of a structure of international relations which the three great Allied powers (or four if China was included) would dominate and thereby preserve the peace.[34] Churchill had made this clear during a meeting with Dominion representatives at the British Embassy in Washington on 10 September, arguing that it would be 'a mistake to regard the future of civil aviation in terms of a competition between the United States and the United Kingdom'.[35]

Nevertheless, competition over civil aviation continued to disrupt the relationship between London and Washington. Lord Beaverbrook, as Lord Privy Seal, was placed in charge of civil aviation discussions and pushed British interests hard. In December 1943 he rejected plans for Britain to develop a superior civil aircraft for trans-Atlantic traffic. Such a plane, he argued, would take six years to develop which would be far too late to beat the Americans. Instead, Beaverbrook proposed that existing bomber aircraft be redesigned for civil use. The choice, he claimed, was clear:

> It is between having or abandoning British Civil Aviation after the war. . . If we fail to provide British aircraft and British engines for the Dominions at the end of the War, then the leadership of air routes in the Empire must pass to the United States.[36]

Beaverbrook's fears were well-founded but the solution was not so simple. If new types of civil aircraft were developed, America would have the advantage for the crucial first years of peacetime, but if resources were ploughed into converting British bombers for civil use, the results would be inferior to the American planes and Britain would still be behind in the aviation race. If her Dominions would take the inferior aircraft, her problems might be reduced, but this could no longer be guaranteed, particularly as the Dominions had their own aviation ambitions as Britain's relative power declined.

Curtin failed to recognize that a weaker Britain would inevitably mean weaker Imperial bonds; he believed that the Dominions would simply share the leadership. For her part, Britain failed to recognize that the genie of Imperial disintegration could not be put back in the bottle. Thus, Leo Amery urged Beaverbrook to stifle the Dominions' ambitions by declaring all air traffic between Britain, Newfoundland, and the colonies as internal traffic. This would effectively prevent the Dominions from operating international airlines. Beaverbrook was more realistic. He told Amery that the Canadians were already doing big business via Newfoundland and could not be stopped, and added dryly that if he stopped the Canadians flying into Newfoundland, he [Beaverbrook] would 'never be able to go home again' to his native Canada.[37]

Britain's predicament was illustrated graphically when her High Commissioner in New Zealand sent a dispatch to London on civil aviation similar to that of Cross, calling for Britain to increase her aviation presence in the south Pacific to be prepared for post-war American competition. The proposal was rejected for a number of reasons, including the claim that aircraft were simply not available for 'a purely prestige service'. As for establishing a British service between New Zealand and North America, the Dominions Office admitted that 'no existing British type could carry a worthwhile payload on this route'.[38]

Britain's relative disinterest in the Pacific air route expressed her diminished power to influence developments in that part of the world. This had been reflected in Beaverbrook's recommendations to the War Cabinet, that Britain cede her aviation interests in South America in return for America giving Britain a free hand in Europe, the Middle East and Africa. As for the Pacific, Beaverbrook had displayed for Dominion representatives a map of the world showing a British air route from Sydney via Manila and Vladivostok to Edmonton, noting that Australia wished to establish a trans-Pacific route via Honolulu to America.[39] Beaverbrook's route would allow Britain to 'girdle the globe'.

Beaverbrook's plan depended on the Americans being prepared to trade, but this they would not do while Britain maintained its territorial advantages and refused to acknowledge that America now controlled their relationship. Britain was placed in the awkward position of trying to compete with America and hold back the ambitions of her Dominions.

On 22 December Bruce tried to talk sense before it was too late, pressing Beaverbrook not to abandon his efforts to achieve a meeting with Washington; otherwise, Bruce warned, the American Government would be under political pressure to claim rights unilaterally over the military airfields it had constructed. This would tie up the Pacific and even lead to claims within Australia, where American construction corps had built airfields across the far north. But Britain was already moving to confront the Americans. On 9 December Whitehall approved New Zealand's surreptitious plan to use her weather reporting station on Pitcairn Island as a cover to ascertain if nearby Henderson Island were suitable for an airfield linking a possible post-war air route from New Zealand to Britain via South America.[40]

America also sought to re-establish her interests in the Pacific. Roosevelt was taking a close personal interest in the construction of a web of bases across the Pacific that would deter any future expansion by Japan. He also tried to appropriate the plan of Australian aviator P. G. Taylor and make it his own, dispatching the Antarctic explorer, Admiral Byrd, on an expedition to assess the potential of Clipperton Island, the key to Taylor's alternative trans-Pacific route that would bypass American Hawaii and link Australia to Britain.

Byrd took with him representatives of six American airlines aboard the *USS Concord* when he sailed from Panama to the Galapagos and Clipperton Islands on 5 September 1943. Upon his return, he wrote to Roosevelt to confirm that the President had been 'entirely correct about Clipperton and the strategic and commercial value of certain key islands in the Tuamotu, Marquesas and Society Groups, as well as about the feasibility of new air routes to the South Pacific'. Byrd urged Roosevelt not to allow the sovereignty of any Pacific island to be transferred without American permission. 'We have ahead of us', he wrote, 'unparalleled opportunities which may never come again that this war has developed for the United States'. The American Chiefs of Staff had also drawn up a schedule of Pacific bases from Alaska through the Philippines, south to the Solomon Islands then east to the Marquesas which they considered Washington should control after the war. They planned to avoid any stigma of imperialism by leasing the bases rather than annexing them and using the cover of the United Nations to do it. Roosevelt pronounced it 'excellent'.[41]

Meanwhile, Taylor continued his lonely campaign to survey the route via Clipperton, apparently unaware of the American survey. He attached himself to the RAF transport command, shuffling bombers from North American factories across the Atlantic to Britain. This allowed him to press his case in both Washington and London, which he did unremittingly. Eventually he came into contact with some RAF officers from the Air Route Planning section and, as Taylor subsequently described it,

> over a few beers in the local pub under Westminster Bridge they were very soon inspired with enthusiasm for the R.A.F route to the Pacific. In this tavern, rich with the smell of good ale, we drafted a signal which, with a smile in our hearts, we felt would produce the necessary results in Washington.[42]

Britain had gradually become determined to challenge the Americans over aviation, and soon Taylor would get his chance to prove the viability of his trans-Pacific air route.

Meanwhile, Curtin wrestled with the problem of Australian manpower while trying to place the Dominion's economy so that it could benefit from the resumption of peacetime trade. The Government had already decided in July 1943 to discontinue the attempt to produce tanks in Australia. Tanks might have been crucial to Australia's defence in 1942 had they been available, but they were of little use in 1943 against the Japanese in the islands; besides the Australian army already had more than enough American and British tanks. As Evatt found during his visit to Washington, better tanks could be secured from American factories and Australian manpower could be diverted to more essential tasks. On 14 October Blamey informed the Advisory War Council that Australian armoured divisions had been broken up into brigades, and it was being considered whether the tanks could be put into storage and the troops transferred to the infantry.[43]

It had been an expensive lesson. The three year programme, begun by Menzies in 1940 for political as much as military reasons, had absorbed the labour of some eight thousand skilled people without producing one worthwhile tank for the Australian army. The Labor Government persisted with it as much for post-war industrial development and future defence as for wartime needs. The Dominion's abandonment in 1942 had determined Australian leaders not to be found wanting again in defence. The tank programme was dropped when the Americans refused to support it any longer and manpower could no longer be found for all Australia's projects. The equivalent aircraft programme continued for similar post-war reasons and with little wartime justification.[44]

In October 1943 the Government decided to increase the supply of food for the United Kingdom and reduce the supply for MacArthur's forces. When MacArthur objected, Curtin at first justified his move by claiming that it was to make up for a cutback in American supplies for Britain. More importantly, Britain was Australia's main customer for primary produce and if she did not recapture her traditional share of that market before the end of the Pacific war, Britain might turn to alternative, and perhaps cheaper, suppliers. As a sign of the changing relationship with MacArthur, Curtin bluntly informed him that the primary reason for the Australian decision was 'the maintenance to some degree of markets which would be important to the Australian export trade in the post-war period'.[45]

Canberra was beginning to realize that Australian forces might not accompany MacArthur on his triumphant march back to the Philippines and beyond. If the troops were to be relegated to a secondary theatre or to garrison duties where they could neither win glory nor guarantee a voice for Australia at the peace table, they might as well be back on the farm or in the burgeoning factories. During a press conference on 17 November Curtin revealed that 'the trend of things showed that despite the Government's objections, the role which Australia would play [in future Pacific operations] was that of the "hewer of wood and carrier of water"'.[46]

Despite MacArthur's 'strongest protest' at the Australian decision, which amounted to a partial demobilization, the American general agreed verbally with Curtin that the final decision was Australia's and that he would comply with it. He also agreed to do what he could to help the Australian production of aircraft after Curtin insisted that the Dominion was pressing ahead with the programme in the face of his advice to the contrary. In a draft press statement describing their talks and meant for release by Curtin, it was claimed that Australia was entering a new phase of the war in which her role would be 'a mighty one' since the continent would be used as a base from which would be 'launched in due time one of the major offensives of the war'.[47]

The statement implied that the war effort was being built up rather than scaled down, with Curtin claiming that MacArthur was

> a vital factor in all decisions to be taken not only along military lines but those of commerce, industry and economics. In this broad field we have endeavoured to fix firmly general principles which will ensure not only a maximum war effort but the future transition to an equally successful after war effort. I am indebted to General MacArthur for the high statesmanship and breadth of world vision he has contributed to the discussion.

High praise indeed, and no wonder considering the statement was drawn up by MacArthur's chief of Intelligence, General Willoughby, before the talks and then handed to Shedden for Curtin to issue on their completion.[48]

In 1942 the statement would have been issued to the press without question. But now Shedden warned Curtin that it was dangerous to suggest that the scale of the Australian war effort had been 'fixed' with MacArthur rather than, as it had been, simply communicated to him by Curtin. Moreover, wrote Shedden, the reference to a 'mighty' Australian effort in the future 'might prove embarrassing at some time', while describing MacArthur as a 'vital factor' in Australian manpower decisions was 'imprudent in view of the widespread comment at present being aroused about American imperialism being furthered under the cloak of the Military Command in the South West Pacific Area'. Shedden submitted a completely re-drafted statement that made the 'fullest use' of MacArthur's verbal agreement to the Australian switch of resources in order to formally and publicly commit him to it. When it was finally issued by

Curtin on 3 December it announced that 'General MacArthur has expressed his full agreement with the general principles laid down by the Government' and that Curtin had 'assured General MacArthur that Australia's war effort, whatever shape it may take by this process of re-adjustment, will be the maximum of which Australia is capable'.[49]

Meanwhile, in Cairo, Churchill and Roosevelt were meeting with the Chinese leader Chiang-Kai-shek at a conference code-named Sextant. To Churchill's horror, when the Chinese pressed for an amphibious operation across the Bay of Bengal to Burma, Roosevelt agreed without consulting the British. Landing craft had already been diverted from Mountbatten's forces to the operations in Italy, and Churchill was obsessed with ensuring that the cross-Channel invasion of France would have sufficient resources to overwhelm the Germans and secure a lodgement on European soil. He also remained adamant that his cherished operation against Sumatra was preferable to foot-slogging through Burma. However, he was unable to prevent Chiang-Kai-shek leaving Cairo with the commitment from Roosevelt.

Following their meeting, Churchill and Roosevelt flew to Teheran for a conference with Stalin, who promised to join the war against Japan once Germany was defeated. The British and the Americans, both daunted by the prospect of defeating the Japanese, were relieved. Before missiles, distance was a formidable barrier to the successful prosecution of war. There was no guarantee that the American atomic bomb would be ready in time or that it would work in the manner the physicists predicted. The Allies had to assume that Japan could only be defeated by an invasion of the Japanese home islands with the millions of resultant casualties. So Stalin's offer was gratefully accepted. To Churchill's relief, it allowed him to convince Roosevelt upon their return to Cairo to cancel the commitment to China. As Eden reported to the War Cabinet, with the removal of this 'great obstacle', the conference 'ended with complete agreement and general satisfaction'.[50]

At Cairo, Churchill also tried to convince a delegation from Turkey to join the war, offering them substantial British forces if they agreed. This dubious plan to enlist the

Turks was probably a legacy of Churchill's thwarted ambition during the First World War to defeat the German empire from the south-east, a strategy that came to grief at Gallipoli and cost him his post at the Admiralty, blighting his career and his reputation. Churchill, according to a close observer, 'persuaded himself that he can bring Turkey into the war and keeps turning over in his mind the consequences, just as if it had already happened. He does not stop to ask what the Turks themselves are thinking'.[51] Once again, his hopes were dashed when Turkey, still fearful of Germany, refused to co-operate.[52] It was further evidence that Churchill would consider the wildest schemes in the Mediterranean in preference to any commitment in the Far East.[53]

As at Quebec, there was much rhetoric about the importance of the Far East but little of substance was announced. Australia was angry that the Allies had met and decided issues of importance to the Dominion without any attempt at consultation and little information after the event. At a meeting with journalists on 6 December, Curtin made it quite clear that Australia had been left out in the cold by her Allies, remarking that 'They don't tell us anything'. Australian newspaper offices received the public communiqué from the conference before the Government had sighted it.[54]

What most alarmed the Australians were the terms of the public communiqué, in which the Allies undertook to strip Japan of all her conquests and restore to China those territories taken from her by the Japanese. Moreover, it proclaimed that they 'covet no gain for themselves and have no thought of territorial expansion'.[55] This was not true of Washington or Moscow, and was certainly not true of Canberra, which was particularly angered by the news that Churchill had pressured Roosevelt to agree that the European empires in the Pacific would also have their territories restored to them. As Eden assured the British War Cabinet, Churchill 'had on several occasions said that we asked for no increase of territory for ourselves at the end of the war, but likewise we were not going to give any up'.[56]

Australia had hoped to create a regional empire from the detritus of the European empires that would shield her from future threats and allow for economic expansion. Even as the Allied leaders were meeting, Evatt had been planning his own conference with New Zealand leaders at which he intended to stake out Australia's territorial claims. Australia was also resisting Dutch attempts to establish an administrative structure within Australia, before re-capturing the Netherlands East Indies.[57] All this was now threatened by the conclusions of the Cairo conference.

13

Dreaming of a White Empire

January to February 1944

THE CAIRO CONFERENCE between Roosevelt and Churchill confirmed that the world was faced with two wars, not one, and that Allied strategy would concentrate on fighting the European conflict first. As General Ismay claimed with relief, Britain had resisted '*any* resources of any kind that were required to beat the Bosche being diverted at this juncture against the Yellow Man'.[1] Australians anxious to see a larger British presence in their region could take cold comfort from British plans to shift the weight of her naval effort from the Indian Ocean to the Pacific in early 1944. On 19 December 1943 the First Sea Lord, Admiral Cunningham, wrote to Somerville to inform him that 'towards the end of March a large slice of your fleet will go to the Pacific'.[2]

Three battleships and four destroyers left Scapa Flow off the far north coast of Scotland on 30 December. They were joined by two aircraft-carriers and three more destroyers and the whole squadron sailed for Colombo, arriving at the end of January 1944.[3] Even before it left British waters, Churchill had queried the cost of sending a fleet all the way to the Pacific. To sustain a modern fleet so far from any British base required a whole fleet of ships, known as a fleet train, shuttling supplies from Britain and elsewhere. At Cairo Churchill had managed to postpone Mountbatten's planned amphibious operations in Burma but now found that the

British naval squadron planned for the Pacific might be an equally heavy drain on limited British resources.[4]

At the beginning of January 1944, while recovering in Marrakesh from an almost fatal bout of pneumonia, Churchill was asked to approve a cable to the Dominions informing them of the tentative plans reached at Cairo with the Americans. He agreed, providing that these plans would not preclude an operation against Sumatra, his much-favoured operation Culverin, the first step back towards Singapore. The Chiefs of Staff admitted that dispatching forces to the Pacific would make Culverin more unlikely but argued that the Pacific strategy should take precedence over South-East Asia. As an added inducement they informed him of 'one final point for your private ear which we have mentioned to no-one'. That was to use Dominion land forces, buttressed with British forces and 'all based on British possessions', to demand that the command of this area also become British. Thus Britain would appear to be making a mightier effort against Japan than she was in fact preparing to do. First, though, the Dominions would have to be advised so that logistical problems could be investigated in Australia.[5]

Churchill was not convinced. He agreed with the Chiefs that Stalin's intention to enter the Pacific war once Germany had been defeated meant that Allied plans for the Pacific would have to be revised. The Chiefs wanted to shift the British effort from the Indian Ocean to the Pacific and join America and Russia in a joint thrust at the Japanese heartland, but Churchill wanted to use the Russians as a cover for Britain's diversionary operations such as the one against Sumatra. He instructed his Chiefs not to inform the Dominions until he had convalesced and been able to discuss the whole issue of British strategy against Japan.[6]

Meanwhile, the British admiral in command of the Australian Navy, Sir Guy Royle, had been informed unofficially by the First Sea Lord, Admiral Cunningham, that a British naval squadron would arrive in Australia in March 1944. Ever since Pearl Harbor, the Australian Navy had been awaiting the British Fleet and had made provision for berthing and dry-docking. The Navy was being squeezed by the manpower crisis, falling behind the Army and Air

Force in the scale of Australian priorities. Royle was therefore very keen to receive the much-vaunted British Pacific Fleet and restore the status of the Australian Navy.

At the end of 1943 Royle informed Somerville that a dry dock suitable for capital ships was being built at a cost of seven million pounds in Sydney Harbour and would be ready by July 1944. 'We should be flattered and pleased', wrote Royle, 'if anyone would make use of that' as well as the facilities developed at Darwin and Fremantle. Royle was coming to the end of his appointment and wanted the custom of appointing British officers to lead the Australian Navy to continue. The Australian Government, however, restricted the intake of sailors and appointed the Australian officer, Captain Collins, to command the Australian squadron at sea with a view to promoting him to Royle's post in twelve months' time.[7]

The Australian Government was still assimilating the lessons of the war and trying to formulate a scheme of post-war defence that would guarantee Australian security. Curtin was hamstrung by a labour movement mistrust of the pre-war system of Imperial defence. Curtin, with Shedden's help, set out to soften this attitude and re-capture for Britain the allegiance of most Australians.

Australia's post-war security, Shedden argued, would have to be based on three things: Australia's own defence effort; co-operation with the Commonwealth; and collective security under the United Nations once it was established. Shedden recognized that the Labor Party as a whole favoured the first and third options and that if Curtin was to mobilize support for the Commonwealth, he would have to stress the contribution that Commonwealth co-operation would make to a new world order. As for the pre-war isolationists within the party who had helped Labor lose the 1937 election, their arguments should be repudiated and the party persuaded to include support for Imperial defence within its policy documents.[8]

On 14 December 1943, in a major speech on foreign policy at a federal conference of the Labor Party, Curtain took Shedden's advice, artfully blending together an appeal to delegates to accept their new citizenship responsibilities in the prospective postwar world. 'The full expression of these responsibilities', Curtin argued, 'is to be a good Australian, a good British subject and a good world citizen'. To clinch it, Curtin reminded the delegates of the 'teeming millions of coloured races to the north of Australia' and that the Dominion must therefore be 'harnessed to other nations'. Under the aegis of Commonwealth co-operation and regional co-operation with the United Nations, Australia could establish her 'pre-eminent position to speak with authority on the problems of the Pacific and have a primary interest in their solution'.[9]

Curtin captured the isolationists, the imperialists and the internationalists within the party. His appeal to racism was an expression of a deeply-felt fear common to most Australians. It was based upon a fear of dispossession and an underlying feeling that they had yet to enjoy the sort of total control of their land that most nations take for granted.[10] In order to secure their possession and meet the future possibility of a hostile 'Asiatic bloc', Curtin planned a massive immigration programme to bring settlers from northern Europe and Italy. He rejected suggestions of a Jewish colony being established somewhere in Australia, although he stressed that he had nothing against Jews. He did, however, have something against Asians and these would be rigorously barred from the continent because of their 'antagonism to the white man'.[11] Curtin realized, though, that it would be difficult for Australia to sell manufactured goods to 'Asiatics and other colored people, [while] at the same time strenuously refusing them access to an empty Australia'.[12] To prevent any embarrassment with the Chinese allies, Curtin pleaded the exigencies of wartime to suppress debate.

The racial dimension of the war rarely rates a mention in most histories,[13] but it was a very real consideration for the men directing its course and shaping the post-war world. Churchill was as susceptible to racism as any Australian. As Lord Moran observed, when Churchill thinks of the Chinese he 'thinks only of the colour of their skin; it is when he talks of India or China that you remember he is a Victorian'.[14] Similarly, Admiral Somerville was 'disgusted' to find that British nurses at a hospital in Ceylon were treating wounded Indian servicemen. He immediately ordered it to stop and for the 'segregation of asiatics from Europeans'.[15] In 1941, Anthony Eden had objected to a plan for settling the problem in Palestine, observing to his secretary that 'the arabs are neither as black as you paint them . . . nor a Jew-filled Palestine so white a panorama. . . If we *must* have preferences, let me murmur in your ear that I prefer Arabs to Jews!'[16] In 1943 his secretary noted that Eden 'loves Arabs and hates Jews'.[17]

In September 1939 Beaverbrook had revealed sympathy with Eden's hatreds when replying to an Australian friend who had likened Jewish refugees in Australia to 'a cancer developing in our midst'. At the time, German armies were brutally invading Poland, where they would establish one of their most notorious extermination camps. Beaverbrook observed that Britain also had her 'share of the Jewish refugee problem' which he regarded as 'extremely difficult and disagreeable'. 'Let us hope', he wrote, 'that, if this war does nothing else, it will relieve us of this cause of anxiety about the future'.[18]

Curtin had more complicated causes for anxiety at the end of 1943. Bureaucratic inertia and resistance from interest groups were subverting the Government's decision on manpower. Instead of releasing forty thousand people from the services and munitions industries, a Cabinet sub-committee passed the matter back to Curtin for further decision. With harvest-time approaching, primary producers were becoming restive about their promised labour supply and, as Shedden warned Curtin, 'after nearly three months' delay, *you are personally going to carry the responsibility*' for deciding on cuts in the war industries. It was time to be 'direct and ruthless', wrote Shedden.[19]

Curtin found that MacArthur began to resist the Australian Government's attempts to limit the Australian manpower servicing his forces. Previously he had supported the construction of Lancaster bombers in Australia, but in February 1944 he informed Curtin that the matter was now out of his hands due to a change in American regulations that ruled such long-term projects out of court.[20] As another side of the cooling relationship between Curtin and MacArthur, and while Australian and American troops were battling with the Japanese for control of the Huon Peninsula in New Guinea, the Australian Government sought to forestall American territorial ambitions in the south-west Pacific and to establish her own power in the region. This quest for recognition of Australian interests had gone unrequited, but would soon demand a reaction when Australia and New Zealand combined to proclaim predominance over their region of the Pacific.

The Australia–New Zealand Agreement of 21 January 1944 was reached after a conference in Canberra between Australian and New Zealand Ministers. As the fighting ebbed away from their region, the Dominions feared that the post-war Pacific order was being determined in their absence, so they unilaterally declared their right to share in discussions. In direct contradiction to the Allied decisions in Cairo, the two Dominions declared at the ANZAC conference that the disposal of enemy islands in the Pacific 'should be effected only with their agreement and as part of a general Pacific settlement'. Moreover, no other Pacific territory should change hands without the concurrence of Australia and New Zealand, and the United States should not be able to retain control over those bases built on foreign soil during the course of the war.[21]

The Dominions would decide their own immigration policies whatever was decided by the United Nations. In 1919 Billy Hughes had resisted Japanese attempts at the Versailles peace conference to abolish immigration barriers based upon colour, and now the Australian Government was keen to pre-empt any Chinese attempts to challenge the 'White Australia' policy.[22] Australia knew that Roosevelt and Churchill had made promises to Stalin and Chiang Kai-shek at Cairo about the carve-up of Japanese territories at the peace table. The agreement in Canberra was a belated attempt to assert Australia's right to share in the shaping of the Pacific.

However, Washington and London heard only another shrill scream for attention. Britain's Dominions Secretary, Lord 'Bobbety' Cranborne, tried to put the Australian–New Zealand Agreement in the best possible light for Britain. On 2 February he informed his colleagues that the agreement had been reached without any consultation with Britain but that Evatt had claimed its *raison d'être* to be

> derived from anxiety . . . concerning United States attempts at infiltrating in non-American Pacific Islands south of the Equator and 'anxiety concerning similar tendencies in Australia and New

> Zealand'. A second motive, Dr Evatt said, derived from the view . . . that the United Kingdom Government tended 'to concede too easily proposals made by the United States of America in relation to the Pacific'.

Britain too, was worried by American encroachments in the Pacific where they impinged upon British sovereignty, as in Fiji. Cranborne was prepared to overlook Evatt's criticism of Britain and the risk to the Empire of such unilateral declarations, arguing that the assertion of defence responsibility by Australia and New Zealand 'may be extremely valuable when we come to arrangements for the post-war period'. Nevertheless, he claimed, the international conference envisaged by the ANZAC agreement threatened the prerogatives of the Great Powers in settling the peace, and must be discouraged. This must be done carefully since Evatt would 'resent anything that he may regard as grandmotherly restraint by the mother country'.[23]

The War Cabinet welcomed the Dominions' 'offer to share in defence responsibilities in the Pacific [as] a notable landmark' and instructed that the British response to it 'be made more cordial'.[24] It was, and it worked. The New Zealand Prime Minister, Peter Fraser, expressed his 'warm gratitude'. 'To have shown any sign of soreness', Fraser said, 'would have been a great mistake. It was most important to do nothing to antagonise Evatt, who was now "on our side"'.[25] However, both Britain and America opposed the international conference, thereby neutralizing a crucial part of the agreement.

In a message on 3 February, the American Secretary of State, Cordell Hull, asked Curtin not to press for a conference of Pacific powers for fear of causing disunity among the Allies if they clashed over the post-war allocation of Pacific territories. Hull suggested that Curtin discuss the issues personally in Washington *en route* to the forthcoming Prime Ministers meeting in London. Of course, Washington did not really fear disunity, since the various informal talks between America, Britain, Russia, and China would soon settle all the issues of sovereignty without Australia being involved. What America feared was that the two Dominions might disturb Washington's plans.

As Evatt realized, the battle for control of the Pacific was

After you: Prime Minister John Curtin escorts General Douglas MacArthur into Parliament House for an official dinner in his honour in March 1944.

Saluting the saviour: Curtin and MacArthur exchange pleasantries at a Parliament House dinner in March 1944 to mark the second anniversary of MacArthur's arrival in Australia. Seated from left to right: Ben Chifley (Treasurer), James Scullin (former Prime Minister), Frank Forde (Army Minister), Douglas MacArthur, John Curtin (Prime Minister), Nelson Johnson (US Minister to Australia) and Bert Evatt (External Affairs Minister).

Calm before the storm: Churchill interrogates General Ramsden, commanding Britain's XXX Corps, while the AIF Commander, General Sir Leslie Morshead, guards the morning tea on 5 August 1942.

In memory of the fallen: Dr Evatt prepares to broadcast from the Anzac garden at the Rockefeller Centre in New York during his visit in June 1943. Mrs Roosevelt, with posy, waits patiently.

Life on the officers' deck: a doubles match between officers and nurses aboard HMT *Nieuw Amsterdam* in February 1943 as the 9th Division crosses the Indian Ocean in the stately presence of the *Queen Mary*.

A diversion for the troops: a boxing match aboard ship as the 9th Division returns home from the Middle East in February 1943.

Calling the odds: a 'race meeting' aboard HMT *Nieuw Amsterdam* during the voyage home from the Middle East by the 9th Division in February 1943.

Brave new world: the mushroom cloud above Nagasaki, 9 August 1945.

Counting the cost: the charred bodies of a mother and her daughter caught by the blast of Nagasaki.

A cast of thousands: the Japanese Foreign Minister, Mr Shigemitsu (with walking stick), awaits his walk-on part aboard USS *Missouri* prior to the surrender ceremony in Tokyo Bay, 2 September 1945.

Where do I sign?: Japanese Foreign Minister, Mr Shigemitsu, signs the Instrument of Surrender aboard USS *Missouri* on 2 September 1945.

Making his mark at last: General Blamey signs the Japanese surrender on behalf of Australia while other Australian representatives and the American master of ceremonies, General MacArthur, look on.

being fought on two fronts—between the Japanese and the Allies on the ground, and between the Allies themselves in various secret discussions. Australia was being side-tracked by MacArthur in the military battle and was not involved at all in the secret discussions. Without consulting Evatt, Curtin assured Hull that the Americans 'need have no disturbance of mind as it did not appear reasonable that there could be an early conference arising out of the Australian–New Zealand discussions'.[26]

When Curtin met Roosevelt in April he blamed the ANZAC agreement on Evatt who had been, according to Curtin, trying to secure the 'future of the white man in the Pacific' and had drawn up an agreement 'in what may well prove to be an excess of enthusiasm'.[27] It is likely that Curtin had only agreed to the declaration in order to appease the anti-American elements within his party and was now denouncing it to Roosevelt to appease the Americans and thereby ensure success for his trip.

In February, Curtin had instructed Evatt to compose without delay a suitable reply to Hull's protest. Three weeks later Evatt submitted a draft reply in which he defended the Pacific conference as 'a helpful contribution to the maintenance of harmonious relations among the United Nations'. He reminded Hull that Australia had not even been informed of the Cairo decisions affecting the sovereignty of Pacific territories and that Roosevelt had been deciding these questions for some time. For instance, at the Pacific War Council meeting in Washington on 12 January 1944, the President had said that France 'should not have New Caledonia back under any conditions and that he believed that in this view Australia and New Zealand would back him up'. Far from it, wrote Evatt, both Australia and New Zealand now wanted it returned to France. This would interpose a European power between Australia and the Americans, thereby preventing the stars and stripes from flying over all the Pacific.[28]

Evatt insisted, in vain, that the conference should still be held, if necessary after the talks in London between the Commonwealth Prime Ministers, arguing that it was not to refute the conclusions of the Cairo conference. As Evatt told the American Minister in Canberra, he had proposed the conference nearly two months before the Cairo conference. The fate of the ANZAC agreement taught the Dominions a lesson in politics: they could not unilaterally claim the power to decide the future, even of their own region.[29]

In fact, General Blamey had prepared a paper for the ANZAC conference in which he disputed the recommendations of the Australian Defence Committee, which had urged that the 'best means of securing Australia from invasion is by taking strong offensive action from established and well defended forward bases'. These bases were to be established in territories stretching in an arc from Java through Timor and the Solomons to New Caledonia and Fiji. The committee, composed of Shedden and the Chiefs of Staff, ignored the central lesson of the war—that Australia could not rely on a distant great power to protect her from invasion. It rejected the United States as a replacement for Britain's protection and reaffirmed that the 'best assurance' of Australia's security was provided by a 'scheme of Imperial defence formulated and carried out by the members of the British Commonwealth in co-operation'.[30]

So Imperial defence remained the cornerstone of Australian defence policy, except that Australia would play a much more forthright role within it. This was a relief to Whitehall. As the British admiral in charge of New Zealand's tiny navy observed, 'the Dominions must play a far larger part in protecting British interests throughout the world after the war'. Dominion forces in the Pacific islands would be defending British territories as much as their own. It was with this that Blamey took issue.[31]

Blamey realized from bitter experience in New Guinea that it was no good the External Affairs department drawing Australian flags on nearby territories if the Dominion did not have the armed strength to hold them. Australian forces garrisoning the screen of Pacific islands considered necessary to hold for her defence could be by-passed by a future invader in the same way as the Allies were presently by-passing the Japanese island strongholds. According to Blamey,

> Australia should seek its protection by a means which may be anticipated will stand the test of time, and will allow

> considerable development of strength and the continued recognition of our place as a member of the British Empire. This envisages a closer degree of co-operation, and a closer alignment of common interests in the Empire than ever before in our history. This aspect is so paramount as to require no further elaboration.[32]

As for American bases in the Pacific, these should be encouraged, but only where they could not pose a future threat to Australia in the event of American hostility. Moreover, the Dutch should be 'wholeheartedly' supported in retrieving the Netherlands East Indies which, although of 'vital interest' to Australia, was 'beyond the capacity of Australia to exert any direct influence'. Lastly, Blamey disparaged the idea of an alliance with New Zealand, claiming that there was little of common interest between the two Dominions, despite their position in the Pacific. They would not stand or fall together but one after the other, with Australia being first in line. Thus, Australia's defence effort had to be concentrated at home rather than in the islands or in New Zealand. Blamey urged that 'a very close study of this position should be made before any commitment is made by Australia on behalf of New Zealand, or before we commit ourselves to common action or agreement with that country'. Curtin read this advice on 19 January. Two days later, he signed the ANZAC Agreement.[33]

In fact, Blamey and Curtin were in broad agreement about the need for Imperial defence and Australia's inability to defend the island screen to her north without help. In a statement on defence to the Canberra conference, Curtin had proclaimed the need for a strategic naval base such as Singapore and a fleet that could command the seas of the region. Since neither Dominion could provide such a base, or the fleet to go with it, 'Co-operation with the United Kingdom', Curtin said, 'is therefore essential'. Since it would be easier to reach agreement on Imperial defence than it would be to establish a system of collective security, which must await the creation of the United Nations organization, it was essential 'that an understanding should be reached as quickly as possible in regard to closer co-operation in Empire Defence'.[34]

The Australia–New Zealand agreement was not, as is

sometimes suggested, a trick pulled out of the hat by Evatt at the final meeting of the conference and signed by delegates too tired to notice its significance. Despite Curtin's attempts to disown the agreement, its terms accorded with statements by both Evatt and Curtin stretching back for many months. In that sense, the agreement was not an aberration in Australian foreign policy but its logical outcome. In any case, it was a resounding failure, since it depended on the willingness of Washington and London to accept its conditions. It also presumed that Australia and New Zealand had common interests, although as Evatt pointed out, the two Dominions were both primary producing countries,[35] but produced the same products and tended to compete for business within the same markets.

Meanwhile, there was such widespread ill-feeling in Washington about British motives that the British naval representative, Admiral Sir Percy Noble, warned Whitehall on 12 January 1944 that the Americans were beginning to argue: 'What is the use of giving Great Britain this or that when they don't seem to want to use them'. This anti-British hostility had arisen out of the Cairo conference. Noble reported that Admiral King 'quite openly said that it was his opinion that we were not trying our hardest . . . in South-East Asia'. Another American admiral had claimed that British was going slow in the Far East to cripple China and prevent its post-war recognition as a great power. This admiral had welcomed Mountbatten's appointment but now 'felt like other Americans that the appointment was a "piece of window dressing" on the part of the British'.[36]

Noble's report would not have surprised Admiral Cunningham—King's hostility to Britain was well known and both admirals were eager for Britain to get moving in the Far East—but Cunningham would have been taken aback by the report of the Admiralty's director of plans, Captain Charles Lambe, to whom he had referred Noble's messages. Lambe now informed Cunningham that the American view was 'understandable', and urged Cunningham to order the immediate use of the Eastern Fleet's aircraft-carriers in an operation 'against any target regardless of its importance'. Subject to King's agreement, a British naval task force should operate in the SWPA regardless of whether it had sufficient logistical support. Lambe argued that without some such dramatic action, the Americans and Churchill might combine to force the navy into the operation against Sumatra just 'because we *must do something* to save our faces'.[37]

Australia received news of this proposed British naval squadron when Admiral Royle informed the Advisory War Council that a task force comprising a battle-cruiser, two aircraft-carriers, four cruisers and twelve destroyers would move to the SWPA where it would be based at Sydney and operate under the command of American Admiral Nimitz.[38] On 5 February Curtin informed Churchill that Australia 'looks forward with great pleasure and keenest anticipation to the arrival here of the first increment of the Royal Navy' which would 'provide a strong uplift to feelings of Empire solidarity'.[39] Churchill, however, had already ensured at a late night meeting of his defence committee in London that the plans would be quashed. The previous day Churchill had returned to London from his convalescence at Marrakesh, and his military advisers brought him up to date on their plans for the Pacific war, observing that the naval commitment to the Pacific 'would . . . restrict operations in the South-East Asia theatre'.[40]

Churchill was furious. He denied that he and Roosevelt had agreed to these Pacific plans at the Sextant conference in Cairo, thundering that it 'was the first time he had heard of these proposals' and that he was 'dismayed at the thought that a large British army and air force would stand inactive in India during the whole of 1944'. Switching the British effort against Japan from the Indian Ocean to the Pacific, Churchill claimed, would be 'casting away the substance for the shadow'. The 'substance' was the operation against Sumatra which he claimed 'would provide an important diversion and contain substantial Japanese forces in the Malayan area'.[41]

Now that Stalin had agreed to enter the war against Japan as soon as Germany was conquered, the Allies could attack Japan from that direction far earlier than they could from a naval advance across the Pacific. Churchill instructed the planners to recast their proposals, based on the assumption that Russia would enter the war on VE Day, which could be any time after autumn 1944. When the Chiefs of Staff had planned to transfer resources to the Far East based on the assumption of a quick victory in Europe, Churchill had scotched the idea; now he argued *against* any such transfer of resources from the same assumption of early victory.[42]

This time, Churchill was not alone. The Labour leader and Deputy Prime Minister, Clement Attlee, wrote to Churchill on the day after the meeting, supporting what he described as the better strategy of an attack against Sumatra. A deep rift began to open between Downing Street and Whitehall as the amateur and professional soldiers battled over Pacific strategy. In Attlee's simplistic assessment, the Sumatra operation would be doing to the Japanese what the Japanese had done to the British in 1942. 'I cannot see', wrote Attlee, 'why we cannot play the same game as effectively, provided we act with the same ruthless vigour'. With political support from Attlee and other War Cabinet colleagues, Churchill tried to remove the issue from the control of his military advisers and make it an issue for debate 'as between Governments'.[43]

Churchill suggested that the naval task force should go temporarily to the south-west Pacific in mid-1944, rather than be the first echelon of an eventual British Pacific Fleet, but that Britain should demand in return American logistical support for the operation against Sumatra.[44] Despite an assurance from Admiral King on 23 January, that 'he personally had never gone back on the Sextant agreement, and that he contemplated the force being placed under Admiral Nimitz', Churchill informed his Chiefs of Staff that he refused to consider himself bound by the Sextant agreement, claiming that he had not been consulted upon it and though he had initialled it, he 'was not even aware at all of what had taken place'. It was 'pretty clear', Churchill wrote, 'that Admiral King is by no means anxious to have the force we offer and we ought not to press it upon him unduly in view of the logistic difficulties'.[45]

It was not, as most historians have suggested, that Britain was prevented from dispatching forces to the Pacific by implacable opposition from Admiral King. Certainly King was none too keen to have them if he thought he could defeat the Japanese by himself. But the real stumbling block was in Downing Street, where Churchill consistently refused to support a powerful British contribution for the Pacific.

Meanwhile, British forces languished in the Indian Ocean, receiving conflicting signals from different quarters in London and insufficient forces for any major operation. In Delhi General Pownall made the acerbic comment that 'there was no need to set up this command at all, with all the emphasis and publicity which, for political reasons, was put on it last September'. Although Pownall supported the Sumatra plan, he realized that it was a side-show to the main thrust against Japan and that it was 'most unlikely' that Mountbatten would ever be given the resources to do it. As far as Admiral Somerville, Commander of the Eastern Fleet, was concerned, the position was one of

> complete chaos . . . because no-one appears to know what the policy is to be. I get signals from A.B.C. [Admiral Cunningham] asking me to reduce personnel in view of our reduced commitments and at the same time get proposals from Dickie [Mountbatten] at Delhi in connection with operations on a considerable scale. When I tax Dickie with these obvious conflictions of policy he tells me that he has received instructions direct from the P.M.[46]

Even Mountbatten eventually realized that Churchill's bold words meant little. On 4 February Mountbatten wrote to Beaverbrook expressing profound disappointment at the collapse of all his plans that had been approved at Cairo. His letter was in vain. Beaverbrook simply instructed his secretary to 'answer at length and most agreeably because of course I am only too willing to give him the opportunity of engaging in correspondence with me in order to work out his disappointments'.[47] Realizing that he was heading a command that might never mount a significant operation, Mountbatten retreated from Delhi to the more comfortable climate of Kandy in Ceylon where he planned to establish his headquarters with some seven thousand staff, complete with a private band. The local commander in Ceylon was aghast, noting that the headquarters increased in size as its prospective operations decreased, and exclaimed in exasperation, 'what the general effect on prices and servants' wages here is going to be I tremble to think!'[48]

When Churchill's envoy to MacArthur's headquarters in Australia, General Lumsden, arrived in London with stories of disunity within the American ranks, Churchill trumpeted to his Chiefs of Staff on 13 February that there were two competing strategies put forward respectively by MacArthur and King. His talks with Lumsden, wrote Churchill, had made him

> more than ever doubtful of Admiral King's scheme. At the Marshall Islands the largest armada ever assembled carried 100,000 men to attack 4,000 Japanese. The waste of effort involved in this kind of operation is indescribable. I certainly have the feeling that Admiral King, although a most agreeable person, is the evil genius of this war.

The following day, Churchill jibbed at Admiralty proposals for the Pacific 'fleet train' of supply ships, instructing that 'no ships can be set aside now from merchant traffic for the purposes of the Japanese naval war'. He instructed the Defence Committee to re-examine the proposals in the light of the 'more promising plan put forward by General MacArthur' and involving the British forces in India.[49]

That evening, Churchill and Eden met with the Chiefs of Staff, Lumsden, and a representative from Mountbatten, to thrash out a strategy. Mountbatten's envoy brought with him from Delhi a plan to mount an amphibious assault on Sumatra, Operation Culverin, in the knowledge that this was Churchill's preference, but made the mistake of requiring more men and resources than Churchill was prepared to concede. Churchill wanted Culverin to proceed but not at such a cost. Still, Churchill stubbornly argued its merits, claiming that it would be a way of forcing Japan to make terms with the Allies. Although there was no question of Churchill making terms with Hitler, he never excluded this possibility with Hirohito. The Chiefs of Staff supported the Pacific plan and justified it partly on the grounds that it would foster good relations with Australia and New Zealand and convert the SWPA into a British theatre of operations. They regarded Mountbatten's South-East Asian Command as a military backwater where no decisive result could be achieved. The Chiefs urged that, with Australia as a base, British forces should strike north towards the heart of Japan in tandem with MacArthur.[50]

British inaction in the Pacific was damaging Anglo–American relations. Apart from the hostility within the American navy, the American Chiefs of Staff now alleged that Mountbatten was fighting a guerrilla war in Burma to conserve his main forces for an attack on Sumatra. They urged their British counterparts to order Mountbatten to extend his operations in north Burma. But, with Churchill's support, the British Chiefs refused. London regarded it as a conspiracy by the American representative in Chungking, the buccaneering General Stilwell. Meanwhile, Churchill harboured his own suspicions about American strategy in the Pacific. He had ordered a report on the American capture of the Marshall Islands and when it arrived in Downing Street he denounced the strategy as one 'of using a steam hammer to crack a nut'. When the First Lord of the Admiralty, A. V. Alexander, pointed out that the overwhelming American force had saved both time and lives, Churchill observed laconically that: 'There is no doubt that a steam hammer will crack a nut'.[51]

On 21 February the British Chiefs of Staff met yet again to decide on convincing arguments to support the Pacific strategy. Apart from Churchill, the Chiefs faced formidable opposition from Mountbatten, the American President and Chiefs of Staff, and the entire British War Cabinet. Two days later, in a memorandum of their arguments, the Chiefs dismissed the Culverin operation as a diversion from the main effort in the Pacific that would only lengthen the war. Britain could not mount the Culverin operation before the defeat of Germany, so British forces might be idle for another eighteen months. 'On purely military grounds', concluded the Chiefs, the Pacific strategy 'offers the best opportunities for the earliest possible defeat of Japan'.[52]

Anthony Eden now weighed in on Churchill's side, supporting the Sumatran operation and sending him a dispatch from one of his officials, Esler Dening, who was acting as Mountbatten's political adviser. Dening argued that Britain would gain little by joining America's central Pacific thrust and had instead to defeat the Japanese army in South-East Asia if it were to recover its standing with colonial peoples and the Dominions. 'The record of disaster', Dening wrote, 'must be wiped out' and the 'British part in the Far Eastern war should be a principal and not a subsidiary one'. Anything less than this, he argued, and it would be 'no exaggeration to say that the solidarity of the British Commonwealth and its influence in the maintenance of peace in the Far East will be irretrievably damaged'. Eden backed this up, warning Churchill that if the British were 'merely dragged along at the tail of the Americans in the Pacific, we shall get no credit whatever for our share in the joint operations'.[53]

Churchill sent Eden's and Dening's comments to his service advisers, observing that he was 'increasingly convinced that CULVERIN is the only step of importance which it is open to us to take, and I think we should take it even if we have to wait till April 1945'. Churchill could dismiss the convincing military arguments of his advisers and instruct them, on political grounds, to plan for an operation against Sumatra, but he was loath to do so for fear they would resign. Also, the Dardenelles in 1915 and Norway in 1940, both warned against mounting a hazardous operation on his own responsibility. So the paper continued to fly between the combatants in Whitehall while bullets flew in the Pacific.[54]

In fact, that same day, news was received of a Japanese naval concentration at Singapore that had the British Navy preparing to scurry for cover, much as it had done two years previously. After two years of war against Japan and with the defeat of Italy and the virtual elimination of German surface ships, Britain was still unable to confront a force from the hard-pressed Japanese navy, as it recoiled from the American assault in the Pacific. Somerville wrote in exasperation to Mountbatten about finding himself 'in the same position now as I was two years ago, i.e. with a quite inadequate force which would be a gift for the Japanese if they came out in full strength'.[55] One of Somerville's officers wrote in his diary that Britain had 'always taken the most appalling Risks in this Ocean and they're now terrified that their Bluff is to be called'.[56]

At Mountbatten's headquarters, General Pownall viewed the Japanese move with relative equanimity, believing that the Japanese would probably attack Darwin or Western

Australia, or move against shipping in the Indian Ocean.[57] There was little to stop them. Only the caution engendered by repeated bloody battles with the Americans, and the need to conserve the formidable remains of their Fleet for the climactic battle for control of Japan itself, prevented the Japanese commanders from risking their ships in an attack against Australia.

14

Struggle Over Strategy

March to April 1944

THE CONCENTRATION OF Japanese warships in the former British base of Singapore sent a tremor of apprehension through Australia. At a briefing for reporters on 2 March, Curtin tried to calm public disquiet by observing that invasion was unlikely because of the lack of any transport ships among the Japanese Fleet. The Japanese were probably using Singapore as a 'funkhole', pulling back their mauled ships from the Pacific 'rather than send a defeated armada back to Japan'. Australia could be confident, he claimed, that 'there would now never be any danger to the eastern side of Australia'. Western Australia was at risk, however, so long as the Japanese ships remained at Singapore, since the fleet of seven battleships and two aircraft-carriers was 'much bigger than anything we have in the Indian Ocean or are likely to have for some time'.[1]

Much to his chagrin, Admiral Somerville was forced to adopt the same 'cat and mouse' tactics he had used in 1942, ordering the Eastern Fleet out of its base at Trincomalee in Ceylon and into the distant reaches of the Indian Ocean, west of the Maldive Islands. If the Fleet returned to its base it would have to be 'haphazard and occasional in order that the enemy should remain in doubt concerning [the] movements and composition of the Fleet'.[2]

Meanwhile, Curtin asked Churchill for the Admiralty's

opinion on the 'probability of Japanese incursions into the Indian Ocean', whether they were 'likely to attempt anything more than raids', and whether Britain had the 'capacity to repel such attacks'.[3] It seems that London anticipated Australian nervousness. Curtin's querulous cable crossed with a steadying one from Churchill, sent at the instigation of the Dominions Secretary, Lord Cranborne, which dismissed fears that there was any serious danger to Western Australia. 'Our battleship squadron in Ceylon is well posted', claimed Churchill, and 'our shore-based aircraft are strong'.[4]

The British Fleet was certainly well posted—far from the reach of the Japanese—while the shore-based aircraft in India and Ceylon were not suited for attacking an enemy fleet. On 25 February Mountbatten had sent out an urgent call for torpedo bombers to correct this deficiency in his defences. Churchill was on firmer ground in claiming that Japan would not risk its Fleet in a major naval offensive but would conserve its strength for the final stages of the war. All Australia had to fear, Churchill cabled, was 'the possibility of occasional offensive sorties'.[5]

Churchill's claims had the required effect; Curtin confided to journalists on 7 March that Churchill, too, believed that the Dominion 'has not much to fear from the Japanese concentration'. There was a brief scare in Perth when reports came in of the Japanese fleet moving southward. 'We got everything in readiness for an attack', said Curtin, 'but the Jap ships turned back'.[6] In fact, they had never left; the report had been a furphy.

Curtin's bravado hid serious inadequacies in the defence of Western Australia that persisted more than two years after the attack on Pearl Harbor. As a British naval intelligence officer passing through Perth at the time observed, 'it was—and is—a fact that Western Australia is pretty wide open'. The only immediate protection against a naval attack was from any American submarines at their base at Fremantle, and from a squadron of Liberator bombers in far-off Darwin. In practical terms, this capital city was defenceless if the Japanese chose to bombard it. Moreover, reinforcements from eastern Australia were either totally inadequate or never arrived. The British naval officer noted that, 'Of 86 [aircraft] that started to fly across from Eastern

Australia, only 23 had arrived after 48 hours . . . No really offensive aircraft were sent at all'.[7] MacArthur had scotched the notorious 'Brisbane line', behind which Australian defences were concentrated, on his arrival in Australia in 1942; but the 'Adelaide line' seems to have remained in place throughout the war. Australia was still only able to defend the populous south-eastern corner of the continent.

A nation of seven million could not defend a continent the size of Europe, and Western Australia was cut off by three thousand miles of featureless desert crossed by a single-track railway. The reinforcement fiasco, however, emphasized Britain's failure to protect Australia, with Somerville providing only a vague and distant menace to the Japanese fleet. In vain did the American submarine commander at Fremantle urge that the Eastern Fleet, or part of it at least, be stationed at Fremantle.[8] But the move of the Japanese Fleet had jeopardized British naval plans for the Far East and postponed yet again the arrival of British ships in Australian waters.[9]

Meanwhile, Japanese troops in Burma were mounting a major offensive against the British defences around Imphal on the Indian border. This move was primarily to protect their position in Burma rather than to presage a move into India. In a series of desperate and prolonged battles over several months, the Japanese attack was blunted and then turned back into Burma. So ended what one American historian has recently hailed as the 'decisive battle of the war for Southeast Asia'.[10] Further south, Australian troops were completing their own hard-fought campaign for control of the Huon Peninsula in New Guinea.

Meanwhile, the Chinese and Japanese armies, far larger than the British or American forces being deployed against Japan, were battling for control of the Chinese countryside. For a time in early 1944 it seemed that the Chinese might collapse and allow the Japanese army to break free, but they held on, steadily bleeding the Japanese nation. Although not formidable in battle, the Chinese armies were numerically strong and had to be countered with strong Japanese forces. Most importantly though, once Chinese territory passed into Japanese hands it had to be held and administered. All this took troops that might have been used in Burma or to bolster the hard-pressed garrisons of the Pacific islands.

In the central Pacific, the American Navy had grabbed the Gilbert and Marshall islands from Japan with much loss of life on both sides. The Japanese stronghold in the central Pacific, the island of Truk in the Carolinas group, had been robbed of its naval and air strength in one devastating raid by an American carrier task force at the beginning of 1944. At the same time, Rabaul, Japan's stronghold in the south-west Pacific, was also neutralized by repeated Allied air attacks, although its 100 000 strong garrison was deemed too powerful to capture. Instead, and much against MacArthur's wish, it was by-passed by American troops landing in the Admiralty Islands to the north of Rabaul. The accelerating Allied effort was directed by military leaders confident of richer prizes as they hammered at the inner ring of Japan's defences.

As the Axis floundered, preparations were under way for the peace that would follow, although it was difficult to make conclusive plans when the post-war strength of the various powers was not known. It was also unclear whether there would be genuine international co-operation or a return to the pre-war system of imperial autarky, in which the world was divided into closed-off systems of economic and political power, of which the British Empire was but one example.[11]

As part of the Lend-Lease negotiations with Britain, the United States had extracted a much-disputed commitment from Churchill for lower tariffs after the war that would allow greater American access to world markets. This was enshrined in the Atlantic Charter concluded with Roosevelt at their ship-board meeting in August 1941 off the coast of Newfoundland.[12] Churchill had always favoured a policy of free trade, but he was acutely conscious of the political risks of forcing such a commitment on his national government composed of conservatives such as Leo Amery who were strongly in favour of retaining Imperial preferences. In order to concentrate his energies on the war, he performed a precarious balancing act between the Americans, particularly the Secretary of State, Cordell Hull, who demanded the abolition of Imperial preference, and some of his own supporters who refused to countenance such a prospect. By

1944 he was fast losing control as Washington intensified the pressure; a decision could not be delayed much longer.

When the issue came up for discussion at a ministerial meeting in February 1944, Amery sent a message from his sick-bed to Eden asking that it be read to his colleagues. In it he blasted the Americans with their 'quasi religious free trade outlook' that was 'blended with an American economic Imperialism' that sought to shatter the British Empire into 'separate fields for American exploitation'. As a sign of his desperation, Amery called for British agriculture to be boosted in the post-war period, mollifying a substantial section of the Conservative Party tied to the rural vote and suspicious of preferences for dominion agricultural surpluses. This contravened the whole concept of Imperial preference, which depended upon the Dominions being the 'bread basket of Britain'. Amery tried in vain to buttress his arguments by painting a pre-war picture of selfless Dominions loyally buying British manufactured goods. 'How anyone can imagine', he intoned, 'that we can afford to drop that substantial bone for its shadow in the water I cannot conceive'.[13]

Britain's tough talking Home Secretary, Herbert Morrison, expressed the brutal truth: neither Britain nor the Dominions could live merely on Empire trade.[14] Moreover, as Cranborne correctly pointed out, the Dominions would much prefer a general expansion in world trade rather than a return to the restrictive pre-war system.[15] Bruce tried to press this point home with Lord Croft, assuring him that Australia no longer favoured Imperial preference and could not support the Empire lobby in its efforts to have it preserved.[16]

Australia had not, however, abandoned her protectionist tariff policies which were an insurmountable barrier to any agreement with the United States. A general lowering of tariffs might devastate Australia's carefully nurtured industries, so the Government pushed for an international commitment to full employment that would, it was hoped, lead to sufficiently high consumption levels to guarantee world markets for Australian surpluses sold through orderly marketing mechanisms.[17] The Depression was still vivid in the minds of Labor leaders who sought to prevent the widely anticipated post-war recession and cement a political alliance between the industrial working class and various rural interests.

The second issue bedevilling Churchill was American anxiety about oil. Washington was pressing Britain to accommodate their post-war oil needs after reports that America might face an oil shortage after the war. Congressmen pressed Roosevelt to gain access to alternative supplies, particularly in the Middle East where Britain enjoyed a sphere of influence. When Washington called for a conference with Britain on the issue, Beaverbrook asked Churchill to 'pigeon hole' the American request. 'Oil', wrote Beaverbrook, 'is the greatest single post-war asset remaining to us. We should refuse to divide our last asset with the Americans'.[18] In a paper prepared for Beaverbrook on the world oil reserve situation, it was suggested that America could be provided with 'some of the less attractive areas', such as 'large areas in Australia and New Zealand and parts of Africa'.[19]

The third contentious issue was civil aviation, and all three issues combined in the minds of some leading British politicians into general opposition towards the United States. During a War Cabinet meeting on 11 February 1944, Beaverbrook said the choice facing Britain was clear—either to be 'a senior partner in the Empire or a junior partner in an association with the USA'.[20] It was the takeover battle of the century in which Great Britain PLC was eventually to become the branch office of Wall Street. However, Beaverbrook and others were unwilling to give in without a struggle. While supporting the idea of collaboration with Washington, Beaverbrook nevertheless argued that Britain 'should aim, with an equal constancy, at maintaining and strengthening our own position as a world power'.[21]

Civil aviation was considered by many to be the key to post-war prosperity and security. Accordingly, British and American companies, backed by their governments, busily competed for routes across South America and Africa, while in Britain the aircraft companies surreptitiously designed new post-war civil aircraft under the cover of war work. The British ambassador in Washington, Lord Halifax, warned London in January 1944 that this intensifying

Anglo–American competition was beginning to affect the wider relationship with Washington. He predicted that it would soon degenerate into 'a competitive struggle which will become increasingly naked and which we must confess fills us with concern'. Several weeks later, the British War Cabinet approved the development of a modern airfield at Heath Row for the use of military transport aircraft but also to provide 'an airfield which could be developed as an international civil airport for the capital of the Empire after the War'.[22]

In the context of stiffening Anglo–American competition over post-war aviation, Britain suddenly decided that P. G. Taylor's alternative route across the Pacific could be justified by its plans for the war against Japan. Officials in London dusted off Taylor's proposals of 1942 and suggested that he be part of any survey party. Taylor was ecstatic, proclaiming that it would fit into place 'the last section of a round the world route on which I have worked for years'.[23]

At the same time, Evatt was pressing for British naval forces in the Pacific, not so much to defeat the Japanese but as a counterweight to the Americans.[24] The Australian Navy also hoped for an enhanced role once the Royal Navy arrived. Up until now, Blamey's troops had taken centre stage, recapturing New Guinea and saving Australia from attack. But Blamey's troops would soon be left behind as MacArthur plotted his return to the Philippines with an all-American cast. Here was a chance for the manpower-starved Navy to have its moment of glory in the Pacific and to re-establish itself as the senior Australian service.

The Navy was hamstrung by having at its head the British Admiral, Guy Royle, who had a limited grasp of political manoeuvring. Royle called for an extra four to five thousand sailors to be enlisted to crew a mini-fleet of surplus ships that he hoped to secure from his colleagues at the Admiralty. This fleet of modern ships would comprise an aircraft-carrier, one or two cruisers and six destroyers. In his eagerness to achieve this proposal, Royle made the question a partisan football to be kicked around in the politically-charged atmosphere of the Advisory War Council.[25]

Putting Curtin on the spot during a meeting of the Advisory War Council on 21 March was not the way for Royle to achieve his aims. When he suggested that Curtin take up the question of naval expansion during his stay in London, the Prime Minister adroitly side-stepped the trap Royle had so clumsily laid and passed the question back to the Chiefs of Staff for resolution, leaving Royle and his colleagues to fight over the limited amount of manpower. Following the meeting, Royle tried to re-write the record of the discussion to suggest that Curtin would support the Navy's expansion. He was firmly informed by Shedden's assistant that this was not so and that the battle would have to be fought among the Chiefs of Staff before Curtin would give it his approval. As a parting shot, Curtin apparently ordered Royle never again to abuse his access to the Advisory War Council.[26]

Juggling Australia's limited manpower was an almost impossible task, particularly when the army resisted Curtin's orders to release men from its ranks for food production. As Shedden warned Curtin in February 1944, manpower was inadequate to satisfy all claimants and the resulting political fallout 'will inevitably alienate some of [the government's] present support'. The Government's targets for food production were not being met because people in lucrative industrial employment were reluctant to transfer back to rural industry, while the army and the munitions industry were mainly releasing unfit workers who were consequently unfit for the needs of the farms. The unemployment lines were growing while angry farmers watched as their crops lay unharvested. According to Shedden the choice was clear: either the food targets must be reduced or the manpower found to fulfil them and 'to stop the growing criticism'.[27]

In this atmosphere of mounting panic, Royle's proposal for an expanded Australian Navy was received with dismay by defence planners. The proposal had its attractions, particularly if the ships could be obtained from Britain at little or no cost and if they included an aircraft-carrier; Australia had never before possessed one, but in the context of the Pacific war, it seemed to be a necessary part of any modern navy.

Despite these attractions, Shedden was determined to prevent Royle railroading the proposal past Curtin, and counselled that Royle's proposal would only exacerbate the problem of food production and threaten Australia's capacity to send her army forward with MacArthur's advance. Shedden argued that the arrival of British forces in the Pacific would drain Australian food stocks, although he claimed that the Americans were unlikely to let the British into the Pacific. Most important, in Shedden's view, was that 'from the aspect of post-war policy, we cannot overlook the fact that the best service we can render to the future of Australia is to build up the R.A.A.F. to the maximum degree of our capacity'. This was an argument calculated to appeal to Curtin since it had been Labor policy during the 1930s and seemed to have been vindicated by the events following the Japanese attack on Pearl Harbor.[28]

Shedden's counselling worked. Although Royle made a last attempt in a meeting with Curtin, the Prime Minister refused to be moved. Curtin was angry that Churchill had not replied to his cable of October 1943, requesting Britain's agreement to the re-balancing of Australia's war effort. Shedden triumphantly recorded that Curtin 'went so far as to say that if Mr. Churchill did not choose to reply to his representations . . ., he certainly was not going to adopt such a humble attitude as to offer him gifts by manning additional ships'. Curtin was coming to the view that food, together with a limited military effort, would be Australia's best contribution to the Allied war effort.[29] During their trip to London, Shedden and Blamey could ensure that Curtin did not deviate from this view.

Britain shared Australia's manpower problems. The demands of the forthcoming invasion of France had to be balanced with the demands of her civilian economy as it prepared to satisfy pent-up domestic demand and to capture post-war foreign markets for British goods. The British Government agreed to a proposal from the powerful Minister for Labour, Ernest Bevin, to maintain control over the labour market and, following the defeat of Germany, to allocate men straight from the armed forces to designated factories.[30] However, if this scheme were to work, Britain had to control her manpower obligations for the war against Japan. As the Americans pushed the Japanese back towards their home islands, the lifeline between the British Pacific Fleet and its supply bases lengthened. The 'fleet train' of ships considered necessary to supply the Fleet gradually crept up from 134 to 158 ships by March 1944 and, for planning purposes, the period between the end of the war against Germany and that against Japan was set at three years. The implications were awesome for a government gearing up to satisfy civilian needs and to face its first election since 1935.[31]

For Churchill, the choices were simple. The first priority was to allow for twenty-four million tons of imports to be shipped into Britain during 1944 and 1945. The second priority was to allow any left-over shipping to be used for the fleet train. The third priority was the size of the fighting fleet that could be used against Japan.[32] It was a funny way to fight a war, but Churchill hoped for a short fight, relying on the promised Russian entry into the war against Japan to bring it to a quick end, perhaps through a compromise peace. His Cabinet agreed almost unanimously that Germany's defeat should signal better times for the war-battered British civilians. Hugh Dalton, head of the Board of Trade, joined his ministerial colleagues during a War Cabinet discussion on 13 April to 'stress the need to lift civilian standards and push export trade as soon as Germany is beaten'. Dalton observed that Churchill was 'always very responsive to this, though also very insistent that we must "do our utmost", whatever that might turn out to mean, to help the Americans to beat the Japs'. According to Churchill, once Russia joined the war against Japan, the Japanese 'might offer terms, short of unconditional surrender, which it might be well worth our while to examine'.[33]

While Churchill was planning a parsimonious British effort against Japan, there was still no agreement, either in London or with Washington, about the form this effort should take. Churchill was doggedly committed to the proposed Culverin operation against Sumatra, an operation which 'gravely concerned' Roosevelt, who wanted the British to secure and develop the links to China in upper Burma and establish Allied air control in the Pacific triangle between China, Formosa and Luzon.[34] Churchill had the support of his War Cabinet and the Foreign Office in seeking to recapture Singapore rather than play junior partner to the United States in the Pacific, but his Chiefs of Staff threatened to resign *en masse* if Churchill insisted on the operation against their combined advice.[35]

The British Army was fixated on the forthcoming invasion of France and quaked at the thought of fighting in Malaya with the limited resources allowed to them by the overriding priority that the European war still enjoyed. The Navy wanted to prove its mettle in the annihilation of the Japanese Fleet in the Pacific where it could play the leading British role and establish its claims to post-war seniority. Although Churchill refused to concede defeat, his cherished operation against Sumatra was impossible to mount without American logistical support in the form of landing ships. These Roosevelt refused to provide.[36]

This refusal should have been the end of the matter, but Churchill was not ready to drop his pet project. Realizing that the Americans were lukewarm about British naval forces operating in the Pacific, Churchill went behind the backs of his Chiefs of Staff and sent a cable to Roosevelt, inviting him to acknowledge that the British Pacific Fleet would not be needed until the summer of 1945 at the earliest and that, in the interim, British forces would be better occupied pursuing an Indian Ocean strategy. This Roosevelt gladly did.[37] The result was that planning, rather than military operations, continued, with Churchill hoping that time would provide him with the wherewithal to mount the Culverin operation.

In the meantime, Churchill ordered a mission to Australia to examine the logistical problems of basing the British effort there instead of in India. He was confident that the investigation would show that the costs of supplying forces at such a distance would finish the Pacific plan. In fact his Shipping Minister, Lord Leathers, had already advised him that the shortage of shipping would preclude the use of Australia as a base, while Curtin's oft-repeated complaints of a manpower shortage suggested further problems. As Admiral Somerville reflected from his vantage point in Ceylon, 'the question of manpower is important since . . . none is likely to be available in Australia and what there is is usually on strike'.[38]

On 11 March Churchill cabled to Curtin seeking permission to dispatch the mission, although it was to be 'on the strict understanding that we are engaging upon no commitments and reaching no firm decisions'.[39] Churchill did not intend to base sizeable British forces in Australia, assuring Mountbatten that he remained

> entirely opposed to shifting the British centre of gravity against Japan from the Indian Ocean to the Pacific, at any rate for the next eighteen months. I have agreed to a Mission going out to reconnoitre Australian bases etc. in order that these possibilities may be explored. This in no way implies any decision to change the existing policy.

Even the Chiefs of Staff were turning increasingly to the idea of using Indian and Dominion forces as a large part of the projected 'British presence' for the Pacific.[40]

British forces arriving in Australia so late in the war would have no defensive purpose and their support and supply services would only soak up Australian manpower that would otherwise be directed into civilian industry and post-war reconstruction. Curtin and the Australian Chiefs of Staff took fright, fearing also that this would compromise Australia's own plans to forge an independent military presence in the region. MacArthur also was worried that the injection of British forces into the SWPA might compromise his supreme command of Allied forces. Curtin's Chiefs advised that Britain already knew of Australia's potential and that anything else could be communicated by cable or through the British military mission already in Australia.[41]

Although Whitehall was taken aback by the negative Australian reaction, Churchill was relieved, instructing that the study now be made from material available in London. The Admiralty deemed this insufficient and eventually received permission to send a team to Australia under Admiral Daniels to study facilities, particularly in the north and west of Australia, that might be used by the British Pacific Fleet. Churchill was contemptuous: Daniels 'was a maker of difficulties and on many points a defeatist and certainly he does not carry my confidence'.[42] In Ceylon, Somerville could only wonder at the 'extraordinary cross currents at work in connection with the general strategical policy for the East' as he chafed at the further delay.[43]

At a staff meeting at Chequers on 8 April, Churchill agreed to a middle route. If MacArthur resisted the idea of major British forces being based within his command, why not slice off north-western Australia from his command and base sufficient British forces there to be able to strike north at Timor, Borneo, and the Celebes? Intelligence reports had already suggested that the Australian Army would be able to contribute several divisions to such a 'British' force. Once in place, such forces would have the choice of tightening the stranglehold on Singapore and Malaya or striking northward with MacArthur towards Japan. Such an island-hopping strategy was also very economical in terms of troops, an important consideration in both London and Washington, where the prospect of fighting major land battles with the Japanese Army was a daunting one. The fear of such battles spurred the development of the atomic bomb in America, while on a tropical island off the northern coast of Australia, unsuspecting Australian soldiers volunteered for a chemical warfare unit that was used by scientists, including the future Nobel Laureate, Sir Macfarlane Burnett, in a British-inspired experiment to test the effects of mustard gas on unprotected troops in jungle conditions.[44]

Curtin left for Washington and London on 5 April 1944, hoping to establish a new post-war pattern of co-operation between Australia and Britain and win Allied agreement to Australia's shifting war effort. Unlike Evatt who had confronted his fear of flying, Curtin declined MacArthur's offer of a special aeroplane and chose to sail to San Francisco in an Australian cruiser. When operational reasons prevented this, he opted to travel in a crowded and blacked-out American troop-ship on this his first trip as Prime Minister outside of Australia.[45]

The British High Commissioner, Sir Ronald Cross, and his wife went on board to wish Curtin and his wife *bon voyage*, taking them a bunch of flowers for their hot and stuffy cabin. They found Curtin shut away and 'very depressed'. Cross ascribed it to feelings of inadequacy at the prospect of dealing face to face with a procession of world leaders. Not even Cross's wife could cheer Curtin up so they quickly made their excuses and departed 'with all possible speed!'

As Cross confided to the Dominions Office, Curtin led a reclusive existence, having a

> small home where his wife always does all the housework without, so I hear, even the occasional help of a char. He has

> probably not habitually thought much about the sort of things he will be called upon to discuss in London, and he probably lacks the confidence that comes from knowledge and definite opinions. All this may make it difficult to set him at his ease, the more so as he is sensitive and will be on the look-out for people trying to influence him by showing him great attention and flattery.

Cross claimed that Curtin was particularly apprehensive of Churchill and that he was 'anxious to justify himself and to stand well in Churchill's regard'.[46] There was an element of truth in this. Curtin certainly did want to rebuild the bridges between Canberra and London. However, he would find it beyond his power and persuasion.

15

A Question of Manpower

April to June 1944

AN EXHAUSTED AND very relieved Australian Prime Minister walked cautiously down the gangway at San Francisco. He had spent the voyage pacing the deck at night, terrified that he might at any moment have to abandon ship. During the long, torpid days, he was often shut away in his humid cabin as the returning American troops crowded the decks and scanned the horizon for some sign of home. Once safely on American soil, Curtin opted to travel by train to Washington rather than risk his life in an aircraft.[1]

Australian officials in Washington were hoping that Curtin's visit would smooth American–Australian relations. Australia's First Secretary in Washington, Alan Watt, cautioned that American officials were having 'to be handled with rather more care than usual', and warned Canberra against allowing the formerly warm relationship with Washington to cool too much since, 'if one looks twenty to fifty years ahead, Australia needs the support of the United States in the Pacific'. Watt worried that Curtin would not have time to restore the previous warmth and to raise Australia's standing in the discussions about post-war problems. Watt lamented that Australia would continue to find herself 'out in the cold so far as current world problems are concerned'.[2]

Curtin was simply not the man to pursue these questions.

He had no one from the External Affairs Department in his entourage, no doubt partly for fear that they would report his moves to Evatt, but also from disinterest in these matters. Curtin had rejected Evatt's suggestions that the head of the External Affairs Department should accompany him to provide advice on foreign policy, arguing that the London meeting was not meant to be a policy conference but an exploratory meeting that would require follow-up conferences between the responsible ministers.[3]

Apart from his wife, Curtin was accompanied by his loyal secretary of the Defence Department, Frederick Shedden, and the increasingly Anglophilic General Blamey. His travelling companions reflected his limited overseas agenda —to gain Allied agreement to the partial demobilization of Australian forces, to secure a respectable place for Australia in the remaining operations against Japan, and to realign Australia firmly within the British Imperial camp, but on a basis closer to partnership than the former subservience. Curtin wanted to break down the wall of British hostility towards Australia that had been built up during the war, and to ensure that the Dominion's post-war destiny within the Commonwealth would not be compromised by its fleeting dalliance with America. On a domestic political level, Curtin wanted to build up his personal prestige for a forthcoming referendum campaign designed to give the Federal Government wide-ranging powers over the life of the nation in order, so it was argued, to ensure that reconstruction went smoothly.[4]

MacArthur had indicated that there was nothing he wanted raised in Washington on his behalf. Curtin did call on Roosevelt in South Carolina where he tried to smooth over the rough patch in Australian–American relations caused by the Anzac agreement in January. This agreement had marked the high point of Australian assertiveness, when its traditional dependence upon Britain was temporarily reduced and anything seemed possible. The agreement was designed to restrict American post-war influence in the Pacific to the northern hemisphere and to establish Australian and New Zealand influence in the southern hemisphere under the protective cover of the British and Americans. Just five days before Curtin met Roosevelt, the agreement was described by the powerful Democrat Senator for South Carolina, James Richards, as the 'unkindest, most disturbing cut of all' which 'in no way recognises or appreciates the dependence of Australia and New Zealand upon the U.S. Army, Navy, and Air Force'.[5]

Curtin had been alarmed by the damage the agreement had done to Australia's relationship with Washington and was more than willing to see it hang like a dead albatross round Evatt's political neck. So he was relieved when Roosevelt guessed that Curtin 'had had very little to do with the drafting, but that Evatt had done most of it and the others had merely agreed'. Curtin admitted that the agreement 'was made and carried in what may well prove to be an excess of enthusiasm' following a discussion in Canberra on 'the future of the white man in the Pacific' in which, Curtin claimed, fears had been expressed about a future Asiatic bloc of India and China turning against all white men. Roosevelt advised that 'it will be best to forget the whole incident'.[6]

The abstemious Curtin then spent several desultory days in Washington, blind to the charms of the capital in its springtime plumage and increasingly distracted by General Blamey's hard-drinking and womanizing. At a meeting with the Combined Chiefs of Staff, during which Blamey outlined Australia's military plans, the jaded general fell asleep as his Allied counterparts discussed the points he raised. Then on 28 April Blamey dossed down in the flying boat's Prime Ministerial bunk while Curtin sat tight-lipped and stiffly upright, enduring the long flight across the dark and hostile Atlantic to London.[7]

The British High Commissioner in Australia, Sir Ronald Cross, had secretly advised Whitehall that Curtin would be so keen to score a personal success that he may not be 'very insistent on anything if his insistence might result in his appearing publicly to have had a failure'. Cross warned the Dominions Office not to overwhelm Curtin with a whirl of social engagements as they usually did with Dominion leaders. Curtin had 'simple tastes' and 'would not like being entertained in a social way and the simpler his surroundings, the more likely he is to enjoy himself'. Cross suggested that Churchill show Curtin 'personal attention' since the 'success he made of Dr. Evatt in 1942 had a transforming influence on Anglo–Australian relations'.[8]

For some time Curtin had intended to raise the issue of a Commonwealth secretariat that could formalize the relationship between members of the Commonwealth. Curtin was ahead of his time in his vision of a Commonwealth in which Britain did not rule but merely participated like any other member. He was defeated before he began. Neither the other members of the Commonwealth nor Britain shared Curtin's desire for closer integration of their foreign policies or the establishment of formal machinery to oversee it. Canada's position was well-known and should have given Curtin pause before suggesting such a proposal. The British Government decided to give Curtin time to raise the proposal during the Prime Ministers' meeting, with Britain taking a back seat while Canada's Prime Minister, Mackenzie King, knocked the plan on the head.

Such was the artful design of the trap set by the Dominion Secretary, Lord Cranborne, and laid out for the approval of the War Cabinet some ten days before Curtin's aircraft lumbered into view. In the event, Churchill did not even attend the meeting at which Curtin's plan was finished off by the Canadians. Curtin was upset by Churchill's absence, but it allowed Britain to escape the odium of quashing a proposal designed to strengthen the Commonwealth, and then permitted her to propose a compromise that would provide the form, but not the substance, of Curtin's proposal.[9]

Canada was the stumbling block not only for Curtin, but also for Churchill who wanted to resurrect the pre-war British Empire to inflate Britain alongside the other great powers. Canada, too, had her own ideas for the future of the Commonwealth, wishing to build upon her close relationships with both Britain and America to construct a quasi-independent place for herself at the head of the 'middle' powers.[10]

When the London conference assembled in the Cabinet room of 10 Downing Street on 1 May, even Churchill realized that the days of whipping the Empire into line were long past and that the prospects of reaching a joint agreement on anything were slender. Accordingly, Churchill agreed with his War Cabinet that the discussions should be purely exploratory and that there should be no attempt to reach conclusions. To help ensure this and to minimize dissension, the various Dominion High Commissioners in London were to be excluded from the talks, other than the opening and closing sessions where they would be included in the general photograph.[11] This would help soothe their ruffled feelings and prevent them pushing for resolution of difficult issues.

Britain was involved in discussions, mainly with Washington, of fairly intractable issues and she did not want this conference to disturb them. More embarrassingly, there were a number of issues upon which Britain herself had not been able to reach agreement, the prime example being post-war economic policy; British ministers were at loggerheads over the direction that Britain should take in her talks with the United States. On this issue, Churchill proposed that he make only general statements about the desirability of fostering trade within the Empire and set out in a non-committal way the state of the Anglo-American economic talks.[12] This would appease all sides, those who wanted a general boost to world trade and those who wanted to see an Imperial trading bloc.

The economic Imperialists within the British Government were left out in the cold. When Lord Croft sought Beaverbrook's help to distribute copies of the Empire Industries Association's bulletin to the four Dominion Prime Ministers, Beaverbrook declined, warning that the 'perils of direct propaganda of this type are too great'. Another vexed question was a future world organization. Churchill was adamant that peace and security would depend on the organization being effectively controlled by three or four great powers (depending upon whether America's favourite, China, was included), whereas Eden proposed ten to twelve nations for the council of such an organization, with room for the Dominions sitting in rotation.[13]

Curtin's political agenda was more immediate, but there were still pitfalls. On 8 April Australia's acting Prime Minister, Frank Forde, had announced that ninety thousand men would be released from the Australian Army during the course of the year. This was seized upon by the local and overseas press as Australia's attempt to take a back seat in the Pacific war, reaping rewards as a food supplier and reducing the risks of fighting. Forde's news was old news and was designed to screen a lamentable failure to discharge men back on to the farms. MacArthur had agreed to the numbers in 1943, and most men were discharged for injury, illness or indiscipline. Only twenty thousand were to be specially discharged to cope with the needs of civil industry.[14]

Forde had issued the statement to assuage growing domestic concern about sluggish discharges into industry; instead, he had blown up an international storm that threatened to buffet Curtin during his visit to the Allied capitals. Curtin immediately instructed Forde not to announce any change in the strength of the armed forces before first clearing it with him. But the damage was already done. Churchill was aghast at the idea of Australia directing men from the forces back on to the farms so as to profit from the war. Such calculations were considered permissible in British planning, but not where Australia was concerned.

Australia's intention was even worse than Churchill feared. So concerned was Forde about the needs of industry, particularly the low paid and otherwise unattractive dairying industry, that he teamed his press announcement with a secret suggestion to the Army department that farmers' requests for the release of particular troops be granted regardless of the troops' situation, thereby 'assisting in increasing production of butter for export to England'. In other words, troops would be taken from the front line to butter Britain's bread. The Army resisted this suggestion, maintaining that most of the troops requested for release were 'key personnel' in operational divisions of the AIF, awaiting orders from MacArthur to take the offensive against Japan.[15] So the manpower problem remained unresolved.

When Churchill read American press criticism of Australia over Forde's announcement, it confirmed all his own prejudices about the Dominion. He instructed his Chiefs of Staff to report on the Australian war effort, which he personally felt to be 'a very poor show'. Five rather than three army divisions was the least that Australia should be maintaining outside the country, and he hoped his Chiefs would back up this opinion with statistics that might be useful in his forthcoming talks with Curtin.[16]

The Chiefs of Staff duly reported back on 1 May, the day that the Dominion Prime Ministers were assembling for their first meeting, but not as Churchill had hoped. Instead, they exonerated Australia. Considering the terrain and the tropical diseases with which they had to cope in New Guinea, the maintenance of two to three divisions throughout the campaign was 'a very remarkable achievement'. The Navy was similarly absolved, while the Chiefs considered that the RAAF had been hampered by inadequate aircraft rather than by any failure of political or military will. In fact, the Chiefs looked forward to integrating the Australian forces with those Britain was committing to the Pacific and accepted some reduction in Australian forces to allow for logistical support for the combined Commonwealth force.[17] It is doubtful whether these comments affected Churchill's prejudices.

The following day, Churchill asked his office to brief him on the subject of Australian manpower so that he would have ammunition for his discussions with Curtin. He had planned weekend talks with Curtin at Chequers, where affairs of state could be determined round the fireplace with brandy and cigars. As an ex-alcoholic, this would have been very painful for Curtin, and may explain his aversion to social situations in London. Apart from resisting liquid temptation, Curtin was also suspicious of being 'duchessed' by the British so, instead of Chequers, he went with Shedden to visit the elderly Lord Hankey, the retired secretary of the defunct Committee of Imperial Defence, who found Curtin to be a 'particularly nice fellow with no "side"'.[18]

On 4 May, in the face of Curtin's request to re-shape the Australian war effort, Churchill instructed Ismay to have

> a small body of competent officers forthwith begin to examine, in cold blood, what really is Australia's contribution. A separate, subsequent note might be written about New Zealand. They both say they are forced to maintain a great number of men under arms and are now compelled to withdraw them to grow everything we shall require in the Pacific etc. Let us have a look at it without the slightest desire to prove a case one way or the other.

Once again, Whitehall exonerated the Australian war effort, and Ismay demonstrated that 'the degree of mobilisation for the war is greater in Australia than in the United States in both armed forces and munitions', although only if manpower employed in Allied food production were counted as part of the total mobilized manpower. Ismay tried to mollify Churchill, buttressing his favourable assessment with statistics and assurances from Shedden that Australian troops would only be switched to agricultural production if Australia had to support greater American and British forces.[19]

Churchill was concerned that Australia's partial demobilization might force Britain to play a much larger role in the Pacific than he intended. He and his Chiefs wanted the Australian troops to camouflage the relatively small British component in the contemplated British Commonwealth force. Similar motives lay behind Britain's wish that Australia man a British aircraft-carrier and associated warships. That way some of the manpower burden of the Japanese war could be shifted from Britain to Australia. The nucleus of a strong post-war Australian naval force would be formed that could patrol the Pacific at Britain's behest, protecting the scattered pearls of her once great empire.

Curtin was suspicious of British motives, particularly about the offer of warships. Churchill, however, assured the assembled Prime Ministers that his forces would begin to move to the Pacific as soon as ports could be secured in Europe by the forthcoming D-Day landings. He reiterated his pledge that 'Britain had promised to do all she could against Japan and that that pledge would be carried out'. He added that the war against Japan would not long outlast the war against Hitler, though this was a personal feeling for which Churchill 'had nothing in the way of facts to prove it by'.[20] In fact, Churchill knew of the secret and feverish development of the atomic bomb, and Stalin's secret undertaking to join the Allied effort against Japan as soon as Germany collapsed.

On the day Curtin left Australia, Forde had called a conference of the Chiefs of Staff to discuss boosting the monthly manpower intake for the Navy. Not surprisingly, there was stiff opposition, although the Army was prepared to concede if an increase were needed to man additional ships. This would have given heart to Admiral Royle's campaign to have Australia accept British warships.[21] Despite explicit instructions from Curtin, on 20 April Royle took his campaign back to the Advisory War Council where he pressed the Navy's manpower needs on the sympathetic conservative politicians. Although Forde criticized him for stirring up the political waters, a week later Forde confided to a night-time press conference at Parliament House, that Curtin was planning to 'seek the transfer of some cruisers or destroyers from the British fleet to Australia'. One journalist observed that Forde was 'rather cautious in giving this information but intimated that they might be gift ships and would be given to Australia on the condition that we manned them'.[22]

Such a transfer was supported by the conservative members of the Advisory War Council at their meeting on 2 May, and the following day Royle rounded off his campaign with a statement to the War Cabinet arguing for the ships. It would not only recoup the wartime losses of Australian ships but, argued Royle, 'ships steaming and fighting side by side with those of the British Forces will be highly tangible evidence of our active participation in the war to the very end'.[23] Royle won converts on both sides of the political fence, and the Government agreed to increase the naval intake at the expense of the Army and Air force.

Curtin was notified of Royle's attempted policy coup on 4 May in the midst of his discussions to reduce commitments for Australian manpower rather than increase them. He was disturbed that his colleagues agreed, in principle, with the transfer of British warships to the Australian Navy provided they were free and new.[24] Although Curtin agreed to an increase in the naval intake to man various small ships being built in Australia, he was angry that Britain had not deigned to reply to a seven-month-old cable from Canberra requesting the return of Australian Navy personnel serving with the Royal Navy. Curtin contrasted the tardy handling of this question in London with the 'urgency with which additional commitments have been pressed'.[25] Unsure of himself in the high-powered world of wartime Whitehall, and anxious to avoid further commitments, Curtin played a cat-and-mouse game with the Admiralty in which he adroitly sidestepped their attempts to corner him for discussions.

On 12 May Churchill notified Curtin of Admiral Daniel's mission to Australia and of the need for his party to be reinforced with 'a few more sailors and soldiers'. He asked

Curtin to meet the First Lord of the Admiralty, A. V. Alexander, and his naval chief, Admiral Cunningham, so that they might explain the need for extra men despite Curtin's reluctance to accept them. A week later, the Admiralty still had not met Curtin who was soon to depart for Washington. Alexander begged Churchill to help collar the reluctant colonial since, he complained, Curtin 'does not seem anxious to come to the Admiralty!'.[26]

Alexander neglected to inform Churchill that one of the matters they wanted to discuss with Curtin was the possible transfer of British ships to the Australian Navy. This was part of the Admiralty's secret agenda inspired by Royle, who had informed them that Curtin would raise the question of his own accord. The Admiralty could then avoid the risk of raising it themselves and appearing to pressure Australia into further commitments. However, this plan nearly went awry during a meeting at Chequers on Sunday 21 May when Curtin raised the issue with the unprepared Churchill as he sought the recall of some three thousand Australian naval personnel on attachment to the Royal Navy. These men would be almost sufficient to man the so-called 'gift' ships.

Churchill hid his confusion with a non-committal response then demanded an explanation from Alexander who filled in the background to the issue and Royle's part in it. The Australian newspaper boss, Sir Keith Murdoch, was also in London, and had been briefed on the ships proposal. When Alexander confessed to Murdoch that Curtin had not even raised the question, Murdoch had offered to remind the Prime Minister about it. Murdoch's prompting may well have caused Curtin to mention the matter to Churchill. Whatever the reason, Alexander was ecstatic that the matter was now on the table for discussion. As he informed Churchill, the offer to man the ships would not only 'show a welcome revival of Australian interest in the importance of a Navy but, if made and accepted, it would be a real contribution to our own manpower problem and ensure a satisfactory foundation for Australia's post-war fleet'. Alexander warned Churchill not to mention Royle's role and to pretend that the proposal was purely Australia's.[27] This Churchill did.

On 27 May Churchill responded to Curtin's claim that Australia was being asked to find three or four thousand extra seamen to man the British ships. 'I think there must have been some misunderstanding on this', wrote Churchill, 'since I am assured by the First Sea Lord [Admiral Cunningham] that no such proposal emanated from the Admiralty'. However, if Australia were willing to provide the men, Churchill was willing to provide an aircraft-carrier and one or more cruisers.[28] Curtin's suspicions turned to anger when Murdoch inadvertently told him that Royle had been secretly conveying Australian moves to the Admiralty. The matter was finally sealed when Churchill declared that Australian sailors and airmen seconded to British forces would not be transferred home to ease the manpower shortage. Curtin was unlikely to aggravate the shortage by accepting the 'gift' of the British ships. On 29 May, just before leaving London, Curtin cabled to Forde recommending that a decision on manning the extra ships be delayed 'until all aspects of the war effort can be looked at together'.[29]

Curtin assumed that the British Government's failure to reply to the cable of October 1943 about Australian manpower problems, was rudeness. Certainly Churchill often ignored Australian needs and feelings, but this case had as much to do with Britain's failure to settle upon an acceptable Pacific strategy. Taking Churchill at his word, Curtin believed that Britain's plans for the Pacific were settled and that the British Pacific Fleet would arrive promptly, placing sudden and unbearable pressure on Australian manpower.

During the Dominion Prime Ministers' meeting on the morning of 3 May, Curtin informed Churchill that Australia would not be able to provide food and supplies for additional Allied forces without reducing the armed forces. Curtin's caution was interpreted by at least one British observer as hostility, with Admiral Cunningham concluding that Curtin and Blamey were 'in MacArthur's pocket' and that Curtin was 'rather rude and gave the impression that he did not want UK forces in Australia'. When the meeting reconvened after lunch, Curtin dispelled this impression, at least in Cunningham's mind, when he made an 'excellent statement' claiming that Australia was 'anxious for UK forces but do not know how many they can support and are most anxious not to have to be relegated to producers only but to keep forces in the field commensurate with the importance at the peace table of their views on the Pacific being taken into account'.[30]

With Australia's position clarified, it was possible to discuss the level of British forces likely to be based in Australia if Churchill approved the Pacific strategy. In a meeting with Blamey on 5 May the British Chiefs of Staff set the maximum force at six army divisions, sixty squadrons of aircraft, four battleships, five fleet carriers and associated escort ships.[31] These were figures that Australia could plan for, but they meant nothing until Britain's strategy for the Pacific war was determined. When the Chiefs handed Blamey a memorandum on strategy, Churchill rounded on them, fearing they might win the strategic argument by default. Although Churchill admitted that Britain's indecision was 'not unfolding a creditable picture to history', he continued to oppose any strategy based upon southern Australia.[32]

Curtin seemed unaware of the debate still raging within Whitehall. On the penultimate day of the conference, he and Mackenzie King hotly disputed whether Britain and the Dominions had finalized plans for the war against Japan. Mackenzie King sarcastically recorded in his diary that,

> Curtin began to speak at length saying that he thought we had worked them out, and I asked him to give me the outline of what they were. In doing that, he used the expression . . . 'the British Government is to consider what it is going to do'. I took these words down in pencil and kept them so as to have that record. This was part of his argument that we were all agreed on plans.[33]

Whitehall could not decide whether India or Australia would provide the base support for the British effort. Churchill and several of his ministers were pressing the Indian case, stressing the abundant and generally pliant Indian workforce, and the savings in shipping. Amery, who had favoured Australia despite being Secretary of State for India, changed sides after hearing Curtin speak of the manpower difficulties in Australia. In a letter to Churchill, Amery asked whether 'anything could be made of Australia without so many off-sets as to more than countervail the advantages'.[34] Amery was writing to the converted.

Meanwhile, Churchill was alarmed that Mountbatten's Indian forces would be drawn into a jungle war in Burma that would preclude the bold amphibious strokes he envisaged along the Burmese coast and, later, against Sumatra.[35] He was determined to revive the Sumatran operation and recapture Britain's Imperial territories in South-East Asia.

Nothing could be done, however, until the great gulf was bridged between Britain's political and military leaders. As the link between the two groups, General Ismay was determined to find some common ground since, as he confided to Mountbatten's Chief of Staff, 'It would be far better to agree to some plan, even though it may not be the best, and to push it forward unitedly and whole-heartedly, than to go on as we are doing at the moment'.

The obvious solution was to opt for what Ismay termed a middle course: a combined British–Dominion effort based upon north-west Australia and striking northward towards Borneo, from where it could either link up with the American Pacific effort or turn southward to descend upon Singapore.[36] It was this compromise that the Chiefs finally adopted, with an 'Empire Force i.e. Australian Divisions and British Fleet to go for Amboina [Ambon Island in the Netherland East Indies] late 1944 or early 1945'. They also conceded Curtin's point that, at least initially, such a force would operate under MacArthur's command.[37] However, Churchill was far from joining them to present a united front to the world.

One of the advantages of the middle course was that Britain could demobilize more of her forces once Germany was defeated rather than dispatch them against Japan. When the Dominion Prime Ministers heard of Britain's demobilization plans, Curtin immediately suspected that Britain wanted to shirk the all-out effort against Japan that Churchill had promised. Attlee assured him that in the demobilization programme, 'military requirements would over-ride all other considerations and the primary objective in view was to make the maximum contribution possible to the defeat of Japan'.[38]

Ernest Bevin added his support, claiming that it was

physically impossible to ship all Britain's three million fighting men to the Far East, and only the surplus would be demobilized.[39] This was all nonsense. Military considerations in the Pacific were the last priority for British manpower. As the Chiefs revealed in choosing the middle course, Britain expected the Dominions to provide much of the manpower in what was now called the British Empire effort against Japan.[40]

After three weeks in London, Curtin had still not achieved his objectives, partly because he avoided informal discussions. Curtin had dodged Churchill's invitation to visit Chequers and had not joined the other Prime Ministers when they accompanied Churchill on a weekend train journey to inspect preparations for the forthcoming D-Day landings. On 21 May, when Curtin finally did visit Churchill at Chequers, it was ostensibly to meet Churchill's daughters. To Curtin's consternation, 'he was ushered into a room where the Prime Minister had [Lord] Leathers, [Lord] Cherwell and [Major General] Hollis armed with a large note book'. Curtin peered through Churchill's cigar smoke for a sign of his daughters bearing an afternoon tea tray, but found himself surrounded and dragooned into discussing the issue of British forces going to Australia. As he later told S. M. Bruce, 'he was taken completely by surprise but had to go in to bat'.[41]

Curtin had to accept that Churchill would not release any of the Australians still fighting in Europe and the Middle East, or the Australian seamen manning some six destroyers of Britain's Eastern Fleet. Curtin's request for the transfer of some Australian squadrons serving in Britain was also refused. Churchill agreed to release some of the Australian ground crew serving in the Middle East if they were replaced from Australia, man for man, in batches of one hundred per month. Many had been overseas for four years, twice the normal tour of duty, and were anxious to return home.[42] Australia would welcome them, but there was still no relief for the Dominion's manpower problem.

However, Churchill did agree to Australia limiting its war effort to the maintenance of six divisions for active operations, with three divisions being kept in the field at any one time; the Australian navy would be kept at its present strength plus new construction; and the RAAF would maintain a strength of fifty-three squadrons. On top of this, Australian food exports to Britain would be kept at the 1944 level. Curtin was keen for Australia to send even more food to Britain.[43]

Despite Curtin's recalcitrance in London, he still wished to reintegrate the Dominion into the Imperial framework. In briefing Curtin, Australian defence chiefs had reaffirmed the principle of Imperial defence, despite the disaster of 1942, with 'each portion to accept a primary responsibility for its own local defence and to co-operate fully with the other portions for the defence of the whole'. The role of the Australian navy would be, as before, 'to deter and delay invasion until the arrival of British Naval reinforcements enables us to assume the offensive'.[44]

Curtin reassured British MPs at a meeting of the Empire Parliamentary Association in the House of Commons, that Australia remained a microcosm of Britain. 'We carry on out there', he said,

> as a British community in the South Seas, and we regard ourselves as the trustees for the British way of life in a part of the world where it is of the utmost significance to the British Commonwealth and to the British nation and to the British Empire—call it by any name that you will—that this land should have in the Antipodes a people and a territory corresponding in purpose and in outlook and in race to the Motherland itself.[45]

Despite this, the Dominion also had a newfound determination, forged out of bitter experience, not to be caught so completely defenceless in any future conflict. Increasing the population through immigration was one arm of this defence; developing an aircraft industry and a strong Air Force was the other arm.

Australia's Director-General of Civil Aviation, Daniel McVey, had preceded Curtin to London via the United States to clinch a proposal to build four-engined, British Lancaster bombers in Australia. In a meeting on 18 April, McVey appealed to Beaverbrook's well-founded fears about the United States, claiming to have

> seen in Washington a map of the Pacific as America hopes to see it after the War. All the Japanese mandated islands are shown annexed to the United States with American air lines flying to China, Malaya, the Philippines, the Dutch East Indies, Australia and New Zealand along entirely American owned island bases.

Beaverbrook realized that McVey's appeal was not disinterested, that the Australian 'naturally wants some of these islands for Australia'. However, he was impressed by McVey's graphic description of American aircraft factories preparing to produce the post-war generation of civil aircraft and by his warning that, unless Britain moved quickly, 'America will scoop the World'.[46]

Although Beaverbrook was sympathetic to McVey's plea, he wanted to avoid discussing civil aviation with the Dominion Prime Ministers, since a Commonwealth approved document had already been submitted to the United States. If discussions were re-opened 'Our agreed document would probably be thrown into the lion's den', warned Beaverbrook, 'and it might not have Daniel's luck'. However, Curtin had demanded that civil aviation be placed on the agenda. The British Government hoped that the discussion could be contained to Australia's idealistic proposal for the complete internationalization of civil aviation and then finished off.[47]

In fact, as McVey confided to Beaverbrook over lunch on 3 May, internationalization was already finished as far as he was concerned. Only Evatt persisted with this ideal, which had been one of the pillars of his Anzac agreement in January, and which he regarded as necessary to safeguard the post-war peace. What the Dominions now wanted, according to McVey, was a scheme of 'Empire Co-operation in Air Transport' in which Britain and the Dominions would form a joint Imperial aviation company that would straddle the globe with a fast, prestige air service. Separate British and Dominion airlines would then operate the slower trunk routes of the Empire. Australian airlines would thus gain access to Imperial airfields and become a major force across the Pacific and perhaps the Indian Ocean. Beaverbrook took up this suggestion, proposing it as the best way of gaining Commonwealth co-operation in the air, although

British officials tried to ensure that Britain had an effective veto power in the proposed joint company.[48]

Britain also wished to meet the American challenge over the supply of civil aircraft. Once Australia began to use American aircraft in civil airlines, British manufacturers would find it difficult to alter that commitment. Wishing to avoid such an outcome, McVey suggested to Curtin that the services order British transport aircraft for war work which would 'permit the design and development of aircraft for post-war commercial purposes, but [that] during the war the aircraft [would be] delivered to the Services stripped of the internal furnishings and fittings required for post-war civil use'.[49]

Curtin wanted the Commonwealth to formulate a united approach on civil aviation so that the Americans could not pick the Dominions off one at a time as they had been attempting to do. The American Ambassador in London, John Winant, had indicated to Curtin on 4 May that the State Department was willing to discuss civil aviation with him on his way back from London. As a result, Curtin and Beaverbrook agreed that a British air route across the Pacific needed to be settled and that landing rights at Honolulu should be sought in return for the already-conceded American landing rights in Australia. As for alternative routes across the Pacific that might avoid an American stop-over, Curtin agreed that they could be explored but felt they would not prove economic.[50]

At MacArthur's prompting, Curtin had refused for the last two years to back P. G. Taylor's survey of an alternative trans-Pacific air route. While Curtin was in London, Taylor was assuming command of a British-organized expedition to assess the viability of his route, when last minute objections from the American naval chief, Admiral King, prevented him. Taylor was furious but returned to the fray, urging Britain to ignore the American objections and alleging that 'U.S. commercial airline people, who dominate both the Navy and Army air transport services, will, and have put every obstacle in the way of this route, which they cannot control in the future because it does not go through any U.S. territory'. Taylor pointed to the damage to Australia and New Zealand if Britain submitted to this

American 'bullying'. The two Dominions, he claimed, were 'very ripe just now for strengthening ties with Britain' and the effect of his survey would 'give a warm feeling to a lot of British people in the S.W. Pacific'.[51]

Although Britain was wary of upsetting the Americans over a peripheral issue, Whitehall wished to survey this route for Britain's effort in the Pacific war, and also in case Anglo–American talks on civil aviation broke down and there was a 'free-for-all'. With British backing reaffirmed, Taylor returned to Washington at the end of June 1944 to try to overcome Admiral King's objections.[52]

As Beaverbrook had intended, there was little progress on post-war civil aviation at the London conference. On the eve of his departure to Canada with Curtin, McVey expressed their disappointment to Beaverbrook. He charged Beaverbrook with deliberately avoiding the issue so that no discussion could take place at the conference. According to McVey's account of the meeting, Beaverbrook admitted to this but then mollified him by promising to back McVey's various proposals and ensure that he received the credit for their success.[53]

In fact, Beaverbrook was hamstrung. Australian disappointment counted for little compared with the Canadian threat to strike out on their own using their strategic position as a stopover *en route* from Europe to America. There was also considerable opposition from the British Air Ministry which wanted resources concentrated solely on winning the war; and there was Churchill who deprecated the likely post-war importance of civil aviation and wished to accommodate the Americans on such contentious issues. With so little progress in London, Curtin declined the American invitation for talks on civil aviation, promising to examine the matter upon his return to Australia.[54]

Once the Dominion Prime Ministers had left London, Churchill was free to concentrate on the forthcoming D-Day operation that would make or break his role in history. Timed for 5 June, the operation was postponed by bad weather in the English Channel. That day at lunch in Downing Street, Churchill was 'in almost a hysterical state' as he waited to see if two years of careful planning would founder on the whim of the English weather.[55] He was banking upon a quick victory in Europe as the Allied armies converged on a cornered Germany. The date for the end of the European war was optimistically set at October 1944, then it would be time to turn against the Japanese in the Pacific. Exactly how this would be done remained to be seen.

16

Pacific Cut-backs

June to September 1944

AT THE BEGINNING of June 1944 Allied troops massing in southern England for the invasion of France waited impatiently for an unseasonal storm to abate. On 6 June conditions had calmed sufficiently for the invasion to begin. Hundreds of aircraft spread like angry locusts across the darkened sky, dropping the first waves of parachute troops on the sleeping countryside of Normandy. Then, as the sun rose on a settling sea, a massive Allied armada forged forward across the English Channel. Behind a murderous curtain of naval artillery and aerial bombing, thousands of Allied troops stumbled ashore among a jumble of amphibious tanks, jeeps and armoured cars. By the time that the soft summer twilight was snuffing out the last vestiges of that blood-stained day, some three thousand Allied troops, uncounted German defenders and hapless civilians lay contorted on the beaches and scattered among the trampled crops.

Despite overwhelming Allied superiority in men and arms, particularly in the air, the invasion force was vulnerable during those first crucial hours and days until they had secured their supply line across the Channel and carved out sufficient countryside to defend in depth against the expected counter-attack. By 8 June the Allied armies had captured Bayeux with its celebrated tapestry of another historic invasion. Four days earlier, American troops had entered

Rome, which had been declared an open city by the retreating Germans. It seemed that German forces would soon be expelled from Italian soil. In Russia, the front line writhed across several thousand miles from Finland south to the Crimea, with the debilitated but still formidable German armies retreating to the homeland from whence they had sprung with such vigour three years previously. It would now just be a matter of time before the Third Reich crumbled, although millions more would be dead by then.

From his distant Bavarian eyrie at Berchtesgaden, Hitler remembered the Armistice of 1918 and hoped that history might repeat itself, perhaps after his secret weapon, the flying bomb, rained terror on England. The Allies also had a secret weapon. Atomic bombs would put Hitler's flying bombs in the shade of their mushroom cloud, provided they worked and did not set off a chain reaction that would destroy the whole world as some scientists speculated. Australian uranium was one of the keys to this awesome secret, and Curtin had agreed in London to hand over this key to the British, promising to develop Australian supplies of uranium and reserve all the production for Britain's use.[1]

The Japanese were still stoutly defending their Pacific empire. Between June and August 1944 Saipan, Tinian and Guam in the Mariana Islands fell in quick and bloody succession to amphibious attacks by American marines while the Japanese Fleet tried unsuccessfully to ward off the blows. With the capture of Guam on 8 August, the United States were within bomber range of Japan's vulnerable cities and, as they revealed twenty-five years later, also of Vietnam's. Meanwhile, the island of Tinian became the base for B29 bombers that would eventually unlock, Pandora-like, the power of the atom.

Curtin left London and travelled via Canada and the United States, where he met the Combined Chiefs of Staff and received their imprimatur to adjust Australia's war effort along the lines worked out with Churchill. This was a victory of sorts, and he announced it as such to his expectant Cabinet upon his arrival in Canberra.[2] More important was the knowledge gained during his visit; he was convinced that Australia would probably not have to support a huge Allied force in the Pacific after Germany's defeat. In a briefing to journalists on 3 July, Curtin dismissed as 'poppycock' the 'idea that millions of men might be based on Australia'. There was insufficient shipping, and the British were planning for a 'great contraction' of their armed forces once Germany was beaten. European armies would not switch to the Pacific, but he assured the disappointed reporters that the British would not welch on their promises, that the 'British flag will fly' and that some British troops would be sent to Australia.[3]

Evatt had been galled to see Curtin fall in uncritically with British foreign policy. As he realized, decisions were being made, or about to be made, that would determine the shape of the post-war world. Australia's wartime contribution gave her a degree of influence in these decisions that would decline markedly once hostilities ended. It would be regrettable if that influence were sacrificed for the sake of superficially restoring better relations between Curtin, Roosevelt and Churchill. Yet this is what Curtin had done, boasting that he had 'got on handsomely with Mr Churchill'[4], thereby satisfying the Australian hunger for acceptance.

Although the Allies had approved changes to the Australian war effort, there was still no agreement about what burdens would be placed upon Australia by the arrival of British forces. American successes in the Pacific were threatening to make redundant the British 'middle' strategy that would strike at Borneo from north-western Australia. Blamey confidently expected to command this strike; it would allow him to escape MacArthur's shadow and lead a Commonwealth force in a combined attack that might well earn him his field marshal's baton. Curtin was as keen as Blamey to see British forces in the Australian theatre but he wanted them on MacArthur's terms.

It was MacArthur, rather than the Australian cabinet or parliament, to whom Curtin reported on his return to Australia. The American general was planning to push on to the Philippines without Australian forces and opposed any Commonwealth force being established on his flank that might diminish the effect of his campaign. As a result, Curtin warned Churchill that the proposed middle strategy was being overtaken by the speed of the American advance and that Britain should simply allocate its naval forces to MacArthur, supplying the maritime muscle to beat his American naval colleagues in the race to Tokyo.[5]

Curtin's timely warning to Churchill reinforced the views of the American Chiefs of Staff who were in London in the wake of the D-Day operation; they claimed that the British timetable for dispatching forces to the Pacific was increasingly out of phase with the progress of the Pacific war. The British Chiefs of Staff welcomed Curtin's proposal as being 'in general accord with our present view of the most effective initial contribution we could make'. It also presented Britain with the 'effective means for restoring Empire prestige in the Far East'. The only 'essential' change to Curtin's plan was for the Combined Chiefs of Staff rather than the American Chiefs to take control of MacArthur's command so that Britain would have a 'proper share in the direction of operations to which we will be so largely contributing'. This was something to which neither MacArthur nor Curtin was prepared to agree.[6]

MacArthur wanted the Allied land effort in the Philippines to be an American one, so he made an offer that he knew Blamey would not accept, but which might satisfy Curtin and maintain the fiction of his readiness to carry the Australians forward with him. MacArthur suggested that he use two Australian divisions, each linked to an American division and part of an American corps. Australian commanders could only operate as individual commanders of separate divisions, and perhaps more to the point, there was no place in such a plan for Blamey, whom MacArthur was now trying to have removed from his position as Allied Land Force Commander. Such an Australian effort might prove militarily valuable but it would sink politically beneath the weight of the American command structure.[7]

Anyway, Churchill had shied away from the middle strategy and, encouraged by the American Chiefs of Staff who were keen to sidetrack Britain's Pacific effort, was again embracing what Britain's naval chief called his 'island idiocy': seizing northern Sumatra as a launching pad for an attack on Singapore. The British Chiefs were angered by Churchill's change of tack, but knew they had broken the solid political support Churchill had enjoyed for his position.[8]

On 12 June Eden advised Churchill that the Foreign Office supported the middle strategy as the one most likely to serve

British interests. These they defined as the complete defeat of Japan; the re-establishment of British prestige, especially in British territories and China; and greater collaboration with the United States and the Soviet Union. Once in Borneo, British forces could join the attack against Japan itself, or descend upon Malaya as the Japanese had done in 1941. However, Eden cautioned Churchill to ensure that Britain could not subsequently be charged with 'having wilfully withheld our support from the United States operations now in progress in the Pacific in order to pursue objectives of our own having no relation to the main business in hand'. This was precisely what Churchill was guilty of, and what he intended to do again.[9]

Although the Chiefs of Staff had conceded that the Commonwealth effort might be under MacArthur's command rather than Mountbatten's, Churchill stipulated that the upstart Royal must have control of it.[10] Churchill wanted the South-East Asian Command (SEAC) area extended to cover north-western Australia so that Britain could avenge the humiliating defeat at Singapore; it was more important to recapture British territories than forge an ongoing alliance with the Americans. As Churchill argued on 24 June, 'Rangoon and Singapore are great names in the British eastern world, and it will be an ill day for Britain if the War ends without our having made a stroke to regain these places'. His even greater desire was to participate in the Pacific on the cheap. He indicated that, if Australian and American forces would pursue the middle strategy under British command, he might support it. However, in a late night meeting of British military and political leaders on 6 July, Churchill continued to stall for time rather than relinquish his Sumatra plan, despite the united opposition of his Chiefs and now his colleagues.[11]

Any strategy would have to accord with Britain's available manpower. The War Cabinet had asked the Chiefs of Staff whether they could reduce their service levels to three million men and women within one year of the defeat of Germany; that would be a reduction from an estimated 4 768 000 at the end of 1944 when the European war was confidently expected to be over. Such a massive reduction was impossible according to the Chiefs. The war against

Japan would need about one million men, Germany must be occupied and the Middle East secured for Britain. All in all, they claimed, 3 404 500 would be necessary, of whom perhaps one million would be Indian and colonial troops. They reminded Churchill of his repeated pledge to join in the war against Japan 'to the limit of our power'. Anyway, they argued, 'for reasons of our prestige in the Far East and in America, it would be inexpedient for us not to do so to a substantial extent'.[12]

As in Australia, the manpower figures were just not adding up. The Chancellor of the Exchequer, Sir John Anderson, pointed out on 14 July that all the demands for manpower could not be met; the War Cabinet would have to cut them, starting with the services and the supply departments. In curtailing military commitments, the War Cabinet should remember that 'the effect on the requirements for munitions labour of a cut in Forces of occupation will be far less than the effect of a cut in Forces taking part in the war against Japan'.[13] The implication was clear: substantial cuts could best be made from the forces planned for the Far East.

At a War Cabinet meeting on 26 July, Churchill described the projected manpower levels for the war against Japan as 'ridiculous'. He suggested that Britain should ignore her pledges over the previous two years and simply ask Washington what support she required against Japan. Britain would therefore be called upon to make a minimum effort, since America was anxious that she be kept out of the Pacific. Churchill's attack on service levels was supported by Lyttelton and Cherwell who, according to Admiral Cunningham, 'both wished to ration the Services for the Japanese war in the interests of (a) getting exports going, (b) raising the standard of living of the Civil population and (c) rebuilding the knocked down edifices and houses'. Cunningham was dismayed, particularly as the Navy was set to play a leading role against Japan. He blasted ministers for trying to squirm out of their pledge and 'fight the war on a limited liability basis', and was relieved when Churchill called a truce, suggesting that he would write a paper setting out a solution to the problem.[14]

Cunningham's relief was short-lived. On 30 July Churchill enquired what sort of Pacific Fleet Britain could have with just 400 000 men instead of the 518 300 planned by the Admiralty. Cunningham was determined 'to bring him down to earth', but it was Cunningham who had to face political realities.[15] On 4 August Churchill issued the Chiefs with a directive, cutting manpower in the services and supply departments by 1 100 000. The Navy was to be cut by 200 000 which Churchill justified by pointing to the 'overpowering [American] Naval force which will be available for the war against Japan'. The Air Force faced similar cuts, with Churchill disparaging the possibility of a large-scale British Air Force effort against Japan. As far as the Army was concerned, the Chiefs of Staff were planning to draw mainly on Indian manpower for their effort against Japan. Of thirteen Indian divisions being used outside India, all but one would be retained for use against Japan.[16]

Australia was also rationalizing its manpower. Primary produce had to be increased to recapture the country's traditional export markets, as well as to feed the Allied armies accumulating in the Pacific. In May 1944 the Government lifted health regulations that restricted the numbers of prisoners of war in rural industries.[17] However, MacArthur refused to reduce his use of civilian labour.[18] There was also a strong push to capitalize on Australia's recent industrial development under the pressure of war. An export committee of the War Cabinet predicted that Australia would be left behind in the post-war export race, due to surplus capacity in Britain and America that would capture local export markets that might otherwise have been won by Australia. The committee urged the promotion of those exports 'which will give us a sound trade foundation and which are so closely linked with the conduct of the war and rehabilitation of re-occupied areas that our effort in that direction would in every way be a direct contribution to the war'. So the development of export trade was added to the manpower equation, throwing out the calculations even further.[19]

To make matters worse, the services resisted all attempts to cut back their numbers. When the Defence Committee reported in August that their monthly intake of men must increase, Curtin ordered that they aim to cut the Army by thirty thousand and the Air Force by fifteen thousand by the end of June 1945.[20] The Navy alone was safe from cuts, although Royle was censured for his overactive pursuit of its interests. Royle's term of office was over, and the Government bowed to British pressure not to promote an Australian officer to the post. Instead, Curtin agreed to a British officer, along with a British Air Force officer to head the RAAF.

In pressing their case for a British naval chief, the Admiralty emphasized the need to appoint an officer with recent experience of using aircraft in naval warfare. With such an officer in place, Australia could be persuaded to accept aircraft-carriers as part of her Navy and thereby play a larger post-war part in the protection of Britain's scattered Pacific possessions. Australia did object when Britain, in a sleight of hand, raised the rank of the officer nominated as naval chief after he had been accepted. This made him the senior service chief and thus automatically chairman of the Defence Committee. The Government refused to accept this, stipulating that the Australian chief must be chairman. Despite this partial demurrer, these two British appointments still left the Defence Committee with a majority of British officers.[21]

When the Chiefs met with Churchill on 14 July for what they hoped would be the final meeting on Pacific strategy, they faced the same mixture of bluster, bombast and delay that they had experienced for the past year.[22] At a meeting in his underground bunker, Churchill once more brought the Culverin operation back to life to the horror of his Ministers and the Chiefs of Staff. Churchill was frustrated that so many servicemen in India were tied up supporting operations in Burma, where the British counter-offensive was pushing back the exhausted Japanese, and objected to contributing 'driblets' to the Pacific as and when the Americans wanted. Britain had the men and she had an Eastern Fleet that was growing in strength but she lacked the amphibious shipping to transfer her army in India across the Bay of Bengal to avenge what Churchill called the 'shameful disaster' of Singapore. He dismissed the meeting with another promise to reach a decision, intending to call in Mountbatten to shore up his embattled position.[23]

So Mountbatten was called to London from his mountain

retreat at Kandy in Ceylon, where life was reported to be so luxurious that junior officers were forced into debt to survive and 2400 tons of shipping was needed each month to supply it from India.[24] Summoning Mountbatten caused further delay despite the Chiefs' warning that facilities in Australia needed to be started now if the middle strategy were to succeed. Before Mountbatten had even arrived, Churchill shifted ground once more with a suggestion (or a threat) that if Culverin were not mounted he would push for an attack against Rangoon rather than agree to the middle strategy.[25] By 8 August Mountbatten was in London, but much to the Chiefs' chagrin, their exhaustive meetings settled 'absolutely nothing'. Churchill returned again and again to the Culverin operation. 'What a drag on the wheel of war this man is', complained Cunningham.[26]

The following day, a decision of sorts was reached. At a morning meeting, Churchill and the Chiefs each presented papers setting out their preferred options for the Far East. There was a surprising amount of agreement and Ismay was directed to draft a joint paper drawing both documents together. The Army chief, General Brooke, quietly suggested that he use the material in the Chiefs' paper, but embroider it with Churchill's phraseology. The ploy worked. Instead of Borneo or Sumatra, the decision was made to strike across the Bay of Bengal to attack Rangoon before the end of 1944. In the view of General Pownall, both sides regarded the Rangoon operation as 'something different from the old old arguments on which they had committed themselves so deeply'.[27] British forces would not become bogged down in northern Burma and Japanese forces would wither away in the Burmese wilderness. Churchill anticipated that the capture of Rangoon would release seven divisions for use elsewhere. From Rangoon, British forces could strike at the Malay Peninsula and Singapore.

Under the agreement, four Indian and two British divisions were to be moved from the European theatre as soon as they could be spared. Britain would offer its fleet in the 'main operations against the mainland of Japan or Formosa', fully expecting that the American Navy would refuse it. Such a refusal, said Churchill, 'would be of enormous value as a bulwark against any accusation that we had not backed them up in the war against Japan'. As Cunningham caustically observed, Churchill 'wants to be able to have on record that the US refused the assistance of the British Fleet in the Pacific. He will be bitterly disappointed if they don't refuse!!!' If the offer were declined, the fleet would be offered as part of a 'British Empire Task Force under a British Commander, consisting of British, Australian and New Zealand, land, sea and air forces, to operate under General MacArthur's supreme command'.[28]

It was a confident show after months of dithering but it concealed many uncertainties. For one thing, the attack on Rangoon still relied for its success on American landing craft, on the war against Germany finishing according to plan, and on forces being made available beforehand from various sources 'including, if necessary, Australia'.[29] None of these things could be guaranteed.

Despite appearances, Churchill still retained his allegiance to the operation against Sumatra and the Chiefs still preferred to switch their effort to the Pacific. Their fleeting agreement on the Far East was overshadowed by disputes about the course of operations in Europe, where the Americans wanted to mount an invasion of southern France with forces from the Italian campaign. Churchill feared, correctly, that a loss of strength in Italy would threaten the resounding victory he hoped to achieve. He proposed instead that an attack be mounted on Brittany to support the Allied forces steadily gaining the upper hand in Normandy. Churchill was overruled by the Americans, leaving him 'raving' and 'absolutely unbalanced', according to Ismay. In this state of mind, he flew off to Italy to see the situation for himself.[30]

Curtin had reminded Churchill of the 'importance to the British Commonwealth of flying the Union Jack in the impending operations in the Pacific'. He had warned that the Americans wished to appear to win the war against Japan single-handed, and that the effect on Britain's eastern Empire would be grave if America were able to claim that she 'fought a war on principle in the Far East and won it relatively unaided while the other Allies including ourselves did very little towards recovering our lost property'. Upon his return from Italy, Churchill assured Curtin that Britain would play its full part in the Pacific war, informing him of the British offer to Washington.[31]

Once again, arrangements were being made without consulting Australia. Although Curtin had pressed for British forces to partake in Pacific operations, he reminded Churchill that he could not countenance MacArthur's command over Australia being compromised by the arrival of independent Commonwealth forces under a British commander. MacArthur also let Churchill know that the British would only be welcome in a subservient position.[32]

Sir Ronald Cross, in London for consultations, was asked if he could 'throw any light upon the troubles that appear to be stirring Mr. Curtin's mind'. Cross blamed MacArthur and Blamey for influencing Curtin, and Curtin's own propensity to support 'anything that may touch Australian nationhood'. The three of them together were 'hypersensitive on command questions' and tended to 'smell a rat' without justification.[33] Cross also reported to Churchill a conversation he had before his departure for London, in which MacArthur had offered to use the British Pacific Fleet as the 'spearhead of his attack upon the Philippines', claiming it as a 'great thing that an American General should sail into Manilla under the British flag'.[34] With these seemingly conflicting signals from Australia, Churchill was understandably puzzled and more than a little annoyed.

Britain did seek a position for herself in the Pacific at the cost of MacArthur's command and with troops partly supplied by Australia. In replying to Curtin, however, Churchill preferred to believe that Curtin had misunderstood the British proposal which, he assured both Curtin and MacArthur, was not designed to interfere with the existing command set-up in the south-west Pacific. Rather than taking Australian troops from MacArthur, it was intended to add a British naval force to MacArthur's command.[35]

This was not strictly true since Churchill had discussed removing Australian troops for the attack on Rangoon and subsequent operations. The capture of Rangoon would release seven divisions from the SEAC, and Churchill planned to join them to six British and Indian divisions from Europe and two Australian divisions to create a formidable force for use against Malaya and Singapore. When Curtin and MacArthur objected, Churchill denied such intentions and said he was on his way to Quebec to discuss the Pacific strategy with Roosevelt and the American Chiefs of Staff. He justified the lack of consultation with Australia by claiming that it was the established and necessary practice for such plans to be drawn up without consulting every country involved in them. Anyway, as he reminded Curtin, 'when you were in England, we told you and General Blamey how our minds were working'.[36]

Although there were discussions with Curtin in London during May and June, there had been no firm Pacific strategy and consequently no agreement reached between Britain and Australia. There should have been some consultation once Churchill and the Chiefs of Staff had reached their compromise in August, but he refused to be dragged into the twentieth century over this continually contentious issue of consultation between London and Canberra. Churchill refused to take Evatt's hint that the appearance of consultation would satisfy the Australians, blithely treating the Commonwealth as Britain's fiefdom. When the local press criticized the lack of Australian representation at the Quebec conference, Curtin dispatched the head of Australia's military mission in Washington, General Lavarack, to Quebec where he would not be a party to the conference but would be available for discussions.[37]

Churchill and the Chiefs of Staff left for Quebec on 5 September on board the *Queen Mary*. Churchill was gloomy and ill-tempered throughout the humid voyage, suffering the effects of anti-malaria tablets combined with another attack of pneumonia. He must also have realized that his role was coming to an end. With Allied forces occupying Paris and pushing the enemy back over the Belgian frontier, Germany must soon be vanquished. His military advisers forecast an early German collapse, but Churchill insisted that the European war might drag on into 1945. In a series of shipboard meetings, he reneged once more on undertakings to boost the British effort against Japan, and ordered his Chiefs not to discuss the option of basing British forces in the south-west Pacific and to stop any transfer of British and Indian troops from Italy to the Far East. Suddenly, the operation against Rangoon was in doubt, as events in Europe claimed Churchill's attention.[38]

Churchill went to Quebec confident that the Americans would refuse his offer of the British Pacific Fleet. Instead, Admiral Leahy informally advised him on 12 September that the Americans would accept. The following day, during the first plenary session of the conference, Roosevelt responded to Churchill's grand gesture by saying, 'No sooner offered than accepted'. Cunningham was ecstatic, but his American counterpart, Admiral King, was livid. Although he grudgingly accepted the *fait accompli*, King made it clear that the British could not expect any logistical support from the American Navy.[39] Churchill was now locked into a strategy that he had rejected for more than a year. General Brooke expressed relief that Churchill had finally adopted their 'sane strategy' and 'now accepts the naval contingent for the Pacific, a Dominion Task Force with MacArthur, etc'.[40] The recapture of Malaya and Singapore now seemed unlikely.

Churchill intended to do a bit of everything: to contribute naval forces to the central Pacific effort against Japan; to partake in the long-range bombing of the Japanese mainland; to offer a naval task force and air units to serve with MacArthur; and to pursue his aim of recapturing British colonies. Despite this plethora of promises, he told his Chiefs of Staff that Singapore was still 'the supreme British objective in the whole of the Indian and Far Eastern theatres'.[41] The main British effort was therefore to be directed against Singapore. Of the forty squadrons for use against Japan, Churchill intended that the majority come from the Dominions, with a large contingent from Canada. The naval task force that might be attached to MacArthur's command would not be an added dollop of British naval strength, but would be detached from the British Pacific Fleet.[42]

Churchill realized that the Americans could criticize the British effort against Japan, so he had armed himself with statistics to show that Britain was holding down many more Japanese than the Americans. The statistics omitted the Japanese by-passed by the Americans, and included on the British side those Japanese held down by Australian forces under American command.[43]

Nevertheless, the press was alive to British reluctance to effectively engage Japan and, to Churchill's annoyance, published stories to that effect while he was meeting with

Roosevelt. At the conclusion of the conference on 16 September, during an informal talk with journalists, Churchill rejected 'suggestions that the British wish to shirk their obligations in the Japanese war, and to throw the whole burden onto the United States'. Quite the opposite, he claimed. The problem was to get the Americans to allow the British into the Pacific, although the conference had sorted this out so that

> Great Britain with her fleet and her air forces and, according to whatever plans are made, her military forces, all that can be carried by the shipping of the world to the scene of action will be represented in the main struggle against Japan. And we shall go on to the end.[44]

That day, the Washington *Times-Herald* called for the United States to go it alone against Japan rather than allow Britain, France, Holland and Russia to contribute token forces at this late hour. These 'Johnny come-latelies who contributed nothing of real weight to the winning of the war' would then demand an equal place at the peace table. If these countries were excluded now, the United States could dictate the peace settlement and 'restore such of those pieces of empire as we decide upon to the British, Dutch, French and Portuguese on our own terms, and they will hold them henceforth by grace of our sufferance and generosity, and the world will know it'.[45]

Churchill was already back-tracking on plans to transfer forces from Europe for the attack against Rangoon. He ordered the Indian divisions to remain in Italy until at least mid-December, and tank landing craft to be retained in the Mediterranean rather than proceed to Mountbatten's command. By the time Churchill had returned to London, the operation against Rangoon had slipped back to November 1945 to protect operations in Europe.[46] The British Pacific Fleet was still due to arrive in the Pacific by early 1945. Curtin was led to believe that it would largely be based in Australia, when it was intended to link it with American naval forces in the central Pacific. When Churchill informed him of the decisions at Quebec, Curtin promised 'complete co-operation' with British plans, but that Australia's contribution would depend on another review of her manpower resources.[47]

At the same time, the powerful Production Executive, a committee of the Australian War Cabinet, reported that the arrival of British forces would place Australia at an economic disadvantage. The Executive recommended that the 'manpower allocations for indirect war and civilian purposes . . . should not be curtailed in order to make provision for British Forces which may be based on Australia'. In particular, construction projects for the British forces 'should not be allowed to absorb men and materials which would otherwise be available for the housing programme'. The Executive must have assumed that British troops as well as British ships would arrive in Australia, whereas Churchill strongly opposed 'sending any British troops to join the Australians and New Zealanders under General MacArthur'.[48]

General MacArthur was pessimistic about Australia's ability to host such a fleet.[49] In a meeting with Curtin in Canberra on 30 September, MacArthur disclosed that he was planning to push northward from Australia as fast as possible, allowing the Dominion's manpower to return to more normal conditions. He claimed that he still intended to take two Australian divisions with him for the attack on the Philippines, after which they could be used against Borneo and then Java. Apart from that, Australian forces would be limited to garrison duties, watching over pockets of Japanese left behind by MacArthur to waste away in their island 'prisons'. If they found these duties 'irksome', they could fight the Japanese, although it would contribute nothing to the Allied effort.[50]

The decisions at Quebec foreshadowed a quickening of the war in the Pacific, although the British commitment remained lukewarm. As General Pownall cynically observed in Kandy, the news from Quebec had transported Mountbatten to the 'seventh Heaven of delight. He is so very simple minded. For my own part, and in his position, I wouldn't get over-excited, or unduly pleased, until I was plumb certain of the resources.'[51] Australia could have done with some of Pownall's cynicism as they prepared to farewell most of the American forces and await the arrival of the British Pacific Fleet—an affirmation of the Dominion's bonds with the British Empire.

17

Taylor's Hollow Triumph

October to December 1944

ON 20 OCTOBER General Douglas MacArthur stepped from a landing craft into the warm waters of Leyte Island. More than seven hundred American and eight Australian ships converged on the Philippines to batter the Japanese defences before disgorging more than 160 000 American troops. As the soldiers fanned out from the beaches to protect MacArthur's historic moment, the General intoned to the people of the Philippines: 'I have returned!' The invasion of Leyte was an all-American operation with the Australians wallowing in MacArthur's wake and the British Navy still committed to the Indian Ocean.

While MacArthur was preparing for the invasion of the Philippines, Stalin had repeated to Eden his commitment to join the war against Japan two or three months after Germany's defeat. Eden had accompanied Churchill to Moscow in early October to discuss the carve-up of Europe between the victorious powers. Churchill was absent from the discussions about Japan, having succumbed to a violent attack of diarrhoea, but he proclaimed Stalin's promise as 'the most important statement at the Conference'.[1] Russian entry into the Pacific war would prevent a prolonged war against Japan.

Japanese forces in Burma were being hammered by combined American and Chinese forces, and General Slim's

British Indian forces were streaming across the border from India after bedraggled Japanese troops defeated in their last large-scale offensive. The capture of distant Rangoon by land suddenly seemed possible if the monsoon rains held off long enough and China, where the Japanese were mopping up large areas of territory, did not rob Slim of his all-important air support.

Australian Army forces had no role in the initial attacks against the Philippines, but Curtin had committed three divisions to be kept in the field. The Australian Government was under political pressure not to leave Australian forces idle while great battles were being fought by Allied armies, and so began what the Australian journalist, Peter Charlton, has termed the 'unnecessary war'. Charlton blamed Australian army generals, and absolved Australian politicians, for organizing offensives against Japanese garrisons cut off from sustenance by the swift strokes of the American advance.[2] These Japanese forces could have been safely contained until they surrendered with the eventual defeat of Japan. Instead, at a cost of more than one thousand Australian lives, and countless Japanese, a series of attacks was launched from Bougainville to Borneo from October 1944 until August 1945.

Charlton's depiction of these attacks as unnecessary is correct, but absolving the politicians is questionable. On 28 September 1944 Curtin had promised the Advisory War Council that he 'would discuss with General MacArthur the question of the acceptance, as an Allied responsibility, of operations for the neutralisation and ultimate liquidation of Japanese forces in rearward areas'.[3] It is clear from this and from other evidence that both Evatt and Curtin were determined to maintain a substantial Australian Army force in action so as to claim a seat at the Pacific peace conference.

Public morale and support for restrictive war measures had to be maintained, and it was difficult to do this without Australian forces in action somewhere. As Shedden advised Curtin on 28 October, an active Army would help to erode widespread public dissatisfaction with the war effort. The problem facing the Government, warned Shedden, was to 'correct the threatened relaxation of the Nation's war effort, lest any drift should later have political repercussions'.[4] The implication seems clear: men were to kill and be killed to shore up the political position of the Labor Party.

Australia was also determined to recover by force of arms those territories she controlled before the war and to establish a claim on the destiny of those colonial territories that she coveted. Australia had emerged from the First World War with the prize of German New Guinea, now she hoped for more, although not necessarily in terms of a formal empire. From New York in mid-October, Evatt's unofficial ambassador and businessman at large, W. S. Robinson, reported on talks with the British Dominions Secretary, Lord Cranbourne; Information Minister, Brendan Bracken; and Lord Beaverbrook. Robinson assured Evatt that he had emphasized Australia's deep interest in the colonial question since 'our defence as well as our economic development very largely depends on our freedom of access and equality of economic opportunity in the New Hebrides, New Guinea, Timor, N.E.I. [Indonesia] Tonkin [Indochina] etc'.[5]

As Robinson indicated, Australia was interested in creating an 'informal' sub-empire in the south seas, with access to the markets and American-built defence facilities of these European colonies. Evatt had considered taking over some of these strategically placed colonies, but had since abandoned the idea in the face of Anglo–American opposition and the problem of defending such acquisitions. Now he wanted a linked British and American defence shield behind which Australia could exercise power over her immediate neighbourhood. As Shedden put it, the events of the war 'should have convinced all Australians that their future will hinge on the maintenance of the goodwill of our powerful friends'.[6] Trade and Customs Minister, Senator Keane, expressed it more bluntly during a tour of North America: 'Great Britain and [the] United States will have to back Australia both during and after the war' since, 'within a few hundred miles' of Australia's seven million whites were 'millions of colored people'.[7]

Britain's attitude towards Australia and the Pacific remained equivocal. Rather than rebuilding her defence shield in the Pacific, Britain wanted to use Australian and other Dominion forces to conceal the paucity of the British effort against Japan.[8] However, for a few months during late 1944, it seemed that Britain might assert her post-war interests in the Pacific more aggressively. It was the Australian aviator, P. G. Taylor, who prompted this when he finally convinced various officers within the Air Ministry to approve his plan for a survey flight of an alternative trans-Pacific air route, even without American permission.[9]

For some reason Admiral King, who had been blocking American agreement, changed his mind in early August 1944, and agreed in principle to the survey flight. Meanwhile, the Air Ministry representative assured King that the flight would be 'only a preliminary survey to ascertain whether or not it is possible to establish a reinforcing air route over the South Pacific'. Any development of the route and associated bases would be subject to further discussion with Washington.[10] The subterfuge, then, was that the route was being developed for military purposes, to allow the transfer to the Pacific of British forces after the defeat of Germany. No mention was to be made of the post-war civil aviation aspect of the alternative route, although as Taylor privately proclaimed in a letter to an Air Ministry official, Britain's 'survival as a first rate nation in the air is the thing which is really at stake'.[11]

Taylor, who had been lobbying in Washington, now readied his flying boat for the epic flight, but King did not grant final permission. Taylor railed against Britain's prevarication in the face of America's stalling tactics, observing angrily in a diary note that it was

> now *8 months* since Britain decided to do the survey flight. We are still standing by. The moon is waning. Pacific nights will be dark, the weather unknown. Diplomats and politicians and senior R.A.F. officers still quibble about trifles and nothing effective is done . . . Australia must set out to establish herself as a self-contained nation. We must not count upon England at all for the future. We must do all we can to hold and strengthen the British Empire but we must not again be left in a state of weakness because of our belief in our strength coming from outside Australia.[12]

In London, the Air Ministry consulted the Foreign Office about whether the survey should proceed without King's final permission. The Foreign Office raised no objection, but then King suddenly announced that he had referred the matter to Roosevelt.[13] This raised the stakes; it would need Churchill's agreement for the survey to go ahead without

Roosevelt's sanction. At the end of August, Taylor was called back to the British base in Montreal. Several weeks later, in another reversal of policy, American approval was received, and Taylor rushed back to Bermuda where his crew was waiting. On Sunday 27 September with his engines at maximum throttle, Taylor lifted his heavily laden aircraft free from the suction of the surrounding water, turned off his radio to block any last minute messages and headed for Mexico on the first leg of his trail-blazing flight.

His first stop in the Pacific was Clipperton Island, a huge submarine mountain, the peak of which rises only some seventy feet above the Pacific. After an eventful six weeks surveying the uninhabited French island for suitable landing sites, during which an awesome hurricane almost destroyed their aircraft, Taylor completed his flight across the Pacific to Sydney. He was confident that his successful survey was being followed up by two British Dakota aircraft loaded with engineering equipment to prepare an air base on Clipperton that might challenge the supremacy of Hawaii as an aerial staging post.[14]

Taylor was met with a signal from RAF Transport Command, advising him that 'international considerations preclude any action beyond [the] completion of [a] comprehensive survey of [the] route'. He was not to hold any talks 'of an official or an unofficial nature' with American representatives about the development of a permanent air route, nor was he allowed to press Australia to develop the route for fear that it would 'embarrass ourselves and may jeopardise future high policy discussions'.[15] For four years Taylor had fought to prove the viability of this alternative route. Exhausted after his long flight, he now faced failure. Returning to Bermuda, this time via Hawaii, Taylor found the Dakotas grounded at Acapulco in Mexico by renewed conflict between Britain and America.[16]

Taylor's survey flight had awoken French interest in Clipperton island, a territory they hardly knew they owned. Mexican interest was also aroused since sovereignty of the island had long been disputed between France and Mexico. With Britain wanting it as staging post on an all-British route round the world, the prospect loomed of a three-way tussle in America's backyard over an island in which

Roosevelt had taken a close personal interest. Upon being advised that American security was threatened, and that the route would neutralize American control of Pacific air routes, Roosevelt called upon Churchill to abandon plans for the survey of Clipperton.[17] In a strongly worded message, he cited the 'Monroe Doctrine, air agreements now under discussion and American public opinion' in calling upon Churchill to 'cancel any instructions by your people about a further survey of Clipperton until you and I can discuss it'.[18]

Churchill immediately complied. He had been unaware of Taylor's plan and turned on his Chiefs of Staff; it was 'astonishing [that] a step like this should be taken without Cabinet authority being sought'. He assured Roosevelt that it had been done without his knowledge and that it was now suspended. Beaverbrook, who seems to have known of the survey all along, agreed that the 'only wise course here is to lay off', observing disingenuously that he was 'most surprised that the decision was ever taken to lay on'. Although the Foreign Office had been involved in Taylor's survey flight, Eden also recommended that any further survey work should be abandoned 'on political grounds'.[19]

The Chiefs of Staff were far from contrite, urging that the more detailed survey of Clipperton be completed, and berating the Americans for being 'unreasonable in preventing our completing a survey of a route not passing through United States territory and designed to join two parts of the British Empire'. The Dominions had shown 'great interest' in the survey and would be disappointed if it were abandoned.[20] The British Government, however, was not prepared to defy Washington over such an issue, particularly when relations with America were already strained by British moves to restore the Greek monarchy by force of arms and when American support would be needed to reconstruct a devastated Europe and contain Russia's expanding influence.

To safeguard the island from both Britain and France, Roosevelt secretly dispatched an armed 'weather reporting party' to occupy Clipperton and pre-empt any further work by Britain. He informed Britain of the *fait accompli* on 4 January 1945, spuriously claiming that it was 'taken as a matter of military urgency as a result of an increase in

Japanese operations in the waters off the western sea frontier'. Britain was advised not to send any unauthorized parties to the island 'in order that there should be no possibility of incidents through mistaken identity'.[21]

On receipt of Roosevelt's peremptory message, Churchill advised the Chiefs of Staff and the Air Ministry that Britain 'really ought to lay off this'. When he learned that the French were sending a ship from Acapulco to Clipperton, Churchill urged Eden to 'tell the Americans about this'.[22] Taylor vainly appealed to Churchill to press ahead despite the American occupation, suggesting that American commercial and military interests, without Roosevelt's knowledge, were seeking to seize the island from France and that Britain should join France in developing an air base there. Taylor claimed correctly that the Americans intended to 'prevent us developing for war purposes what must later emerge as a peacetime asset to us' but he was wrong about Roosevelt. Roosevelt was very much involved in Clipperton, and was determined to exclude Britain, even suggesting that Mexico be given the island so America might then lease it from them.[23]

The Clipperton affair demonstrated the more subservient attitude that Churchill felt Britain must adopt towards America, and raised serious questions about Britain's ability to assert her power on the other side of the world once the war was won. The lesson should not have been lost on the Australian Government who were still commited to reviving the pre-war system of imperial defence, but they had always regarded the Clipperton survey as a personal venture of Taylor's, with Evatt occasionally cheering from the sidelines. Although an alternative aerial link to Europe would have served Australian interests, the Government did not have the imagination to pursue it. Hamstrung by a colonial past, and by the continued dependence which events of the war seemed to confirm as inevitable, Australia watched as her great power allies tussled over the isolated Pacific outcrop.

In civil aviation generally, Australia was similarly sidelined as the Allies competed for supremacy in the post-war skies. Australia was slow to replace its policy of internationalization and was forced to follow London. When

Washington called for a conference on civil aviation, Australia finally formulated a fresh policy; on 25 September Cabinet agreed to support the quick restoration of pre-war Commonwealth services, and the establishment of a new British service across the Pacific via Hawaii to Canada.[24]

The British War Cabinet accelerated the design and production programme for civil aircraft, taking care not to alert the Americans by developing the aircraft under the cover of 'transport types'. In November the British Overseas Airways Corporation (BOAC) was given the first of twenty-five converted Lancaster bombers to resume the route from London to Sydney in co-operation with Qantas. Meanwhile, Commonwealth delegates gathered in Montreal in October 1944 before an international conference in Chicago which would finally obliterate the push for internationalisation of civil aviation and establish the basis for the airline system that continues to this day.[25]

Questions were now being asked in Washington about whether Britain could or should play any part at all in the Pacific war.[26] The Japanese had risked the remnants of the Fleet during the Philippines battle to protect her all-important oil supplies in Borneo. With the loss of the Philippines and the naval force that Japan had husbanded for so long, the enemy's back was broken and there now seemed less need for the British Pacific fleet. However, the kamikaze planes that had appeared first during the Philippines battle would take such a heavy toll of American ships that the more heavily armoured British aircraft-carriers would be summoned.

Britain planned that Australia would provide the initial base facilities for the British Pacific Fleet, despite the difficulties Cunningham foresaw. There was, complained Cunningham, 'a sort of passive resistance about the Australian Gov[ernmen]t's attitude'. At a meeting in London between Admiralty staff and the commander of the Fleet, Admiral Sir Bruce Fraser, it was decided that Fraser would arrive in Australia by battleship in early December and then fly to Hawaii to consult with the American naval chief in the Pacific, Admiral Nimitz. The rest of the fleet would reach Australia by the end of December.[27]

To appease the service men and women who would have

to fight this unpopular war, Churchill introduced measures to improve their amenities and instil in them the will to fight. Their food and entertainment were to be improved, with the beer ration being increased to a minimum of four pints per week. British women were to be encouraged to go to India to help at leave centres and so boost service morale, and Churchill urged the Viceroy of India to 'do all that you can to encourage the British women now in India to play their full part'.[28]

Just how difficult that period was expected to be was revealed on 31 October, when Churchill announced that the first election since 1935 would be held as soon as Germany was defeated. If the so-called second stage of the war lasted for two or three years, no further prolongation of the parliament could be justified, particularly since the Tory majority no longer represented the sentiments of the people. Churchill also realized that his popularity would peak with victory in Europe. If he were to be re-elected, there would be no better time to face the people; if he lost, the Labour Party would then have the unpopular task of continuing the war in the Pacific, maintaining relatively stringent wartime restrictions at the cost of its present popularity. This should ensure that it would be a one-term government and open the way for Churchill's triumphant return.[29]

As the Allies proceeded to destroy the Japanese empire, and their soldiers souvenired human remains,[30] competition intensified for control of captured Japanese territory. Mountbatten tried to galvanize the British into supporting the SEAC with more forces, warning Beaverbrook that General Hurley, Roosevelt's representative in China, had threatened that 'If the British do not capture Singapore by themselves before the end of the war I doubt whether the Americans will let them have it back, in view of their record in the Far East'.[31] Hurley was known for his bluntness. A Foreign Office official described him as 'extremely vain, sometimes rather foolish and always confident that there are no flies on him'. Despite Hurley's bluster, the Americans were committed to the restoration of the tattered British Empire, although they were more equivocal about the return of French possessions.[32] Still, Britain suspected American motives, sometimes with good reason, and kept a watchful eye on her pre-war territory.

In October 1944 the Foreign Office suspected that America might exclude France from the Pacific war and claim Indo-China as part of the China theatre instead of the SEAC; moves they strenuously opposed. Eden saw it as part of an American conspiracy to restrict Britain to a minor role against Japan. Citing a cable from Curtin in support, Eden argued strongly that Britain 'should play a major rôle in the war against Japan and that our contribution should be not merely effective but spectacular'.[33] At a conference in Cairo between Eden, Churchill and Mountbatten in November, Eden impressed upon a relieved Mountbatten the need to recapture at least Singapore and Rangoon before the Americans forced the Japanese to the peace table.[34]

Eden also pressed for the return of Indo-China to France since:

> Any attempt to interfere with French sovereignty over Indo-China . . . would be passionately resented by France and would have incalculable results not only in the Far East but in Europe. It would also put in question the future of all other Far Eastern colonial possessions which have been overrun by Japan.[35]

In other words, if Indo-China were taken from France, British colonies might be taken away also, and the prospects for a strong and united France in Europe might be complicated. Such self-centred calculations created the conditions for three successive wars that continue to convulse the emerging nations of Indo-China.

At the end of October 1944 Australian and New Zealand Ministers met in Wellington to continue the discussions from the previous January that had so outraged British and American governments. The meeting Evatt had wished for would have included America and the European countries with Pacific interests; but now the Dominions talked in a vacuum since they could not expect to impose their design for the Pacific on their great power allies.

The Wellington meeting, between the Australian Deputy Prime Minister, Frank Forde, Evatt, and New Zealand representatives, called for a South Seas Regional Commission that would involve the Pacific peoples. The meeting was largely intended to buttress Dominion claims to a major voice in the peace settlement with Japan, and reinforce their determination to restrict American post-war control to the northern Pacific. It also challenged European empires by calling for all colonies to be administered by the United Nations as trustees, with the regular international inspections and benevolent development that implied, rather than exploitation and continued subjugation.[36]

Britain took particular exception to this part of the declaration. Although the Dominions were rarely consulted before Britain issued declarations on matters of interest to them, the British Government was incensed that the Dominions might do likewise. Cranborne urged the British War Cabinet to 'make clear to them [Australia and New Zealand] the deplorable results which are likely to ensue if regional discussions lead to public declarations of policy on matters of interest to other members of the Commonwealth without prior consultation'.[37]

At the Prime Ministers' meeting in May, London had plainly objected to the mandate system including colonies of the British Empire. They had understood, complained Cranborne, 'that Mr. Curtin (who was now ill) shared our view'. The call from the Wellington meeting was badly timed, embarrassing Britain when she was about to approach the United States on the matter of post-war colonial policy. In the absence of Churchill, who was visiting Moscow, the War Cabinet approved a cable expressing displeasure at the declaration and advising that Britain would have to issue a public statement on colonial policy that would disagree with the Dominion line.[38]

The Dominion declaration revealed a rift in the Commonwealth that could be exploited by the Americans, and Roosevelt made just such an overture during his first meeting with the new Australian minister in Washington, Frederic Eggleston. Although Roosevelt 'didn't like the wording' of the Australia–New Zealand Agreement, which attempted to limit American influence to north of the equator, he realized that there was room for agreement with Evatt over colonies. Roosevelt claimed that Britain was coveting the Netherlands East Indies but should instead offer independence to Malaya and Burma as America was doing for the Philippines. He confided that 'he wanted to keep China as a friend because in 40 or 50 years China might easily become a very powerful military nation'. Roosevelt 'thought the Americans and the Australians could work together on a liberal policy on these matters'.[39]

With Curtin seriously ill from the heart disease that would eventually kill him, Evatt could reassert Australia's commitment to the ANZAC agreement. When Britain took umbrage at the Wellington declaration, Evatt denied that the Dominions had shifted ground since the Prime Ministers' meeting in London. Australia's position, he argued, had been established by the ANZAC agreement in January and nothing since then had altered. Evatt was on weak ground since Curtin had cabled from London in May his 'agreement with the general principles enunciated by the Colonial Secretary' at the Prime Ministers' meeting. Evatt simply denied such an agreement, and argued that the May meeting was 'merely a personal exchange of views'. He blamed elements within the Colonial Office for the British objections, believing that neither the Labour members of the War Cabinet nor even Cranborne really supported these objections from 'the distant past'. Although Australia desired good relations with Britain, Evatt cautioned Bruce that 'we must not be asked to surrender principles to which we are committed. Still less should we be chided for our views by the Colonial Office'. In fact, Evatt confided that he was prepared to accept the British proposal for regional commissions, rather than an international commission, to oversee colonies.[40]

Britain likewise dismissed Evatt's retort as a minority view sent in the absence of Curtin. Accordingly, the British War Cabinet repeated their objection to the Dominions holding discussions without consulting concerned members of the Commonwealth. Britain approved in principle the South Seas Regional Commission, but its implementation depended on agreement with the United States and assurances that it would not interfere with Britain's colonies.[41]

Bruce counselled Evatt to adopt an attitude of 'sweet reasonableness' on the issue of trusteeship, adding that the Dominions should have consulted London before issuing their declaration. In a private cable to Curtin, which found its way to Evatt, Bruce described Whitehall's dismay at the

Wellington meeting and his embarrassment at Australia changing policy without consulting Britain. Bruce had repeatedly harrassed Churchill over the issue of consultation, and now he stood accused of the same offence. He suggested that, before the declaration was endorsed, the clause relating to mandates be quietly dropped. When Evatt saw Bruce's latest attempt to dilute policies he held dear, he controlled his anger with great difficulty, then redirected it at the antediluvian officials of the Colonial Office, the real villains in his eyes.[42]

Evatt had been misled by idealistic pronouncements from the executive committee of the British Labour Party about the future of British colonies. Parliamentary Labour leaders such as Attlee and Bevan did not share these views, nor, of course, did the Tories. Cranborne had received strong Conservative backing for his stand, with Lord Croft comparing international supervision of British colonies with outside 'supervision of Tasmania', maintaining that the British Parliament would never 'consent to sharing its duties with people who have no concern whatever with the British Colonial Empire, and I hope that fact will be made plain'.[43]

Britain moved quickly to pre-empt any move by Washington to 'put forward schemes which were unsatisfactory from our point of view, but [which] might attract support'; they proposed a system of regional commissions dominated by Imperial powers which would prevent international interference with British colonies. Continued British control would be disguised by fine words about economic and social progress for native peoples. During a War Cabinet discussion of colonial policy, the Chancellor of the Exchequer, Sir John Anderson, warned that care should be taken 'to ensure that such bodies [to promote colonial welfare] were not a means of putting pressure on us to spend money which we had not got'.[44]

Despite the best British efforts to avoid it, Churchill was manoeuvred into signing an American-inspired declaration during the Yalta conference in early February 1945 which made British mandated territories subject to trusteeship. Although he tried to ensure that the matter did not receive detailed discussion during the San Francisco conference in mid-1945, at which the United Nations Organization was established, international supervision of colonial territories was practically a *fait accompli.*[45]

While statesmen were sparring about colonial territories, MacArthur was recapturing the Philippines at great cost to the Filipinos, just as his father had done nearly half a century before. In Burma, the failing Japanese Army was conceding territory to the Allies, as they withdrew before the expected onslaught. Australian troops on Bougainville welcomed in the New Year with a strategically senseless offensive across daunting terrain against the debilitated Japanese garrison, with Blamey's exhortation to 'exterminate these vermin' ringing in their ears.[46] Meanwhile, the people of Perth paid a belated welcome to the advance detachment of the British Pacific fleet.

18

Secret Agreement at Yalta

January to April 1945

THE WORLD WAS still very much at war as 1945 dawned. A German counter-offensive through the 'impassable' mountains of the Ardennes had threatened the military timetable for the invasion of Germany. The German offensive was contained, but it delayed Allied progress. British entry into Athens had sparked an internecine struggle for supremacy between the various political factions, and news of British soldiers shooting Greek civilians provoked widespread protests, and divided the Allies, with Canada, Australia and the United States critical of British action.

In the Pacific, General MacArthur had destroyed Japanese resistance on Leyte Island and in early January swooped onto the main Philippine island of Luzon. By the beginning of February, American troops had entered the suburbs of Manila, although it would take another month and some 100 000 civilian deaths before victory was won. Meanwhile, American Superfortress bombers were raiding Japanese cities and would soon begin dropping the new and deadly napalm that would engulf Tokyo and other cities in fire. In lonely Los Alamos, New Mexico, physicists raced to perfect a means of detonating an atomic explosion; the American Government expected to have eighteen plutonium bombs ready by August 1945.[1]

Further south, Australian troops had taken over from the

Americans in New Guinea, Bougainville and New Britain and soon began offensives against these left-over Japanese troops. For several months, military censors enforced MacArthur's news blackout to confuse the Japanese about Allied troop movements and, perhaps, to emphasize his achievements in the Philippines. Nothing had been heard of Australian troops in action, and although they had been meant to join MacArthur in the second wave of attacks on the Philippines, the invasion of Luzon proceeded without a slouch hat in sight.

By 10 January 1945 the *Canberra Times* implored 'anyone knowing the whereabouts of Australian soldiers in action in the South-West Pacific Area [to] please communicate at once with the Australian Government'.[2] As soon as the whereabouts of the troops became public, the conservative Opposition loudly decried the use of first class fighting forces in 'mopping up' operations of no strategic significance, and the press complained that American forces in the Philippines were pushing Australian operations off the front page.

The Labor government was keen to restore the Empire's prestige and Australia's place in it.[3] Thus, when Britain asked Australia to help provide garrison forces for British island colonies recaptured from the Japanese, Curtin acknowledged Australia's willingness but complained that he was precluded from complying by the commitment to MacArthur. Britain was concerned about American reluctance to leave captured British islands, and feared that Washington might claim them permanently by right of conquest. A week later, Curtin amended his reply to allow Australian troops to garrison the Pacific island of Nauru, with its rich resources of phosphate.[4]

Australian farmers had been starved of phosphates for three years and wheat yields had declined alarmingly as a consequence. Nauru was within the southern part of the Pacific from which Australia wished to exclude the Americans, so offering a garrison for Nauru would not only stake out this claim but also safeguard future phosphate resources for Australian farmers and the rural vote for the Labor Party.

Curtin may have had this in mind when he told the House

of Representatives on 22 February that, while Australia had 'a major political and national interest in the British Empire', she had 'a major political issue nearer home, and that is to clear out the enemy who is still in occupation of territories for which this Government is politically responsible'. He rebutted Opposition charges that the Army was idle, claiming that its whole strength was under MacArthur's command and that Blamey's island campaign was 'not only commensurate with all the ambitions and the duty of this country, but is also clear evidence of the undoubted genuineness of Australia . . . to fight where the fighting is hardest'.[5]

Although British observers in Australia had been predicting for more than a year that MacArthur would leave the Australians behind once the Allied armies had fought their way through the outer ring of the Japanese defences, its confirmation seemed to surprise and disappoint the Australians. On 15 February 1945 Shedden had written to MacArthur asking why Australian troops were not being used in the Philippines and claiming that 'Australian opinion considered it a point of honour' for their troops to be involved in such operations.[6]

After the war, MacArthur justified neglecting the Australians by claiming that Blamey had made it so inconvenient that he used two American divisions instead. He suggested that Blamey wanted to retain the Australian forces to support his bid for the command of the British Commonwealth area, then being mooted to replace MacArthur's command in the SWPA.[7] In fact the ambitions of both generals precluded the use of the Australians in the Philippines.

It was Churchill as much as Curtin who impeded any commitment of Australian forces. Churchill had pencilled in two Australian divisions for the SEAC operations against Burma and Malaya, but he steadfastly opposed Mountbatten visiting Australia even though such a visit might have secured for Britain a commitment of Australian forces to the SEAC. It would certainly have increased political pressure on the Australian Government to do so. Although Mountbatten and MacArthur were Supreme Allied Commanders in adjacent areas, they did not meet until mid-July 1945, and then only because MacArthur had shifted his headquarters from Australia to Manila. In June 1944 Mountbatten had written to Sir Keith Murdoch advising that he would soon ask the British Chiefs of Staff for permission to visit MacArthur in Australia.[8] Although they granted permission, Mountbatten was unable to leave during important operations in Burma.

In February 1945 Sir Ronald Cross suggested that Mountbatten use his postponed visit to confer also with the Governor-General, the Australian Government and its Chiefs of Staff. Churchill reluctantly agreed to a brief visit provided there were no press interviews, fearing perhaps that Mountbatten would use such a public platform to demand a greater British effort in the Far East. The British and American Chiefs of Staff raised no objection to the meeting, but when it came back to Churchill for final ratification on 7 March, the Prime Minister suddenly refused to give it, maintaining that 'there can surely be no need for him to go to Australia' and questioning 'the actual business he has to transact with General MacArthur'. Ismay pointed out the embarrassment of refusing this visit after asking the Americans to approve it. He suggested that Canberra be cut from Mountbatten's itinerary, which would limit the visit to two weeks, and that the ban on press interviews be maintained. But Churchill was adamant that the visit be cancelled.[9]

In ignorance of this, Cross informed London that Canberra should not be omitted from Mountbatten's visit in case it appeared that Australia were not being consulted. The newly arrived Governor General, the Duke of Gloucester, also advised through Buckingham Palace that Mountbatten should talk with Curtin, the all-party Advisory War Council and the Australian press. These messages caused the issue to resurface.[10]

A lull in the SEAC after Rangoon was recaptured prevented Churchill from objecting to the visit for operational reasons, but he still insisted that any visit be as brief as possible and that the meeting not be held in Australia. This curious stipulation might have prevented a meeting, but then Ismay advised that MacArthur's headquarters had shifted to the Philippines. In a note to Churchill on 2 June, Ismay revealed that Mountbatten also wanted to visit London for consultations after the defeat of Germany and presumably to remind the British public of the war against Japan. The last thing Churchill wanted in the flush of European victory and in the midst of an election, was for the British electorate to be reminded about distant battles to be fought and further privation to be endured. Ismay suggested that Mountbatten be kept out of Britain until after the August election, that he be allowed to visit MacArthur in late June, 'but that he must not go on to Australia, or give Press conferences anywhere'. 'Yes, exactly', Churchill scribbled.[11]

With a shipping shortage hampering operations in Europe, Britain regarded Pacific operations as a dangerous diversion from the main effort against Germany.[12] It might have been difficult to resist pressure for a greater British effort against Japan if there were a well-publicized meeting in Canberra between Mountbatten, MacArthur and Curtin. Churchill clearly wanted to avoid such a conference. His continuing hostility to Australia must also have played a part, exacerbated by four years of often bitter disputation.[13]

Part of this hostility sprang from a feeling that Australia was profiting from the war at Britain's expense. In a paper prepared by Lord Keynes on 3 April 1945, it was calculated that Australia had made a profit of £94 million in 1944, after her payments for the war were deducted from overseas payments made to her for war purchases and other services. Keynes claimed that Canada was 'doing her full duty' while Australia was 'scarcely doing a thing'.[14] London also resented the limited extent of rationing in Australia compared with that in Britain. Churchill had ordered a study which found that Australia's traditional and continuing reliance on meat gave the appearance of comparative luxury to the Australian ration scale. In fact, British and Australian diets were nutritionally similar, although the Australians enjoyed a more palatable and varied diet.[15]

The Pacific Fleet provided the main thrust of the British effort in the Pacific. Its first echelons had arrived in Fremantle on 11 December 1944 under the command of Admiral Sir Bruce Fraser. The bulk of the Fleet arrived a month later after making diversionary attacks on Sumatra to cover the American invasion of Luzon. They were the first British forces to arrive in Australia apart from the token Spitfire squadrons. Eventually, the British Pacific Fleet would comprise about a third of British naval strength, some hundred ships of various types, but it was dwarfed by the Americans who managed to maintain four separate task forces in the Pacific, each of the same strength as the British one.[16]

Although Australia had repeatedly stated that her manpower shortage would limit her contribution to the support of the British Pacific Fleet, London was slow to understand.[17] On 7 December 1944 the Australian War Cabinet finally set its contribution at £22 150 000, which was made up of various resources from barracks to canned meat and dental equipment supplied to the Fleet. The civilian economy already had a deficiency of 85 000 men, and the Government tried to switch those Australian civilians supporting American forces to the support of their British brethren.[18]

In Britain, similar manpower and shipping shortages forced the Government to reassess its planned effort against Japan. On 2 December 1944 the Government accepted a plan by the Labour Minister, Ernest Bevin, for the gradual re-allocation of manpower' from the services to industry and reconstruction. Bevin did not use the word demobilization, although that is what it was. The call-up of eighteen-year-olds and those previously deferred from service would provide manpower for the war against Japan. [19] All Britain's plans had been overturned by the prolongation of the European war.

The Fleet had been dispatched to the Pacific on the assumption that victory in Europe was close. When the war dragged on through the winter of 1944–45, the strain on shipping became intense, particularly as Britain was trying to prepare for post-war redevelopment as well as fight two wars. Keynes returned from talks in Washington brandishing the news that the system of Lend-Lease would continue in stage two of the war and that the United States was prepared to give Britain 'complete freedom to export (subject always to the needs of the war against Japan)'.[20]

On 19 January, Cunningham wrote to Fraser of the 'great danger that we may dig ourselves in too deeply in Australia', urging that he base his fleet

> as far forward as possible . . . and ship all our stores, aircraft, etc. direct there from the U.K., using the Torres Straits or

> perhaps later going up the west side of New Guinea. This would mean an immense saving in shipping, which is going to be the bottleneck in all operations this next year, especially if the new enemy submarines get really busy.

Three days later, at a meeting of the Chiefs of Staff, the Shipping Minister attacked the size of the fleet train for the Pacific fleet. Cunningham was forced to admit that 'the build up of the Pacific fleet both physical and logistical must be slowed up'. He ordered that aircraft-carriers destined for the Pacific be retained in the Indian Ocean until sufficient shipping was available to support them.[21]

That same day, Lord Cherwell had advised Churchill to reject an Admiralty request for an increase in the fleet train 'in view of the great shortage of shipping and the consequent threatened fall in our imports and our stocks'. Meanwhile, Churchill was supporting moves to have the New Zealand division in Europe sent to the SEAC after Germany's defeat rather than be repatriated. On 26 January the Admiralty request came before the War Cabinet where it was roundly attacked by Ministers who questioned the need for so much shipping. A decision was deferred for two months, and Cunningham undertook to re-examine the figures.[22]

With the Pacific commitment causing increasing concern, Churchill, with an entourage of Ministers and officials, left for a conference with Roosevelt and Stalin in the Crimean city of Yalta. Bruce had urged Curtin before the meeting to remind Churchill of Australia's vital interest in the discussions, particularly those concerning the Pacific, and had deplored Britain's lack of consultation. Curtin was certainly concerned, but his poor health had sapped the fight out of him. He had, he complained to a visiting British official,

> received the agenda for the Yalta conference; there were several items which vitally affected Australia; apparently he was not to be consulted in advance. On this and similar happenings he remarked: 'The time will come when someone sitting in this chair will say: "I won't put up with it"'.[23]

Curtin was no longer the man to do it and his office still waits for that declaration of Australian independence.

On the eve of their departure from London, Churchill had

informed Eden of his eagerness to have Russia enter the war against Japan as soon as possible after the defeat of Germany. This was despite the fact that Russia, through its capture of a warm water port, might become a Pacific power, and despite his strenuous opposition to Russia entering the war in 1941–2 when the security of Australia was at stake. Now more direct British interests were involved, so the situation was different. As Churchill put it, a quick end to the Pacific war 'such as might be procured by the mere fact of a Russian declaration against Japan, would undoubtedly save us many thousands of millions of pounds'.[24] Such a quick end might also coincide with the British election campaign and win extra kudos for Churchill.

British and American officials *en route* to Yalta met in the British colony of Malta. Cunningham found that the Americans' only firm plan for the Pacific was an invasion of the Tokyo area in December 1945. The aim was to avoid major land battles with the Japanese that would lead to heavy casualties and hamper the flow of the Allied advance. Just as the British intended to use Dominion and other forces, so the Americans intended to use the reconstituted Philippine army, and guerrilla forces, to mop up the remaining Japanese in the Philippines. The Americans finally indicated that the Australians would probably not be needed in the Philippines operation.[25]

The following day, Churchill met Roosevelt in the President's cabin aboard the cruiser *USS Quincy*. Churchill exhibited his usual eagerness to partake in the Pacific fighting. His aim, he claimed, 'was to go where a good opportunity would be presented of heavy fighting with the Japanese'. He singled out air warfare as 'the only way which the British had been able to discover of helping the main American operations in the Pacific'. This was a curious comment given the lack of British Air Force units in the Pacific, other than the token Spitfires in Australia. Twice Churchill offered British troops for deployment in China but, predictably, was rejected by the American Chiefs of Staff.[26] As Churchill well knew, the Americans were not about to share with the British their special place at the right hand of Chiang Kai-shek. But Churchill had achieved his aim: the public record would show that he had offered troops for the Pacific and the Americans had rejected them.[27]

At a meeting in Yalta on 9 February between Churchill, Roosevelt, and their Chiefs of Staff, Churchill argued that Russia should join in issuing 'a Four-Power ultimatum calling upon Japan to surrender unconditionally, or else be subjected to the overwhelming weight of all the forces of the four Powers'. If Japan then suggested terms short of unconditional surrender, it would be for the United States to determine the Allied response, however, Churchill's own view was that 'some mitigation would be worth while if it led to the saving of a year or a year and half in which so much blood and treasure would be poured out'. A frail Roosevelt agreed that the suggestion could be put to Stalin but thought the ultimatum was premature since the Japanese 'still seemed to think that they might get a satisfactory compromise'. More mass bombing attacks on the Japanese home islands were necessary, Roosevelt claimed, before the Japanese would 'wake up to the true state of affairs'.[28]

In Europe, Russian troops were already approaching the outskirts of Berlin and Allied troops were pushing the Germans back across the Rhine. It was not surprising that at Yalta, most of the discussions concerned Europe. When the Pacific war was discussed, Churchill again supported Stalin's desire for the warm water port that had been lost to Japan in 1905. In a secret agreement, Stalin promised to launch his Siberian army against the Japanese as soon as it could be organized after the defeat of Germany. This agreement was omitted from the official communiqué at the end of the conference, and Churchill ordered that there was 'no need whatever to inform the Dominions'.[29]

Britain's strategy in the Pacific was now under attack on several fronts: Churchill wished to negotiate an end to the war before the British Pacific Fleet could show its mettle; Admiral King delayed a decision on when and where the Fleet would see action; the Australian Government restricted the amount of assistance to the Fleet and tried to divert resources to civilian pursuits; and the British War Cabinet was reassessing its manpower resources and fixing priorities for their use.

The British Government had planned to switch 315 000 workers from munitions production to civilian production during the first half of 1945, given an early end to the

European war. The delay caused a one-third cut in the number of workers available for civilian work, and the situation was exacerbated when the Board of Trade called for 375 000 additional workers in civilian industries to produce textiles, clothing, furniture, and house fittings, to begin the re-equipment of industry and to boost exports. While Churchill was in Yalta, Sir John Anderson warned of the 'grave difficulties' if these requirements were not met, and called for the War Cabinet to reconsider the whole issue 'including the strategic issues involved'.[30] The simple fact was that the greater the British effort in the Pacific, the fewer the new houses built for returned service personnel, the fewer the new clothes available in British shops for frustrated, coupon-clutching customers, and the less vigorous Britain's export drive to markets such as China which had been starved of Western consumer goods.

Cunningham reacted with dismay to Anderson's attempt to hamper his ships. Fearing a hostile American reaction, Cunningham advised that it would be most unwise to 'restrict operations against Japan so as to release men for production of Civil goods to better the condition of the public'.[31] But there was an election within sight and Churchill had not been in the House of Commons for most of his life without learning a few basic political lessons.

On 26 February Churchill issued a directive to the War Cabinet designed to give the civilian economy a 'considerable reinforcement'. The Board of Trade was allocated 275 000 additional workers for the first half of 1945, but, as always, the European war retained first place followed by the expansion of civil production. Although Churchill did not cut the level of forces allocated to the war against Japan, he did urge that 'there can be some delay in their build-up and equipment, including reserves'. He ordered that older equipment be used against Japan, and that only essential items be sent out. 'Over-insurance in provision', he wrote, 'is a luxury we cannot afford'.[32]

When new German submarines appeared to threaten Britain's maritime lifeline, the Admiralty retained escort ships in European waters. This meant that some of the aircraft-carriers and battleships *en route* to the Pacific had to be held in the Indian Ocean until escort ships became available to protect them from Japanese submarines. On 1 March Cunningham decided to put the 'greatest effort into holding the U boats for the next 3 months' with everything being 'sacrificed to that object'.[33] In fact, the German submarine menace was not as dangerous as it had seemed, but in combination with German jet-powered aircraft and the V2 rocket, it alarmed British leaders.

Meanwhile, the Americans continued to delay a decision on the deployment of the British Pacific Fleet. America's Admiral Cooke, who had been at the Yalta conference in his new role as Deputy Director of Naval Operations, met Mountbatten to incite him to request the use of the British Pacific Fleet in the SEAC. Cooke, described by one British admiral as a 'horrible little man', succeeded. For two years, Mountbatten had been operating on a series of promises from Whitehall that had been successively withdrawn or amended, and the latest manpower crisis in Britain seemed set to delay his operations against Burma and Malaya. So he readily agreed to take up Cooke's suggestion that the Pacific Fleet return to the SEAC to take part in an operation against Phuket. As Cunningham observed with exasperation, Mountbatten was 'stupid enough' to cable this 'infuriating' suggestion to London, thereby 'allowing himself to be made a catspaw of to help the elements in the Navy Dep[artmen]t. who wish to prevent the fleet operating in the Pacific'.[34]

The Chiefs were worried that Churchill would fall for the American-inspired ploy. In a signal to Washington on 7 March they refuted Mountbatten's suggestion, citing to Churchill Mountbatten's admission that 'adequate air cover can be provided by the East Indies Fleet'. They felt 'sure that you [Churchill] would not entertain any idea of diverting units of the British Pacific Fleet for this operation'. They also sent a strong message to their American counterparts, asking for a decision on the deployment of the Fleet in the Pacific, but the Americans were temporizing. MacArthur wished to retain the Fleet within his command, possibly for use against Borneo, while Admiral Nimitz wished to use it in the forthcoming invasion of Okinawa, the final stepping stone to Japan.[35]

On 8 March Admiral Somerville tackled King in Washington, pressing for a quick decision on the British Fleet.[36]

Meanwhile, Fraser kept his ships at Manus Island, north of New Guinea, 'a dismal place' according to Fraser, where they waited in the tropical heat for their assignment. In early March the fleet stayed in Sydney for two weeks, raising the morale of the sailors but straining facilities to the utmost. Still awaiting a decision, Fraser withdrew his spruced-up ships to Manus for more desultory battle practice. Finally Nimitz was assigned the British Fleet and it was quickly deployed south-west of Okinawa to cover the American invasion, timed for Easter Sunday, 1 April 1945.[37]

By this time, the Japanese Fleet was all but destroyed. In early March, the battleship strength of the combatants was twenty-six American, four British and six Japanese; the aircraft-carriers included twenty-five American and six British against three Japanese. The Americans also had sixty-three escort carriers against one Japanese. There was a similar predominance in cruisers and destroyers.[38] However, the Allies had to operate at the end of extended lines of communication while the Japanese Fleet was concentrated in its home waters. The Allies also had to dodge or destroy the thousands of kamikaze planes thrown up as a final defence. Despite this, Allied power at this stage of the war was overwhelming.

Both Australia and Britain intended to limit their effort against Japan to the extent that it would guarantee influence at the peace talks and return to Britain the colonies she had lost. To reduce the forces dispatched from Britain, Churchill approved a plan to shift the British aircrew from the three Spitfire squadrons in Australia to Mountbatten's SEAC.[39] Meanwhile, the Chiefs of Staff pressed ahead with plans to establish a Very Long Range (VLR) bombing force for use against Japan, calling for Churchill's personal support to prevent it being 'elbowed out' by the Americans. To avoid American claims of logistical problems delaying the deployment of the bombing force, the Chiefs intended to make it self-contained, with its own construction units capable of creating an airfield on the Philippine island of Luzon. The Dominions would provide much of this force, particularly Canada, and the Chiefs asked Churchill to 'enlist Mr. Mackenzie King's interest and support'. When the Canadian Prime Minister's response to the proposal was luke-warm,

Churchill asked whether Britain could do without their assistance. Definitely not, replied the hard-pressed Chiefs.[40]

Australia was also proving difficult: Curtin made any contribution to the VLR force conditional on the return of Australian airmen from the European theatre. This bargain fell on deaf ears in London where Churchill was trying to juggle manpower figures in a way that would be electorally advantageous. The VLR force shrank until an additional Australian contribution was unnecessary, apart from two bomber squadrons and one transport squadron to be based in Britain but available to support the VLR force if required. In the event, the Japanese surrender pre-empted the formation of the VLR force.[41]

Britain's attempts to dodge her responsibilites reflected her waning power. In a report to the War Cabinet on future British policy towards the Suez Canal, Attlee called for its internationalization, and for the United States to be drawn into a post-war defence commitment in the Mediterranean. 'The time has gone', Attlee claimed, 'when Great Britain could afford to police the seas of the world for the benefit of others'. Attlee portrayed the Empire, which had previously been Britain's strength, as an albatross around the neck of the struggling British taxpayer.[42] At the same time, he was imposing these views on the problem of reconciling Britain's manpower resources with her commitments against Japan, and the military chiefs were being forced to submit to the political imperatives of the War Cabinet.

On 22 March the Production Minister, Oliver Lyttelton, pressed Churchill to set a limit to the forces Britain would deploy against Japan. Previous estimates had been based on outdated assumptions about the duration of both wars. Following the Ardennes offensive, it had seemed that the European war could drag on to the end of 1945, but with German armies now collapsing faster than anticipated, the war suddenly seemed likely to end by 31 May. Since the United States was secure in the Philippines and able to interdict the Japanese supply route to and from South-East Asia, that war also would end much sooner and the British contribution be that much less as a consequence.

Lyttelton wanted to reduce these forces to token proportions so that Britain could more quickly 'raise our standards in this country and regain our exports'. He suggested that Churchill remind the Chiefs, when drawing up their estimated forces for the new situation to

> bear in mind that:-
> (i) as regards British Pacific Fleet and V.L.R. Bomber Force, our contribution is but to add (for potent political reasons) to the already dominating strength of which the Americans dispose, and that in consequence there is no need for insurance.
> (ii) in view of the speed and priority of American re-deployment our target forces should be limited to those we can bring into action within a stated period from V.E. Day—say twelve months.[43]

Apart from limiting the Pacific forces to improve British living standards, there was also pressure to maximize the demobilization of British forces in Europe as soon as possible after VE Day.

Whether troops were being sent east to the Pacific or home to Britain, personnel shipping was required. On 24 March Attlee set out the priorities for such shipping. Top priority was given to demobilization; the return of service personnel on leave after extended duty; moves for approved Pacific operations; and the return of British Commonwealth prisoners of war. Second priority was given to moves for approved operations in the SEAC and the return of Dominion personnel for the war against Japan, despite a British commitment at the Yalta conference that 'forces should be reoriented from the European theatre to the Pacific and Far East as a matter of highest priority'.[44]

Under Lyttelton and Attlee's prescription, tokenism would be the order of the day in the Pacific. Churchill issued a directive along these lines on 14 April, instructing that it was of 'the utmost importance that releases from all three Services should begin not more than six weeks after the end of the German war' and that forces to be deployed against Japan 'should be related strictly to what can be brought into action in time to play a part with the assumed duration of the Japanese war'.[45]

Churchill promptly backed a plan to allocate extra shipping to move 520 000 tons of animal foodstuffs from

Argentina to Britain. This would translate, during 1946, into an extra 75 000 tons of pork or bacon and 500 million extra eggs. Churchill observed that such shipments of foodstuffs could begin immediately the European war ended. On 20 April the War Cabinet approved this plan to enhance the British breakfast despite objections that it would interfere with shipping movements for the war against Japan. As the historian, Arthur Bryant, later observed: 'The war in the Far East, though it had brought America into the fight and so ensured ultimate victory, had been for Britain only a consequential, though agonizing, incident in the long German war. . .'[46]

One of the worst moments during this 'agonizing incident' in the Pacific occurred on Easter Sunday when more than one hundred thousand American troops stormed ashore on Okinawa. During the three-month campaign that followed, the Americans blasted their way along the sixty-mile-long island, brandishing flamethrowers while kamikaze planes attempted the impossible defence of Japan.

On 6 April the giant Japanese battleship *Yamato* ventured out in a suicidal attempt to scatter the American and British warships. It was sunk without ceremony but with great loss of life. The British aircraft-carrier *Indefatigable* was hit by the Japanese, to the relief of Admiral Cunningham. While deploring the casualties, he was 'glad that they have had some opposition'. As one of the British naval officers in the Pacific observed, 'It's well called the British "POLITICAL" Fleet'.[47] Meanwhile, in Tokyo, the 'peace party' of Admiral Suzuki took power but could see no easy or honourable path to peace. There was little time left for negotiation.

19

Victory in Europe

April to June 1945

THE AMERICAN INVASION of Okinawa revealed the terrible price that waited to be paid during the invasion of Japan itself. Some 12 000 American service personnel were killed in the fighting for Okinawa while 150 000 Japanese perished. Estimated casualties for the invasion of the Japanese island of Kyushu suggested a figure of some 250 000 American dead and wounded. This would just be the first of the Japanese islands the Allies would have to conquer. In April 1945 MacArthur was appointed to command army forces throughout most of the Pacific and to plan the invasion. He confided to a visiting British intelligence officer that the 'chances of the Japs packing up was 50/50 but he wanted to kill some more first'. According to MacArthur, who had an undeserved reputation for protecting the lives of the men under his command, 'his troops had not yet met the Japanese Army properly, and that when they did they were going to take heavy casualties'.[1]

From Guam, Superfortress bombers continued their deadly flights over Japanese cities. These huge bombers rolled off modern assembly lines in America for the long flight to their hastily constructed bases on the other side of the Pacific, from where they delivered death on a scale that defies the imagination. In one raid on Tokyo, more than sixteen square miles of buildings were destroyed, killing more than eighty thousand people. 'We knew we were going to kill a lot of women and kids when we burned [a] town', wrote the American air force commander, Curtis LeMay, but it 'had to be done'.[2] Similar refrains, justifying similarly horrendous war crimes, would be rejected out of hand by the victorious Allies during trials at Nuremberg and Tokyo.

On 13 February 1945 the German city of Dresden was bombed by British and American bombers, setting off a fire storm that laid waste the city and incinerated as many as one hundred thousand men, women and children. It was meant to show the approaching Russian armies what Bomber Command could do, but it became a byword for infamy in a war where mass murder had become the airmens' lot. Germany was too punch drunk to notice the devastation of Dresden. By mid-April, Vienna was occupied by Russian troops and the Anglo–American armies had crossed to the east bank of the Rhine and were streaming across Germany almost unopposed. Hitler, still hoping to snatch victory from the ashes of defeat, had issued instructions to torch the cities and countryside in advance of the Allied armies.

In Burma the British army under General Slim streaked to Rangoon in a desperate attempt to beat both the onset of the approaching monsoon and the retreating Japanese army. On 1 May a combined parachute and amphibious landing captured the city and linked up with advancing elements of Slim's army. Burma became a killing field where thousands of straggling Japanese soldiers perished, along with the imperial ambitions of their emperor. Australian troops in Bougainville, taking no prisoners, continued the offensive strategy that Blamey justified as being necessary to restore Australia's image in the eyes of the native peoples.[3]

Meanwhile, the key to Japan's defeat was being fashioned in New Mexico. The question of using the new and awesome atomic bomb suddenly fell to Vice President Harry Truman when Roosevelt suffered a massive stroke and died on 12 April. News of the bomb's development had been kept from the recently-elected Truman but he was now informed in guarded terms of this 'explosive great enough to destroy the whole world' which might arm America with the power to 'dictate our own terms at the end of the war'.[4]

The ailing Australian Prime Minister, John Curtin, paid homage to the memory of the dead President. Curtin's own hold on life was failing. Without the strength to lead the nation, Curtin stubbornly held onto power, causing a state of virtual paralysis in Canberra. One official later recalled that: 'Almost—almost—one could be glad he's dead so we can do things'. Chifley, Curtin's eventual successor, described it as 'the worst six months of my life'.[5]

Among the issues waiting to be addressed were the shape of the United Nations Organization, the future of colonial territories and the peace settlement in the Pacific. Evatt had already convinced the Australian War Cabinet to limit to just ten officers the Dominion's participation in the Allied control commission to govern post-war Germany. Australians would only be part of the British contingent, he argued, and Australia had not been consulted on the organization of the commission and the treatment of Germany.[6]

Australia was in a relatively strong position with Britain as they approached discussions on the United Nations. Britain hoped the various Dominions would support her approach, so consultation was suddenly *de rigueur* in Whitehall, with Cranborne urging his ministerial colleagues to have the 'closest liaison' with the Dominions. He accepted a South African suggestion for a meeting of Commonwealth representatives before the international conference and urged his colleagues to be prepared to change British policy as a result of the forthcoming Commonwealth discussions.[7]

At the beginning of April, Commonwealth representatives assembled in London to try to agree on a common policy and resolve the two main problems. One concerned voting in the proposed Security Council of the United Nations: Britain wished to restrict it to the 'Great Powers' but the Dominions wanted to open it to representatives of all the powers. The second was the future of colonial territories and the powers of the United Nations to interfere in the administration of such territories.

With Curtin ill, Australia's views were put by Evatt and Frank Forde, Curtin's deputy. They bickered over who was leading the Australian delegation and appealed to Bruce to adjudicate. On 2 April the bemused High Commissioner recorded how Evatt apologized for his 'discourteous' behaviour during his previous visit and now claimed that he 'had found all my telegrams useful; entirely agreed with my point of view and especially appreciated my frankness with regard to the Colonial question'. Evatt told Bruce that 'no decision had been reached as to who was the leader of the Delegation'. Forde arrived next to complain that Evatt was trying to usurp his position. Although technically leader by right of seniority, Bruce recognized that Forde, 'a decent little man without any very great capacity', was not competent to lead the delegation. He advised him 'very strongly not to attempt, in order to maintain his position as leader, to take on jobs that Evatt was more fitted to do'.[8]

This issue of seniority dogged the discussions in London and the conference in San Francisco. Cranborne told Evatt diplomatically that, as Deputy Prime Minister, Forde must be treated as leader unless Curtin instructed otherwise. Undaunted, Evatt again appealed to Bruce for support, pledging 'all the support and assistance of himself and the Australian Government' in any campaign Bruce might be contemplating to secure a post in the United Nations.[9] Bruce was interested in such a post but he would have realized that he could not rely upon Evatt's support to achieve it.

Despite these problems, Evatt transformed the London meeting from a polite exchange of views into a meaningful discussion of differences between the various Dominions. Evatt clarified in his own mind the direction in which he wished the United Nations to move, while the British Government was put on notice that they would have to 'reckon with a very determined, able and non-compliant Australian delegate'.[10]

During the discussions, Evatt pursued some of the issues that he considered vital to Australia's future as an independent 'middle' power rather than as a subservient member of the British Empire. He was determined to break the great power monopoly over international decision-making that had so frustrated him during the war and which he regarded as a future threat to peace. He successfully sought to dilute the exercise of veto by any of the great powers which, under the draft plan, would prevent any changes to the United Nations charter that did not have the unanimous approval of the great powers. Such a provision might cripple the

Organization's ability to act and would certainly diminish the influence of smaller countries.[11]

Evatt was also determined to ensure that colonial empires be moved closer towards dissolution under a system of international inspection. Evatt was motivated by liberal ideas about uplifting the native peoples, but he also realized that Australia could only advance in the Pacific if the European colonial powers would retreat. Moreover, if they could be stripped of their empires, their status as great powers would be less secure and Australia's influence as a middle power would be that much stronger. However, any suggestion of international inspection was anathema to the British Colonial Office which was equally determined to protect Britain's right to rule over her colonies. The declaration of the Wellington conference warned Whitehall that her Dominions would join with the United States to support the ideal of colonial independence.

Many of the progressive moves Evatt supported at San Francisco seem to have been inspired by position papers drawn up by the organizational wing of the British Labour Party during 1944. For instance, a draft policy paper on the international post-war settlement called for colonial trusteeship in much the same terms as Evatt would use in San Francisco a year later. This is not to imply that he purloined his ideas from his British comrades, although some interchange of ideas did take place, but that he believed he was proposing ideas in general accord with those of the British Labour Party. Since the Labour Party was widely expected to form the post-war government of Britain, Evatt must have felt that he would be part of a post-war concensus on this issue and others. In fact, when Attlee recalled his time at San Francisco as part of the British delegation, Evatt was not mentioned, and proponents of trusteeship were described as people 'whose views were based on admirable sentiments but great lack of practical knowledge of actual conditions'.[12]

Evatt should have read those Labour Party documents more closely. He believed that states should be treated equally, much as individuals are meant to be treated equally before the law. As Evatt's friend the American historian C. Hartley Grattan observed, the Australian foreign minister

> rejected power politics and embraced the idea of the sovereign equality of all nations. He sought to realize Australia's objectives through a United Nations so organized as to curb the great powers' disposition to power politics and tip the weight in decision making more toward the numerous small powers.[13]

However, as a British Labour Party document pointed out, while all states 'must have equal rights to Peace, Freedom, and Security . . . all States cannot hope to have equal influence or equal power, and it is only building a theorist's house of cards to pretend they can'.[14] Perhaps because of his training as a lawyer, Evatt remained committed to his house of cards.

To buttress Britain's case for the return of her conquered territories, she helped to ease the return of both the French and Dutch to their colonies. As the Chiefs of Staff acknowledged on 21 April, the early return of French and Dutch forces to the Far East was, from the British point of view, 'both militarily and politically advantageous'.[15] Britain opposed Chinese entry into Indo-China for fear that it would 'encourage false ideas of expansion and martial glory'. Instead, Indo-China was kept firmly within the SEAC so that, at war's end, it could be handed back to the French by Britain. The French return to Indo-China would help legitimize the British return to Malaya and Hong Kong. Moreover, there were European considerations involved. As the Foreign Office advised Churchill on 28 March 1945, 'France is our nearest neighbour and we have a vital interest in her restoration as a strong and friendly power'.[16]

Hong Kong was a problem for Britain because America viewed the colony with disfavour. When General Hurley, Roosevelt's representative in China, visited London, Churchill left his visitor under no illusions about Britain's vehement drive to regain her Chinese colony. After the meeting, Churchill advised the Foreign Office that he had brushed aside Hurley's 'civil banalities' and taken Hurley 'up with violence about Hong Kong and said that never would we yield an inch of the territory that was under the British Flag'.[17]

Britain had wanted colonial policy to be completely left off the agenda at San Francisco, fearing the hostility of the smaller powers. When that was impossible, she sought to restrict discussion to those territories held under mandate from the defunct League of Nations whose sovereignty now needed to be regularized under the aegis of the United Nations. On 3 April there was a wide-ranging discussion of the issue in the War Cabinet, and although Evatt and other Dominion representatives were in London, they were not privy to the discussions. At the suggestion of the Colonial Secretary, Oliver Stanley, his colleagues agreed that the pre-war system of colonial mandates could be continued under the United Nations but that it should not be extended to include other colonial territories that 'might voluntarily be placed under trusteeship'. The War Cabinet agreed that 'it should be made clear beyond any doubt that in this phrase "voluntarily" meant by the will of the Power exercising sovereignty over the territory concerned and not at the option of the territory itself'. In other words, India could not be placed under United Nations trusteeship at the request of the Indians, only at the behest of the British. To protect the British hold over India, the War Cabinet decided to oppose any move by Australia to have the United Nations charter include a guarantee of the political independence of its members.[18]

The following day, Britain tried to sell her line to the Australians and New Zealanders at the Commonwealth representatives' meeting in London. The acerbic Foreign Office head, Sir Alexander Cadogan, found that the Dominion leaders were not amenable to his salesmanship, confiding to his diary that he could not decide 'whether Evatt and Fraser are more stupid than offensive'. Worse than that, British Ministers left Cadogan to present much of the British case. Oliver Stanley was ill with the mumps, while Eden 'shirked' one meeting leaving Cadogan 'to face the music, which was unharmoniously played by Fraser and Evatt'. As the historian, W. Roger Louis, observed, Evatt 'plunged into the sea of the trusteeship controversy like a hungry shark', devouring the self-interested arguments of Whitehall with a sharp legal and historical mind that established the basis for his later stand at San Francisco. It was sweet revenge for Evatt after Curtin had fallen in with British colonial policy during his visit to London in 1944.[19]

On 10 April Cranborne reported to the War Cabinet that the Australian and New Zealand delegates strongly backed the concept of voluntary trusteeship. It would appease American criticism of the Commonwealth and would pressure other colonial powers to accept trusteeship. In particular, the Dominions singled out the colonies of France and Portugal that were presently occupied by Japan, and presumably they would also have had the Dutch colonies in mind. Under international trusteeship, Australia could help to ensure that they would be adequately defended and thus prevent a repetition of the Pacific war; it might open them up to Australian trade; and it would 'improve the conditions of backward peoples and aid them along the road to self-government'. Once they became independent, these colonies would be more open to Australian trade and influence. The mixture of self-interest and liberal humanitarianism in the Dominion proposal would not appeal to British sensibilities.[20]

Cranborne argued that 'there can be no question of our agreeing to place any of our Colonial territories other than those at present administered under mandate under any form of trusteeship'. Nevertheless, in view of the strong Dominion feeling, he suggested that Britain not oppose the concept of voluntary trusteeship, but make clear that she would not be volunteering any of her own colonies. This view was adopted. Churchill suggested they could leave it to the French and the Dutch who would oppose it just as vehemently, Britain could thereby avoid the odium of leading the opposition and simply lend support from a distance. This solution caused dismay amongst the Commonwealth delegates, with Evatt declaring that 'it would be better for the United Kingdom Government not to accept the principle of trusteeship at all than to accept it and refuse to apply it to their own territories'. The stage was set for a further confrontation.[21]

The conference at San Francisco opened on 25 April, Anzac Day as it happened, and it was Evatt who led the fight for colonial trusteeship. Cranborne led the British team and hoped to have the discussions over in a month despite objections from Evatt who, according to Cranborne, was 'violently opposed to any limitations, I suppose because it will curtail his opportunities of shining on the World Stage'.[22] The conference took two months, with Evatt orchestrating a frenzied campaign, to the great frustration of the British delegates. After a month of Evatt's activities, Cadogan blasted him as the 'most frightful man in the world; he makes long and tiresome speeches on every conceivable subject, always advocating the wrong thing and generally with a view to being inconvenient and offensive to us, and boosting himself'. Cadogan claimed that 'everyone by now hates Evatt so much that his stock has gone down a bit and he matters less'.[23] Certainly the British and Americans resented Evatt's tactics enormously, but the resentment was largely a sign of his success; Britain and America were being pushed much further than they would have wished on the issue of trusteeship.

The United States had moderated its objections to colonialism after Roosevelt's death and the realization that the captured Pacific islands must remain as American strategic possessions after the war. The American Congress was unwilling to contemplate independence for island territories over which so much American blood had been spilled. So Britain and America formed an unlikely alliance against the rest of the Commonwealth. As Cranborne confided to a colleague, the American 'house has in fact come to have so much glass in its constructions that they cannot afford to throw stones at others'.[24]

So a new category of colony was created that combined security with trusteeship and thereby left the United States with a pattern of bases across the Pacific. Churchill was happy to concede American control over islands whose sovereignty Britain and America had hotly disputed between the wars. It suited Britain for America to be compromised as a fellow Imperial power and, by their defensive works, help to protect the adjacent territories of the British Empire.[25] Britain had to concede partial sovereignty over her own territories, and the trusteeship council of the United Nations would be able to inspect colonies and report on whether the controlling powers were fulfilling their international obligations. It would also be a forum to which the native peoples could appeal against injustices. Although it was a diluted version of Evatt's proposal, the changes did mark a watershed in the history of decolonization, recognizing the changes that had taken place during the war and preparing for the slow death of formal western imperialism.

After San Francisco, it became a matter of *when*, not *if*, dependent territories would be given their independence. When Cranborne was eventually ousted by the voters of Britain, he felt only relief after the impossible burden of defending the Empire. In a letter to a colleague, he observed how

> first, Canada, and now, as appeared at San Francisco, Australia and New Zealand, are beginning to show most disturbing signs of moving away from the conception of a Commonwealth acting together to that of independent countries, bound to us and each other only by the most shadowy ties. Dr. Evatt is only a particularly repulsive representative of a not at all uncommon point of view in his own and the other Empire countries.[26]

Cranborne complained that if the Dominions were taken into Britain's confidence, it only served to 'irritate them and strengthen the forces of independence. I really, during the last few months, sometimes felt at my wits' end'. The only solution was for Britain to enhance her diminished great power status relative to America and Russia. 'Herein lies the importance of the Western European Pact,' he concluded, 'which Winston wouldn't look at. One can only hope that Bevin [Britain's Foreign Secretary after the July election] may be wiser'.[27] The Western European Pact became the North Atlantic Treaty Organization which failed to enhance Britain's status since its linchpin was the United States. NATO simply confirmed Britain's decline as a great power.

Despite the rift between Evatt and the British delegates over the trusteeship issue, Evatt enlisted British support to enshrine the principle of domestic jurisdiction in the charter.[28] This clause would protect the 'White Australia' policy from outside interference. Just as Billy Hughes flourished a similar victory after the 1919 Versailles peace conference, so could Evatt proclaim the preservation of 'White Australia' as a victory from the conference. Ironically, it was Hughes who criticized Evatt, fearful that the trusteeship debates might compromise Australia's hold over the phosphate-rich island of Nauru. 'No new principle[s] should be accepted', Hughes argued, 'which have the effect of worsening the authority and responsibility of Australia as existing under the present Mandate for Nauru'.[29]

Across the world from San Francisco, real battles were being fought to end six years of slaughter. By early February the Russian armies had reached the River Oder, less than one hundred miles from Hitler's headquarters in Berlin. Impeded by the weather and their attenuated lines of communication, it was not until mid-April that the Russians reached Berlin, surrounding the city by 25 April, the day the United Nations conference opened in San Francisco. Five days later, Hitler was dead by his own hand. As his body burnt in the rubble of the Chancellery, his loyal soldiers fought a fierce battle street by street. American and British armies had poured across the Rhine in late March, meeting little resistance from the debilitated German armies and joining in a triumphant procession towards Berlin. At midnight on 8 May Europe's torture ended.

In the Pacific, embattled Japanese in the caves and tunnels of their island fortresses put up stubborn resistance. Although MacArthur still wanted to kill more Japanese, American political leaders were counting the likely cost in American dead. Roosevelt's call for unconditional surrender had stiffened resistance in both Germany and Japan. With Roosevelt dead, these terms could be moderated, particularly in view of the growing American desire to transform a defeated Japan into a bulwark against Russian expansion. Like Britain in the Atlantic, it could act for America as an unsinkable aircraft-carrier off the coast of Asia.

Britain was more concerned with rehabilitating her cities and economy after five years of German bombs and rockets than with victory against distant Japan. On 2 May, as the guns fell silent on the Italian front, the British Chiefs of Staff pressed Churchill for ships to transport the VLR bomber force for use against Japan. The War Transport ministry had cited such 'inescapable commitments' as the British import programme, relief for Europe, and operations in the SEAC; now Churchill, too, resisted their pressure, instructing them to scour their own shipping resources first, and berating their grandiose plans for the Pacific. 'Here is the real evil we have to face', he argued, '—the whole world being strangled in its development by demands for the war on Japan which have absolutely no relation to the number of warships or troops or aircraft which can be engaged there'.[30]

However, the Chiefs of Staff persisted, so Churchill referred the matter to the unsympathetic Minister of War Transport, Lord Leathers, reminding him of Britain's pledge to participate in the bombing of Japan 'to the utmost limit which the bases will allow'. 'I would have thought', wrote Churchill, that the end of the European war would 'have provided sufficient easement to enable us to implement our undertaking to bomb Japan'. Leathers conceded that it might be possible to provide cargo shipping for the VLR force, as well as increase imports during the first half of 1946.[31]

On VE Day, Australians had crowded round their wireless sets to hear the British High Commissioner, Sir Ronald Cross, remind them that

> Britain has heavy scores to settle with the Japanese; make no doubt about it Britain purposes to settle her *own* scores; and be assured that in priority over all other interests Britain will employ every man, every gun, ship and aeroplane that can be used in this business.[32]

However the Japanese war was well below demobilization and rehabilitation on Britain's list of priorities. When the Admiralty requested an increase in the fleet train to enable the British Pacific Fleet to operate away from Sydney for longer periods, thereby avoiding American criticism and maintaining 'Imperial prestige', Lord Cherwell advised Churchill to reject it since it would exacerbate the expected shortfall in imports. 'The British people', he wrote, 'short of coal and clothing, cannot be asked to face further cuts in food, building materials, etc. If failure to get the ships restricts operations, this may have to be accepted'.[33]

The matter came before the Conservative British Cabinet on 5 June, the all-party War Cabinet having ended with the European war. Churchill supported Cherwell and Lord Leathers, observing that: 'Important as operational requirements were, full consideration must be given to the United Kingdom import programme and the demands of the civil population in this country'.[34] Churchill probably recalled the unrest following the First World War when the government, of which he was a leading member, had not been so mindful of opinion within the services and the population. He would also have realized that he was about to face an electorate hungry for the fruits of victory.

Incredibly, Whitehall could still not agree on the main thrust of the limited British effort. The first option was to concentrate on Mountbatten's Indian Ocean command where his planners were reassessing their timetable after the sudden fall of Rangoon, and Mountbatten was still pressing for the British Pacific Fleet to support his operations.[35] The second option was to recapture Borneo and the Netherlands East Indies. The final option was to join the Allied onslaught against the Japanese home islands. All these options might be mounted, but dissipating the British effort would just give an impression of British weakness everywhere. One option had to prevail to achieve the desired political effect.

The Americans opposed any operations in the SEAC that would detract from the main effort against Japan, although on 12 May they conceded that an invasion of Malaya, preparatory to an attack on Singapore, could be mounted provided it did not prejudice operations in the Pacific. With American approval, the Chiefs brought forward the Malayan operation to mid-August, hoping to recapture Singapore before the Japanese surrender.[36]

Both the Navy and Air Force chiefs backed the strategy of fighting Japan itself since, according to Admiral Cunningham, Britain would have difficulty taking a 'prominent part in the Eastern settlements if we have devoted ourselves solely to mopping up operations'. The Army chief, Field Marshal Alanbrooke, preferred to concentrate on Malaya with Dominion and Indian troops. There was also, as America's General Marshall informed Anthony Eden on 14 May, the 'rather unexpected problem' of German prisoners volunteering to help fight the Japanese. Marshall suggested that the Germans could clear various passed-over Pacific islands, 'letting them and the Japanese kill each other there'. If Marshall had thought the British would oppose the use of Germans he was presumably surprised when Eden assured him that he was 'not shocked by the proposal'. In fact, the British War Cabinet had already considered a proposal to use sixty thousand German prisoners of war for labour purposes in the war against Japan.[37]

Britain not only preferred a prompt return to peacetime pursuits rather than fulfilling her pledge to fight Japan, but she was also concerned at the possible power vaccuum created in Europe by any hasty withdrawal of the Allied forces. For three and a half years, Churchill had assuaged Australian, Chinese and some American criticism of the 'Germany first' strategy by undertaking to switch resources against Japan as soon as Germany was defeated. Now that Germany was defeated, Churchill saw an even greater danger as Russian armies spread across eastern Europe from Lubeck to Trieste, and Anglo–American armies rapidly melted away through demobilization and transfer to the Pacific.[38] Although he was sanguine towards Russian expansion in the Pacific, Russian influence in Europe was a different matter.

On 12 May, just four days after VE Day, Churchill called on President Truman to keep American forces in Europe until they had reached an understanding with Stalin on the status of Yugoslavia, Austria and Poland. Churchill sought Truman's approval for the possible use of American troops 'in the event of hostilities against Yugoslavia', where the guerrilla leader, Tito, was asserting his control over the country. Warning of another bloodbath if these issues were not settled before the withdrawal of the Anglo-American armies, Churchill conjured up the image of an 'iron curtain' being drawn across the Russian front.[39] Ironically, it was the German master of propaganda, Joseph Goebbels, who had coined the phrase, but Churchill immortalized it in March 1946 during a speech in Fulton, Missouri.[40] This speech laid part of the foundation for the Cold War which would eventually help to carry Churchill back to Downing Street.

As a result of Churchill's concern, Truman agreed to their early meeting with Stalin at Potsdam in July, to settle the outstanding problems of Europe. The withdrawal went on as planned; American personnel in Europe were reduced from 3 270 000 in June 1945 to an estimated 1 530 000 by December 1945. Of those being withdrawn, 1 200 000 were to be transferred to the Pacific. Churchill instructed that

> All these facts [of the withdrawals] should be reported to the whole Cabinet. But I cannot at the moment advise them what to do about it. When I see President Truman, I will warn him that if his armies go too soon, they will probably have to come back pretty quick.[41]

Churchill's milder tone was due to the election, which had

been called following the withdrawal of the Labour and Liberal parties from the ranks of the National Government on 23 May.

While anxious to retain American forces in Europe, Churchill backed moves to withdraw British troops from the SEAC before victory. In a speech in the House of Commons on 8 June, the War Minister, Sir James Grigg, suddenly announced that service personnel in Mountbatten's command would be repatriated four months earlier after serving three years and four months. This politically popular decision played havoc with Mountbatten's plans to invade Malaya. Describing it as 'quite disgraceful' and 'an electioneering dodge', Cunningham claimed that it 'looks to have wrecked all our plans for offensives down the Malay peninsular and the taking of Singapore'. When the men were released from their army formations, there was insufficient shipping to bring them home from India. 'Surely it is very silly', wrote an angry Churchill, 'having faced all the grave military disadvantages of releasing these men, of which I approve, to dawdle them about in India'. Churchill was dismayed to find that there were no compensating political advantages. In a note to Lord Leathers, Churchill minuted tersely, 'I rely on you to fix this'.[42]

Field Marshal Sir Alan Brooke, who was already hard pressed to find sufficient forces for the Far East, was horrified that political considerations dominated military planning: 'Heaven help democracies if they must have elections in wars!', he wrote in his diary. His own planners were discovering all sorts of practical difficulties in the British contribution to the planned invasion of Japan. Eventually, on 28 June, the Chiefs of Staff suggested to Churchill that Britain offer the Americans 'a small land force of some three to five divisions to participate in the main attack against Japan'.[43]

However, the following day the Chancellor of the Exchequer recommended to the Cabinet on behalf of the Man-Power Committee that

> if civil industries and services are to receive the additional man-power they require in order to enable them to make a reasonable start in 1945 with the most urgent tasks of civil reconstruction, the scale of our military effort in this period must be reduced to some extent below the level at present planned.

The conservative backbencher, Sir Harold Nicolson, captured the feeling of the Commons when he wrote that the Pacific war 'arouses no interest at all, but only a nauseated distaste'.[44]

Australia wished to make only that effort against Japan necessary to guarantee a seat at the expected post-war peace conference. But she was anxious to avoid American criticism if she demobilized any further. On 9 February 1945 Shedden had warned Curtin of such criticism while urging him to talk with MacArthur 'regarding his operational plans as they affect the Australian Forces, with a view to determining the stage at which appropriate reductions can be made and the future strengths which should be maintained'. Curtin had accepted this advice, refusing to release more manpower for civilian industry, and reminding Parliament on 28 April that Australia was well-off compared with countries ravaged by war; nevertheless he promised to release some manpower by 30 June.[45]

Just where Australia's symbolic effort would be directed remained unclear. MacArthur had promised to use Australian troops for the invasion of Honshu, planned for early 1946, but in the interim diverted them to the capture of Borneo, an enterprise of little strategic significance once the Americans controlled Okinawa. The ostensible reasons for the operation were to capture the oilfields of Borneo and provide a port for the British Pacific Fleet. The Japanese maritime link to the oilfields had already been severed, and the British were loath to use a base in Borneo due to its great distance from Japan. Cunningham dismissed the idea as another American attempt 'to keep the B.P.F. away to the South'. Although the British Army was keen to take control from the Americans in the south-west Pacific, Cunningham correctly claimed that the Army wished to

> use Australian instead of home divisions in SEAC. They say it is to save shipping but it is all to assist in their damned redeployment . . . If they had SWPA under SEAC they would use the Australian divisions in Malaya and keep their own at home.[46]

MacArthur continued to claim that he wanted the Australians with him when he invaded Japan, describing it as a 'wise political move on [the] part of [a] Pacific power such as Australia'. This was confirmed on 28 May when the Australian War Cabinet, despite pressure from Britain, reaffirmed its commitment to push ahead with MacArthur in the final drive against Japan, while retaining a token force in the SEAC, primarily for use in the recapture of Singapore. The Australian naval squadron was left to operate with the American Seventh fleet rather than being attached to the British Pacific Fleet, and the RAAF remained with MacArthur, apart from three squadrons proposed for Britain's VLR force.[47]

When the demobilization of a further 50 000 Australian service personnel was about to be announced in early June, Evatt and Forde protested from San Francisco, warning that 'publication at this juncture . . . might have an adverse effect on various Conference proposals and also on the part we shall be able to play in the final Pacific settlement'. But the domestic pressures were too intense. On 28 June, the Australian War Cabinet decided to begin the release of the 50 000 men to fuel civilian industry, while refusing to undertake more construction work for the British Pacific Fleet because it would affect the Government's house-building programme.[48]

The operations in New Guinea, New Britain and the Solomon Islands were to be run on the basis of economy of effort. On 6 June the all-party Advisory War Council had approved a report by Blamey, in which only operations with 'relatively light casualties [would be undertaken], so as to free our territory and liberate the native population and thereby progressively reduce our commitments and free personnel from the Army'.[49]

The Government also indefinitely deferred any decision to accept the offer of British warships that had been first made more than a year before. The British Government had decided to ask for payment for these 'gift' ships to offset the costs of basing her Fleet in Australia. Australia refused to pay for ships that were not really necessary, given the age of her existing ships and the overwhelming American naval presence in the Pacific. Interestingly, one of the reasons for refusing the ships was that the aircraft-carrier would have to operate for logistical reasons with the British Pacific Fleet rather than with the Americans. The Australians probably realized that the British wanted Australia to do the fighting under a British flag, allowing Britain to gain the credit and Australia to pay the cost in men and money.[50]

The British Government was distracted by the election campaign, a passionate struggle for power between the former comrades-in-arms. On 5 July the British people delivered their verdict, but no announcement was made for three weeks to allow the votes of service personnel to be counted. Churchill retreated to the south of France for a short painting holiday before flying to Berlin on 15 July for the meeting with Stalin and Truman to settle the organization of post-war Europe. At the same time, an American bomber group was practising for the pin-point delivery of a new weapon that would obliterate two cities and change the face of the world.

20

The Surrender of Japan

July to August 1945

ON 22 JUNE 1945 Emperor Hirohito instructed a sombre conference of Japan's military and political leaders to find a way out of the bloody morass into which the nation had been led. They turned to the Soviet Union, hoping that the still neutral Stalin would intercede with Churchill and Truman to produce a peace short of the unconditional surrender previously demanded by the Allies. But Stalin had his own ambitions in the Pacific and was preparing to seize disputed territory from the Japanese. Soviet power, already spread like a frown across the face of Europe, was turning to the Pacific. In New Mexico, scientists prepared for the first atomic explosion, deliberately timed to coincide with the Potsdam conference of Allied leaders.

Churchill arrived in Potsdam, just outside the ruins of Berlin, on 15 July. With the British election result still undeclared, Churchill took with him the Labour leader, Clement Attlee, as a sign of national unity that he hoped would strengthen his hand with the Russians. Two days later, and just before the first plenary session of the conference, he was handed a brief note by the American Secretary for War, Henry Stimson, announcing 'Babies satisfactorily born'. As Stimson explained over lunch, the 'babies' referred to the atomic bomb that had been tested for the first time in the desert of New Mexico just twenty-four hours before.[1]

The *USS Indianapolis*, with its small but deadly cargo securely stowed, was already on its way to a rendezvous with history at Hiroshima. The American physicist, Robert Oppenheimer, who had supervised the New Mexico test firing, recalled the awestruck scientists as they watched the cauldron of fire boil into the night sky:

> A few people laughed, a few people cried. Most people were silent. I remembered the line from the Hindu scripture, the *Bhagavad-Gitu*: Vishnu is trying to persuade the Prince that he should do his duty and to impress him he takes on his multi-armed form and says, 'Now I am become Death, the destroyer of worlds'.[2]

This terrible bomb was Churchill's saviour, promising a speedy finish to the Pacific war and solving Britain's seemingly intractable manpower problems. On 3 July Churchill had urged that the target figure of 135 000 women to be demobilized by December 1945 be increased to a target of 250 000.[3] He also wanted to ship German prisoners of war to Britain to help with civilian reconstruction, and delay returning Italian prisoners of war even though the Italians were now notionally on the side of the Allies.[4]

Churchill's proposals for releasing servicewomen were rejected by the Chancellor of the Exchequer, Sir John Anderson, who argued for more restrained releases of women, urging that 300 000 additional men be released from the services and the munitions industry. There could be no guarantee that women, once released, would go into industry. Even with all these additional men, there would not be 'any appreciable improvement in civilian standards' nor would there be much 'progress towards the restoration of our export trade'. Once again, Anderson called for further cutbacks in Britain's planned effort against Japan, reminding Churchill to keep this 'prominently in mind' at Potsdam.[5]

Cutting back on the British commitment was not as easy as it seemed. The VLR bomber force being formed to destroy Japanese cities was a case in point. Britain had progressively reduced its plans for this force to just twenty squadrons, many of them to be drawn from the Dominions. The deployment of the first ten squadrons had been accepted in principle by the Americans in early June, with the remaining ten to be accepted when conditions permitted.[6] It was difficult to find makeshift airfields on which to base these squadrons, with aircraft crowding the hastily constructed bases surrounding Japan like crows round a carcass, there was little space for the British planes and few cities left that were worth bombing. When the Air Ministry suggested cutting back the VLR force to just ten squadrons as a way of alleviating the manpower problem, Churchill immediately agreed. However, on 11 July the new Air Minister, Harold Macmillan, reminded the Cabinet that America had accepted Britain's firm offer to send twenty squadrons. Macmillan suggested that it would be 'very inadvisable to go back on this decision in view of our relations with the Americans, especially since our original contribution was planned at a much higher rate'.[7]

Although Churchill was advised at Potsdam of the atomic bomb test, he could not be certain that it would force the Japanese to surrender. One of his aims at Potsdam was to convince the Americans that the Allied demand for unconditional surrender, a demand that they had forced Germany to accept at the cost of needless casualties, should be watered down in the case of Japan. To Britain's surprise, the Americans proved remarkably accommodating; the American naval chief, Admiral Leahy, suggested at a meeting on 17 July that Churchill provide them with a memorandum on the subject, which would shift the responsibility for the change onto Churchill's ample shoulders.[8]

Later that afternoon, in a private meeting with Churchill, Stalin advised the British Prime Minister of Hirohito's overtures for a conditional surrender. Not wanting to appear too eager to grasp this olive branch, Churchill warily maintained that Britain would help America in the Pacific 'to the full'. He did, however, confide to Stalin that the Americans were beginning to question the sense of demanding unconditional surrender if it meant that perhaps one million Allied lives would be forfeit as a result, not to mention another ten million Japanese lives. When Churchill discussed the peace offer the following day with Truman, he was careful not to press him to accept it, although he did emphasize the 'tremendous cost in American life and, to a smaller extent, in British life which would be involved in enforcing "unconditional surrender" upon the Japanese'. Churchill was pleased to see that Truman was clearly aware of the 'terrible responsibilities that rested upon him in regard to the unlimited effusion of American blood'. In fact, Truman already knew of the Japanese moves from the interception of their diplomatic cable traffic.[9]

On 22 July Stimson gave Churchill more details of the atomic bomb test. Churchill later recalled his undisguised relief at the news that a full-scale invasion of Japan might no longer be necessary. He claimed it would have meant the 'loss of a million American lives and half that number of British—or more if we could get them there: for we were resolved to share the agony'. How Britain could suffer half a million dead when the latest plans only committed perhaps 100 000 troops, many of them from the Dominions, Churchill did not explain. Instead of this 'nightmare picture' of slaughter, Churchill envisaged the whole war ending in 'one or two violent shocks'. The Allies were doing the Japanese a favour, he claimed, since the atomic bomb would give them 'an excuse which would save their honour and release them from their obligation of being killed to the last fighting man'. It might also avoid the need for calling the Russians into the war against Japan.[10]

The United States was also anxious to avoid Russian involvement in the Pacific war. Until the successful atomic test, America (and Britain) had wanted to use the Russian armies against Japan like the British used them against Germany—to absorb the impact of the enemy on the ground while they delivered a relatively economical *coup de grace* from the air. With atomic bombs in her armoury, America now wished to delay the possibly unnecessary entry of Russia. Russian involvement might provide her with a springboard into the Pacific, and help to turn a defeated Japan into a communist satellite, whereas America intended to use Japan as a bulwark against Russian expansion.[11]

Churchill also believed the atomic bomb would help Britain and America to contain Russian expansion in Europe. As Cunningham observed after lunching with Churchill on 23 July, Churchill was 'most optimistic' and wanted to tell the Russians of the bomb as 'it may make them a little more humble'. Cunningham disparaged Attlee's 'damned silly' suggestion urging that Britain 'not oppose a great country like Russia having bases anywhere she wants them'. 'What an ass!!', commented Cunningham. That evening, during a banquet at Churchill's villa, Stalin looked forward to Russia joining Britain and America in the struggle against Japan and, according to Cunningham, 'toasted our next meeting in Seoul or Tokyo'.[12] The Allies were advised that the Russians would be ready to launch their attack against the Japanese on 15 August, and that same day, Stimson indicated that the atomic bombs might be ready as early as 1 August and 'almost certain[ly] before August 10'.[13] The trick would be to drop them and force a Japanese surrender before the Russians declared war against Japan.

Although the decision to drop the bomb was mainly the prerogative of the United States, it still required Britain's formal approval since the bomb had been a joint Anglo–American development. Churchill had given his approval on 4 July. The British, wishing to assert a greater presence in the Pacific based upon their token forces there, wanted a say in the strategic direction of the fighting. This the Americans were not prepared to concede. They offered to consult with Whitehall, but reserved the right of final decision. Britain had no alternative but to accept. As Cunningham candidly acknowledged, 'In view of the disparity in the size of the forces to be employed this is I think reasonable. [Admiral] King tried to be rude about it but nobody paid any attention to him.'[14]

On 24 July the three Allied leaders met to settle on a statement to Tokyo calling for her unconditional surrender under threat of 'prompt and utter destruction'.[15] This so-called Potsdam statement was issued two days later, and was promptly rejected by Japan since it did not preserve the position of Hirohito. Although both Truman and Churchill had considered modifying their demands, their joint statement bluntly insisted on unconditional surrender. If they moderated their stance now, the Japanese might well accept and remove the need to use the atomic bomb, but a demonstration of the bomb was needed to ascertain its effect on people and buildings rather than empty desert, and also to overawe Stalin. It would also, as Truman believed, ensure that Japan would 'fold up before Russia comes in'.[16]

The decision to use the bomb was not referred to the

Combined Chiefs of Staff, something that Cunningham thought in hindsight to be 'a great pity'. The Chiefs, wrote Cunningham, may well have found that 'neither the atom bomb nor the invasion were necessary to induce Japan to surrender'.[17] Eisenhower, the American Supreme Commander in Europe, shared this opinion, and told Stimson that the bomb was unnecessary: 'First, the Japanese were ready to surrender and it wasn't necessary to hit them with that awful thing. Second, I hated to see our country be the first to use such a weapon'.[18] But Stimson had spent two billion dollars developing the bomb and he was anxious for Congress to see something for their money.

By the time the Potsdam declaration was issued on 26 July, Churchill was back in London and Attlee's Labour Party was returned with a majority of 146 seats. A disappointed Churchill immediately resigned rather than wait for the confirmation of the new parliament. In a radio message to the British people he regretted that he had 'not been permitted to finish the work against Japan' while claiming that 'all plans and preparations have been made, and the results may come much quicker than we have hitherto been entitled to expect'.[19]

The British ambassador in Washington had been instructed by the Foreign Office to suggest (orally) to the State Department that the constitutional powers of Emperor Hirohito be left intact by the Allies to ease the problem of controlling the vanquished nation. Britain wished to avoid a 'total and protracted military occupation, combined with the assumption of all the functions of Government,' which would result in a 'strain on both manpower and physical resources'.[20]

When a copy of this cable was sent to Australia for the Dominion's information rather than as an invitation to consult on the issue, Bruce was angry. It was unforgivable that Britain, Russia and the United States should decide the issue of the Japanese surrender terms when Russia had not even declared war and the British had played such a lukewarm role throughout the war. Bruce reported to Chifley that he had 'left little unsaid as to my view as to the way in which we have been treated and I think you should also telegraph a protest' before the surrender terms had been finally decided.[21]

Chifley had been sworn in as the new Prime Minister of Australia on 12 July following the death of John Curtin. On 29 April Curtin had been admitted to hospital with congestion of the lungs. Although he later managed to return to the Lodge, he was too ill to leave his bed, and on 5 July he died. The contest for the leadership occurred as Evatt returned by ship from the San Francisco conference. His presence would not have made any difference to the outcome. Chifley took the leadership with a vote of forty-five to Evatt's one or two. The hapless Forde was runner-up with fifteen or sixteen votes. With the dour but competent Chifley in charge, government in Canberra could be given new life and direction.

High on the agenda of the new Government was a protest over the Potsdam declaration. Evatt had been assured when in London, that Australia would be consulted about the proposed peace terms for Japan but, as he now observed to Bruce, 'there is grave danger of our being gradually excluded from all really important discussions preliminary to and involving making of peace settlements in Europe and armistice settlement with Japan'. Evatt naively believed that the new Labour Government in London would 'better appreciate [the] great importance of [the] matter to Australia'.[22]

Evatt responded to the Potsdam declaration with a bitter and public denunciation of the way in which Australia had been treated by her great power allies. He revealed that it had been made without consultation or agreement with Australia, and that the Government had learned of it via the press. Evatt was particularly upset that China was a signatory to the declaration while Australia was ignored. According to Evatt, 'Australia's interest and concern are no less significant than those of China'.[23] Of course, Evatt omitted to acknowledge that China still had a formidable Japanese army occupying much of her countryside and that China was being groomed by Washington for great power status.

Evatt was not prepared to countenance the possibility of Japan receiving milder treatment than Germany, 'having regard to the outrageous cruelties and barbarities systematically practised under the Imperialist regime'. This was a sensitive political issue in Australia where the newspapers had reported various Japanese war crimes involving Australian personnel. Evatt pursued the imposition of a harsh peace until Labor's fall from power in 1949. He criticized people, presumably in Whitehall, who had failed to learn from the 'many early disasters of the Japanese war' and 'do not realise that the post-war security of the peoples of Australia and New Zealand and of India too are integrally bound up with the destiny of South East Asia and are therefore directly dependent upon the terms of the peace settlement with Japan.'[24]

Bruce took up Evatt's complaints during a meeting with the plain-speaking Ernest Bevin, Britain's new Foreign Secretary. Although originally claiming to be angered by the issue himself, Bruce thought it was more important to restore harmony to British–Dominion relations rather than to press a particular point. As he assured Bevin, he 'had merely come to ascertain what his reaction was to Evatt's views and to see if there was any method of handling the matter without friction'. He found that Bevin was 'quite outspoken' about Evatt's bid for inclusion into great power discussions and flatly refused to concede it. Bevin would draft a reply to Evatt that would be 'quite sympathetic towards Australia's desire for consultation, but will be quite frank on the subject of Evatt's specific proposals'.[25]

Although Australia had pressed Britain to increase her presence in the Pacific war for the sake of Imperial prestige, she was unwilling to join forces with Britain in a joint Commonwealth effort, because British forces were likely to arrive too late to play any part in the defeat of Japan. Australia's military attaché in London, Brigadier Rourke, advised Shedden on 26 June that Britain still had not committed ground forces to the Allied attack on Japan and that Whitehall was in considerable disarray mainly because of the elections.[26]

On 4 July, the day before Britain's polling day, an exhausted Churchill finally approved the concept of a land force taking part in the invasion of Japan. At a meeting with the Chiefs of Staff during which he confessed to not having read the relevant paper, Brooke stabbed out the strategy on a map of the Far East. Later he wondered

> How much he [Churchill] understood, and really understood, in his exhausted state it is hard to tell. However, I got him to accept the plan in principle, to authorize our sending the paper

> to the Americans and to pass the telegrams to the Dominion Prime Ministers for their co-operation.[27]

Two days later, Bruce advised Canberra of the British plans for a Commonwealth land force of five divisions to assemble in India and go into action against the Japanese main islands in March 1946. It was hoped that some of the land force and most of the tactical air force of fifteen squadrons would be provided by Australia and New Zealand.[28]

Britain would contribute to the second Allied wave of invasion rather than the first, which the Americans had timed for November 1945. Even without knowledge of the atomic bomb, it was clear the war might be over before the end of 1945 once Russia joined the fray. Although Australia was strongly committed to the Commonwealth, she wanted to have her flag flying when Allied troops stepped ashore in Japan. This was the feeling of the Advisory War Council when it met in Canberra on 19 July. 'The primary consideration', according to Billy Hughes, 'was actual participation in the main offensive against Japan'. Although he would have preferred Australia to participate within a Commonwealth force, 'if it could not be organised in time to take part in the main offensive, then Australia herself should not fail to participate in these operations'.[29]

Such participation would allow Australia to continue to go forward with MacArthur, which would help 'to strengthen future Australian–American relations which are of paramount importance from the aspect of security in the post-war period'. It might also entitle Australia to a say in the peace settlement, although events at Potsdam should have thrown doubt upon this. While wishing to go forward with MacArthur, Australia also wanted to play some kind of joint role with Britain to whom Australia mainly looked for her post-war security. Avenging the defeat of Singapore, at which Australia's 8th Division was captured, also played a part in the Dominion's calculations.[30]

Chifley was careful to specify that British forces would not be welcome in Australia itself since the Dominion could not accept any additional commitments. Moreover, Chifley warned that, if command in the SWPA were ceded from MacArthur to the British, London must guarantee to the

Dominion the same level of control she had enjoyed over strategic decisions under MacArthur and control over her own forces.[31] The War Cabinet had already refused a British request to mount an operation against two islands in the Solomons. The British High Commissioner for the Western Pacific had expressed concern that 'the natives would suffer severely from loss of food supplies and enemy violence and that our inability to protect them would be gravely detrimental to British prestige and create a serious local political situation'. The Australian War Cabinet agreed with Blamey that the operation was 'strategically unsound',[32] which was strange since the High Commissioner's reasons had justified similar strategically unsound operations mounted by Blamey in the past.

Australia was also loath to help with the British Pacific Fleet. After what Cunningham described as a 'most unhelpful' cable from Canberra about support for the Fleet, he complained of the Dominion seeking to play 'less and less part in the war'. On 17 July Admiral Fraser wrote to Cunningham explaining the difficulties he had been having with the Australian Government, which he blamed on political factors. He was particularly upset by Australia's refusal to support the British Fleet. On 11 July the War Cabinet confirmed its decision to release 64 000 men from the services 'in order to provide for housing and other high priority needs of the civil economy' irrespective of the British Fleet. Chifley sharply rebutted Fraser's petulant assertion that the Fleet had been sent at Australia's request and on the understanding that it would be adequately supported.[33]

One reason for recapturing Singapore, with its developed naval base, was Britain's wish that its Fleet no longer be dependent upon Australian logistical support. Meanwhile, Mountbatten was rushing to complete preparations for the invasion of Malaya and the attack on Singapore before Tokyo raised the white flag. Despite Japan's embattled position and the ease with which Singapore had been captured in 1942, Mountbatten still feared a prolonged siege. Only after capturing Singapore would Britain countenance its army sending troops against the Japanese home islands. Cunningham observed that there were 'nearly a million first line troops in Honshu backed up by some seven million

Home Guard: it seems to me that, physically, it will take a long time to kill all these'.[34]

At the Potsdam conference, MacArthur had cut Britain's offer of five army divisions down to three, and stipulated that the Commonwealth force was to use American equipment and logistic support, to be integrated within the American army, and to be kept in reserve until after the initial invasion. MacArthur also stipulated that no British Indian divisions be included because of 'linguistic and administrative complications'. These were humiliating conditions for a great power to accept, but Britain's contribution to the Pacific was token from start to finish; now it would be lost beneath the weight of American armour. The British Pacific Fleet was deliberately excluded from the naval operations that destroyed the remnants of the Japanese Fleet at the end of July, and America's Admiral Halsey was determined to 'forestall a possible post-war claim by Britain that she had delivered even a part of the final blow that demolished the Japanese Fleet'.[35]

Britain accepted her reduced status in the Pacific with apparent equanimity. As Ernest Bevin advised the Defence Committee on 8 August, Britain's 'original offer of up to 5 divisions was made very largely with a political object, i.e. to remove any idea in the United States and the Far East that we were not pulling our weight in the Far Eastern war and to re-establish our position in the eyes of the Far Eastern peoples'. General MacArthur's conditions meant that this would probably not be achieved, with the Commonwealth force becoming 'an unadvertised component of the American machine'. Bevin urged that the British propaganda machine should maximize the force's individuality and try to have it used in the initial assault.[36] But such an assault was already in doubt after the dropping on 6 August of the first atomic bomb on Hiroshima.

During the Defence Committee discussion of Bevin's memorandum, with Mountbatten in attendance, Alanbrooke indicated that the three Commonwealth divisions would comprise one British, one Australian, one Canadian and two New Zealand brigades. It would be more than six months before the force could be available for action in Japan after first training in America with American equipment.[37]

However, the second atomic bomb had already been loaded into a B29 bomber on the airfield on Tinian Island, its target the city of Kokura. The alternative target, if Kokura were obscured by cloud, was Nagasaki. News of the first atomic bomb had electrified the world, but there had been no Japanese surrender, and the British Defence Committee continued to discuss operations against Japan that could not be mounted until 1946.

With MacArthur concentrating on the invasion of Japan, Britain was offered command of the south-west Pacific, except the Philippines. Under the arrangement agreed at Potsdam, Britain would control the region west of the Celebes and Java while Australia took the eastern portion although, as Mountbatten admitted, he could not exercise real control until he had captured Singapore and opened the Straits of Malacca. So the transfer of control was left until the Straits were in Allied hands. Alanbrooke also announced that American General Wedemeyer was planning to recapture Hong Kong with Chinese forces, and suggested that British forces participate in this operation to retake their former colony.[38]

On 9 August, the day after Russian armies launched a wide-scale attack against the Japanese forces in Manchuria, an American B29 bomber searched unsuccessfully for a break in the hazy conditions over Kokura before turning towards Nagasaki. The weather there was also unsuitable until, at the last moment, a twenty-second break in the clouds permitted the bombardier to set his sight on a stadium several miles from the specified dropping point. Exploding at 1500 feet above the city in a blinding burst of light and heat, some 70 000 inhabitants of Nagasaki were killed, some quickly and some agonisingly slowly over the following days. As the world waited for the Japanese reaction, a third atomic bomb was prepared for delivery, possibly on Tokyo itself, while further massed incendiary attacks on other Japanese cities continued the conventional killing.

The day after the destruction of Nagasaki, the Japanese offered to surrender according to the terms of the Potsdam declaration except in so far as it affected the prerogatives of the emperor. Cunningham commented in his diary that the Allies now had to

> decide whether that constitutes unconditional surrender. How often do our rulers miss the salient point. When we were lunching with Winston at Potsdam he showed us the draft declaration and all the COS [Chiefs of Staff] told him that something about the emperor should be included. Now instead of having a formula of our own we have this Japanese formula which no one can really understand.[39]

When the Cabinet discussed the offer they covered their confusion by passing the initiative to the Americans, deciding that 'if they were of the opinion that the clause affecting the Emperor was acceptable, we should agree'.[40]

Such a backdown was anathema to the Australian Government. While news of the surrender offer was being transmitted via Switzerland, Evatt was pressuring London and Washington to rip out Japanese militarism by the roots. He also insisted that Hirohito, 'as head of the State and Commander-in-Chief of the Armed Forces, should be held responsible for Japan's acts of aggression and war crimes and would thus demand his removal'. According to Evatt, the Emperor should be forced to descend from his celestial throne to sign the surrender document and then be thrown into prison to await trial as a war criminal.[41]

When Evatt learnt of the Japanese conditional offer, he reiterated Australia's view that it was 'of the first importance . . . that the present Imperial–Militarist system be not only discredited but completely broken'. The Japanese people could only be made to 'appreciate their defeat', by the 'visible dethronement of the system'. Without this, 'the Japanese will remain unchanged and recrudescence of aggression in the Pacific will only be postponed to a later generation'. Australia resolutely opposed the 'small but influential school of thought in some Allied countries which is prepared to save the face of the Emperor' and appealed to Britain to 'resist any claim of the Emperor . . . to immunity from punishment, to support us in bringing him to justice and to deprive him of any authority to rule from the moment of the surrender'. Evatt repeated his cable to Washington, Chungking and Moscow, once again emphasizing the Dominion's right to propound an alternative foreign policy to that of Britain.[42]

Despite Australia's principled and legalistic stance, the United States and Britain would have to deal with the defeated Japanese nation and the considerable Japanese forces holding out from Rabaul to Singapore and from Canton to Siam. London and Washington feared, not without reason, that only the Emperor could order his forces to surrender. By not calling Hirohito to account for his crimes, the Allies might save thousands of lives that would otherwise be lost in enforcing the surrender. So, in replying to the Japanese surrender offer, the Americans were deliberately obtuse about Hirohito's future, insisting only that the Emperor sign the surrender document and that he be subject to the authority of the Allied Supreme Commander. This satisfied the American electorate but still left the Japanese a straw of honour to grasp.

Evatt had asked that no reply be made to Japan until Australia had studied the offer and been able to comment upon it. Despite hostile comments from Australia, and China, the British did not heed the Dominion's objections. Britain wanted to demand even less of Hirohito, suggesting to Washington that the Emperor not have to sign the surrender but only have to command an end to hostilities. This, Britain believed, would 'secure the immediate surrender of the Japanese in all outlying areas and thereby save American, British and Allied lives'.[43]

Washington accepted the suggestion from London, announcing that the four powers, America, Britain, Russia and China, had agreed on the surrender terms. 'So now it's up to the Japs', wrote Cadogan at the Foreign Office. On 14 August the British Cabinet was advised that Washington was 'going ahead on surrender terms etc., without showing much disposition to consult us. We must accept this, [wrote Cadogan] and if Dominions complain we can say that we, too, were not consulted . . .' Evatt had already instructed Bruce to 'spare no effort to see that on this occasion definite action is not taken without our prior knowledge or consent'. But Washington would brook no interference as it brought the Pacific war to an end on its own terms.[44]

On 15 August, in a masterful understatement, a recorded message from Hirohito crackled over Japanese wireless radios: 'the war situation has developed not necessarily to

Japan's advantage'. Citing the American use of 'a new and most cruel bomb' as his reason, Hirohito called on his subjects to 'pave the way for a grand peace for all generations to come by enduring the unendurable and suffering what is insufferable'.[45] The war was over, but the agony had not ended for many of its victims. Prisoners of war still had to be freed and nursed back to health, and some 200 000 more citizens of Hiroshima and Nagasaki faced a lingering death from radiation sickness. The Pacific war had ended in the devastation and defeat of Japan but it had also violently transformed the western Pacific, setting up reverberations which are still being felt today.

Epilogue

THE SUDDEN END to the Pacific war caught Britain unprepared. She had assumed that she would be able to recapture Singapore, avenging its ignoble loss and rebuilding British prestige in the minds of native peoples. Instead, Mountbatten's planned invasion of Malaya became an unopposed landing while British parachutists dropped among the startled inhabitants of Singapore. Britain and China now competed to assert sovereignty over Hong Kong. Shortly before his death, Roosevelt had become obsessed with the idea of wresting Hong Kong from Churchill's grasp, but Churchill had been equally adamant that Britain would never willingly let Hong Kong go.[1]

Attlee's Labour Government now faced the urgent task of reasserting British sovereignty over the colony. On 10 August Field Marshal Alanbrooke advised the Cabinet that their plan to accompany American forces to Hong Kong had to be replaced with 'emergency arrangements'. He suggested sending a detachment of the British Pacific Fleet to Hong Kong before the Chinese or Americans could interfere. This unilateral action raised serious political problems: Hong Kong lay within the region of American strategic control and the British Pacific Fleet was subject to the orders of America's Admiral Nimitz. Attlee proposed to send a personal cable to Truman to try to smooth the waters.[2]

The closest Commonwealth troops to Hong Kong were the Australians in Borneo and on the island of Morotai. So Attlee appealed to Canberra for the use of their troops in disarming the Japanese and asserting British control. It is of 'great importance', argued Attlee, that Hong Kong should be 'initially occupied by British Commonwealth Forces rather than by Chinese or American troops from the mainland of China'. Britain not only wanted to regain her colony but also to rebuild her influence in China. As the Chiefs of Staff urged the British Cabinet on 13 August, 'it is politically important that [as well as reasserting control in Hong Kong,] we should show the British flag in the main Chinese ports such as Shanghai, Tientsin, Tsingtao and Dairen'. They counted upon Australia to provide a brigade of troops from Borneo to follow up the British warships in Hong Kong. In Indo-China, where Britain was also charged with taking control from the Japanese before the return of the French, the British representatives used the Japanese Army and Air Force to put down Vietnamese rebels making a bid for independence during the temporary hiatus in the territory's sovereignty.[3]

Sending troops to impose colonial control in Hong Kong would surely damage Australia's relations with China and would also, as Shedden advised Chifley on 12 August, impede her demobilization programme. Evatt suggested that Australian participation in Hong Kong might stir up Chinese criticism of the Dominion's 'White Australia' policy in retaliation. Although the Australian Chiefs of Staff and Shedden supported both a naval and army contribution to Hong Kong, this was cut back by the Cabinet on 17 August to a short-term loan of some Australian naval units.[4]

The British Admiralty had ordered the immediate dispatch of British warships to Hong Kong, but the Fleet did not plan to leave its base at Manus Island until 20 August. Cunningham complained that: 'They do not seem to realize the necessity for haste'. Australian minesweepers from the Philippines were rushed in to clear the way for the British warships while the British Government brushed aside Chinese objections to the move. By 20 August Cunningham reported to Admiral Fraser that the 'Hong Kong muddle appears to be clearing up', with the British government being 'quite determined to take possession'. MacArthur managed to complicate the situation by demanding that Hong Kong must not surrender before the formal surrender he was orchestrating for Tokyo, with himself in the starring role. Nevertheless, he ordered the Japanese in Hong Kong to surrender to the British rather than to the Chinese, although he told a visiting British intelligence officer that he was 'very angry' over the high-handed British action.[5]

Now the British propaganda machine was cranked up to create an image of Allied unity. A directive from Britain's Political Warfare Executive on 23 August, instructed official broadcasters that

> in order to avoid raising latent Asiatic sympathy for an Asiatic State overthrown by Europeans you should associate other Asiatics both in the work of bringing Japan to defeat and in our joy at her defeat.
>
> Assume Allied solidarity throughout and let major events crowd out all inter-allied misunderstandings. There is no evidence to warrant newspaper reports of difficulties with the Chinese over the immediate future of Hong Kong.[6]

Britain was eager to regain Hong Kong for practical reasons; it would help to relieve the strain on the British budget and allow the Pacific fleet to leave Australia for Hong Kong and Singapore. Cunningham pressed his naval commander in Ceylon to reduce the number of his admirals, claiming that it was 'essential that we reduce our expenditure in the East' since the 'country's financial position is just frightful'. Although Attlee tentatively questioned the need for Britain to maintain a worldwide network of bases, the Admiralty justified restoration work on the naval bases at Singapore and Hong Kong by pointing to the value such work would have in 'restarting British trade to China and the East Indies'. With Bevin's backing, the British Defence Committee decided on 7 September that the Pacific Fleet would be based at Hong Kong. Bevin argued that the presence of the Fleet would have a 'beneficial and steadying influence on various treaty undertakings involving ourselves, the Russians and the Chinese'. The Army was even planning to keep sufficient forces in the Far East to allow 'one Infantry Brigade spare for use at Shanghai or elsewhere if required'.[7]

Australia was concerned not only about the unwanted request for forces for Hong Kong, but also that the planned handover of territories in the south-west Pacific would mean a heavy manpower burden. After the Japanese handed over power, the Australians would have to care for the native population. The Army Department urged Shedden to ascertain the effect on the Australian economy before taking on such a commitment.[8] At the same time, the government was determined to maximize the effect of the limited Australian contribution. This could not be done if her forces were to be scattered across the Pacific at Britain's whim.

Instead of restricting their commitment, the Australian Government doubled their Chiefs' recommended contribution to the Allied occupation force for Japan, proposing two cruisers, two destroyers, two brigades of troops and three squadrons of Mustang fighter planes. To ensure maximum impact, the Government stipulated that its contribution was being made as a 'principal Pacific Power which has for so long borne the heat and burden of the struggle against Japan' and that its forces would be commanded by an Australian officer responsible to the Supreme Allied Commander, rather than to a British Commonwealth Commander. Australia also demanded separate representation at the surrender ceremony being planned for Tokyo Bay. British planning contemplated a joint Commonwealth occupation force with a British commander, though with an Australian as army commander.[9]

The question of this British Commonwealth Occupation Force (BCOF) provoked a bitter and prolonged dispute between the Labour governments of Britain and Australia following peace in the Pacific. Australia sought to emphasize its independence, while Britain wanted to use Australian and other Dominion forces in a Commonwealth force to increase the impact of its own contribution. In a press statement on 17 August announcing the proposed Australian occupation force, Chifley complained about the lack of recognition given to Australia by the Allies when formulating the surrender terms for Japan. Britain responded by admitting that she had little more influence than Australia in determining the Allied approach to Japan, although the British Government did succeed in excusing Hirohito from signing the surrender document.[10]

Meanwhile, Bruce fulfilled his traditional soothing role and warned Canberra that the British Government was 'hot and bothered' by public Australian criticism which undermined the sympathy of Attlee's colleagues for their Australian comrades. He appealed to Chifley not to 'destroy this favourable atmosphere by overplaying our hand just at the moment when possibilities of achieving what we have been fighting for for years are opening'. Attlee advised that his Government would not respond to public criticism from Australia 'in the interests of good relations' and asked that Chifley 'ensure that a similar restraint from public controversy is exercised in Australia'.[11]

Like Britain, Australia had serious problems reconciling her ambitions with her limited means. Although she offered to send a parachute battalion to participate in the occupation of Singapore, the shortage of shipping and the tight schedule prevented it. The best she could hope for was to rush 120 troops by air as soon as an airfield became available at Singapore.[12] As for Australia's plan to maintain an independent occupation force in Japan, the Chiefs of Staff pointed out with some relish that the Cabinet's unilateral decision to double the forces recommended by the Australian Chiefs for use in Japan had created an impossible situation. It simply could not be done, they said, since 'an independent Australian Force would necessitate the establishment of separate Australian base installations, repair facilities and the provision of maintenance, common technical supplies, stores, fuel and lubricants'.[13]

Britain was anxious to bring Australia back within the Commonwealth fold, ending this 'deplorable example of lack of unity' and using the Dominion as a buttress to support her collapsing Empire. To lessen the strain on British resources, Whitehall readily offered to allow an Australian commander to lead the whole Commonwealth force, not just the army contingent as previously contemplated. When British Ministers objected to the power this would give the Australian commander to determine matters of political policy, Attlee assured them that the proposed command structure would leave the Australian commander subject to control by the Australian and British Chiefs of Staff.[14]

To the surprise and annoyance of the British, Chifley persisted with the independent Australian force, sending what Cunningham described as a 'rude reply' to the British suggestion. Cunningham had been advised earlier by Admiral Royle, returning to London after his stint as chief of the Australian Navy, that 'it was no good handing out soft soap to these Australian politicians. We should be as rude as they are'. Fortunately, in this case, Royle's advice was not heeded. Evatt was in London and was apparently anxious to curry favour with his British Labour counterpart. He offered to 'persuade the Australian Government to reverse their decision and take a more favourable view of the suggested composite British Commonwealth force'.[15]

On 19 September the Australian War Cabinet agreed with Evatt that 'it might be possible to use the occasion to demonstrate Australian leadership in Pacific affairs and the Pacific settlement'. It would implicitly acknowledge Australia as the Commonwealth country with the primary interest in the Pacific. It would also provide 'an opportunity for experience in the joint higher direction of British Commonwealth Forces in the Pacific'. The War Cabinet also cut back the proposed army force from two brigades to one. London was very relieved. On hearing the news, Alanbrooke proclaimed: 'Thank Heaven, for if they had been allowed to refuse our last offer of an Australian Command and a Combined Chief of Staff organisation . . ., it would have been the end of all Imperial co-operation'.[16] It would have also meant a bigger British contribution to the occupation force since, without Australian involvement, Britain could hardly have concealed her national contribution behind the Commonwealth banner.

On 2 September 1945 an armada of Allied warships rode at anchor in Tokyo Bay as General MacArthur held centre stage on the afterdeck of the battleship *USS Missouri*, where he called upon representatives of the Japanese government to sign the instrument of surrender, witnessed by the lines of American sailors and Allied officers. Admiral Sir Bruce Fraser represented Britain while General Blamey signed the document on behalf of Australia. Emperor Hirohito was absent, absolved from attendance at the behest of the British. Across the Pacific, surrender ceremonies took place with local commanders, the Japanese command structure being left intact in many places to ease the Allied burden of control. Producers of British propaganda broadcasts were instructed to 'avoid recriminations and questions of war guilt', to 'ignore reports of possible after-effects of the atomic bomb' and to 'keep off the Japanese treatment of Prisoners of War for the time being', the intention being to gain the co-operation of the local Japanese commanders.[17]

Like the Germans after the First World War, there was widespread feeling among the Japanese armed forces that they had not been defeated in the field. As Mountbatten's political adviser observed, from the psychological viewpoint 'the war has been fought in vain'. British weakness in the Far East, together with the suddenness of the surrender, forced them to allow the Japanese to 'follow their own concept of the surrender'. This meant that the Japanese were allowed to 'look on the whole process as a handover rather than a surrender—an attitude which is encouraged by the methods and procedure adopted by the principal Allied Governments'.[18]

As for Australia, her grandiose plans for a sub-empire in the south-west Pacific had faded, and she fell in with British policies that supported the handing back of colonial territories to their former masters. She looked forward to the revival of the system of Imperial defence, which had proved so flawed in 1942. As Curtin explained to the House of Representatives during a major speech on post-war defence and security policy, Australia's foreign policy 'must always be in harmony with that of the British Commonwealth as a whole'. According to Curtin, the Dominion's role was to

> give advice, to state its view, now and again to criticise, and to make suggestions which, in its view, would strengthen the family relationship. This we have done, remembering always that our articulation in the world would be more impressive as a member of a family than it could ever be if we made it as a separate and distinct entity.

Curtin claimed there was 'no abatement of the sovereignty of this country in making that statement'.[19]

In October 1939 another Australian leader had expressed Australian policy in similar terms: 'The fact is, and it is not a fact for which we need apologise, that we maintain our independent existence primarily because we belong to a family of nations . . .'[20] The leader on that occasion was Robert Menzies, of the conservative United Australia Party.

In the years between these two speeches, the words had changed little, and the sentiments hardly at all, despite the world-shaking events of the war that had almost annihilated Australia. Half a century later, the actors have been replaced but the script remains the same, as Australia continues to cling to the sense of dependence upon great power 'protectors', shrugging off those who point to a possible independent destiny.

Conclusion

TAKING A LONG view of the Pacific war, it is clear that Britain's expulsion from the Far East in 1942, her decision to concentrate on fighting Germany first and to send largely token forces against Japan in 1945, had historical antecedents stretching back for nearly a century. Britain's waning ability to defend all her Imperial possessions was hidden by the bluff and bluster of the Victorian age. The Boer War, at the end of that gilded period, revealed fundamental weaknesses in Britain's armed forces, with the Empire being held to ransom by a ragtag army of guerrillas. At the same time, the rise of Germany and the emergence of Japan as a power in the Pacific challenged Britain's grip on her Empire as never before. Britain responded by forming an alliance of convenience with Japan which allowed her to compete in a debilitating arms race with Germany that ended in the trenches of the First World War.

The system of imperial defence was established as a sign of imperial strength but really emphasized British weakness, with the Dominions and India helping to prop up the crumbling Empire. Britain made the protection of the wider Empire the first priority on the Dominion's defence agenda. On the international stage, she sought to win the ongoing arms race by hobbling the other competitors with a series of disarmament agreements. These strategies worked only so long as everyone else played the game.

Russia was never in the game and, under Stalin's industrialization programme of the 1930s, the weapons of war were mass-produced in ways that would have made Henry Ford proud. Russia's rising power was a threat to both Germany and Japan since Stalin had scores that he wished to settle with both of these powers. As a response to the Russian threat, as well as for their own internal reasons, Japan, Germany and Italy began, during the 1930s, a rapid military build-up combined with an expansionist and destabilizing foreign policy. This raised an obvious threat to Britain's hold over her Empire. With potential foes on both sides of the world, Britain's bluff was about to be called.

The gradual contraction of Britain's ability to defend all her imperial possessions forced her to choose between competing priorities. Despite the rich imperial rhetoric between the wars, the Far East was relegated to a distant third place behind the defence of Britain and the defence of British interests in the Mediterranean and the Middle East. Britain persuaded Australia and New Zealand to adopt British priorities as their own, dispatching forces to the Middle East on the outbreak of war just as they had done in 1914, defending distant British interests at the expense of their own defence. The benefit to Australia of following in Britain's footsteps had been the chance to skimp on defence, hiding behind the tattered Imperial banner while trying to construct, in the south Pacific, a society in Britain's image. At Singapore the Imperial banner was torn aside to reveal an island continent denuded of defences. Australia suddenly faced the prospect of annihilation as a nation.

Britain's survival strategy dictated that Germany be defeated before she turned to the Far East to rescue her Empire. This was cold comfort for the seven million people of Australia who feared that an Asian invasion and occupation would dislodge their frail grip on the continent. Fortunately the Japanese were deterred at first by distance from invading Australia, and the United States were able to step into the breach left by Britain, dealing the Japanese Navy several critical blows in 1942 from which it never recovered. The prolonged and desperate struggle for control of New Guinea, which was primarily an Australian effort, finally dispelled the spectre of invasion.

Once the continent was secure from invasion, the

Australian Government increasingly turned towards post-war reconstruction and away from prosecuting the war with vigour. This suited MacArthur once his own forces had built up to a substantial level. It allowed him to leave Australia behind as he raced towards Tokyo. When the surrender terms for the Pacific were being negotiated, Australia's voice and interests were ignored. For Britain, the Pacific conflict had always been a distant war. Just as she had used Australians to help fight her battles in Europe, she now turned to the United States, the Dominions and India to fight her battles against Japan. By such a strategy, Britain succeeded in recovering all the territory she had lost to the Japanese. She could not know that, within a generation, a nationalist revolution ignited by the war would rob her of almost the entire Empire.

The question of Australian independence and the development of a distinctive Australian foreign policy has bedevilled historians of Australian diplomacy. Some writers have searched as far back as nineteenth century colonial records and proclaimed the existence of a distinctive foreign policy by the various colonial governments. Although the colonies certainly had particular interests and occasionally acted upon them, for example Queensland unilaterally annexed Papua in 1883, it was always done within the structure of the British Empire and usually to serve Imperial interests.

It was only during the Second World War that Australia developed the capacity to formulate a distinctive foreign policy. With the appointment of representatives to a wide range of countries and the creation of a diplomatic service, Australia was able to draw on alternative sources of information instead of relying upon Whitehall and the shortwave broadcasts of the BBC. Curtin's dramatic 'turning to America' message in December 1941 was a sign of change in Canberra. However, its importance has been overrated by writers who have, with the benefit of hindsight, claimed it as the point at which Australia switched her allegiance from Britain to the United States. This simply did not happen. The Dominion regarded the close relationship with America as a temporary expedient. In 1945 Australia sought to reconstruct the Imperial framework in the belief that it could still provide for Australia's defence, supported by a strong United Nations

Organization and with the greater presence in the Pacific of the United States.

In rushing back to the mother country, the Dominion abandoned her attempts to create a sub-empire in the south Pacific. This concept of a sub-empire was defined by the 1944 Anzac agreement, which sought to restrict American expansion to north of the equator, reserving for Australia and New Zealand a zone of influence in the south-west Pacific stretching from Fiji to Malaya. This attempt failed when both Britain and the United States refused to recognize Australia's claim to regional pre-eminence. All the same, Britain found the Australian bid for regional power a convenient way to use Dominion forces to help protect British interests in the Pacific.

Although there are signs of the Dominion attempting to define distinctive Australian interests during the period between 1942 and 1945, the issue is not clear-cut. While Evatt was more forthright in developing such policies, Curtin often managed to restrain him. By leaving Evatt behind in Australia when he travelled to London in May 1944, Curtin ensured that the Anzac declaration was allowed to slip off the Allied agenda and Australia more closely aligned with British foreign policy. It was only as Curtin's health declined that Evatt was able to reassert his control over foreign policy. By this time, however, the overriding preoccupation of both the Australian and British Governments was to limit their contribution to the Pacific war and to rebuild their war-damaged societies within a Commonwealth alliance.

As for Australian independence, it remains only partially achieved and has come piecemeal, almost unnoticed by the Australian people. There was no revolution, no upheaval of the toiling masses yearning to be free from the yoke of an imperial master. If Australians were asked when their nation became independent, most would probably point mistakenly to 1901 when the separate colonies joined together to establish the Federal Government. More informed citizens might suggest 1926, when the Imperial Conference declared that the Dominions and Britain were all 'autonomous communities within the British Empire, equal in status, in no way subordinate one to another in any respect of their domestic or external affairs'[1]; or 1932 when the Statute of Westminster was passed by the House of Commons; or 1942 when it was finally ratified by the Australian Parliament; or perhaps 1986 when the Australia Act was passed. Each of these advances marked a step along the road to an independence that has remained beyond the reach of successive governments.

Australian independence is not simply a constitutional issue. The exercise of independence also requires the ability and strength to formulate policies based on a clear appreciation of national interest and to withstand pressure from 'allies' who seek to impose their views. For too long, Australia has looked at the world with British and then American eyes. What has been good for London or Washington has not necessarily been good for Australia. During the 1950s, Robert Menzies's conservative government offered the Australian continent for the testing of Britain's atomic bombs. During the 1980s, Prime Minister Bob Hawke offered Australia as a testing ground for America's controversial MX missile, continuing a tradition of servile military dependency that belies the gradual progress towards political independence.

It is worth repeating that the experience of the war did not propel Australia from the protective British bosom into the arms of America, as popular mythology would have it. In 1945 as in 1939 the Australian defence policy rested on the bankrupt system of imperial defence. Although Australia's defence in 1942 had been guaranteed by the sacrifice of her troops in New Guinea and by the actions of the American Navy in the Pacific, the Dominion remained ambivalent about the United States. Australia hoped that she would defend her in future conflicts but recognized that the United States was also a competitor, and perhaps even a threat, in the region. Only during the post-war years did Australia gradually slip from Britain's grip as she discovered a new sense of security in the shadow of her wartime ally. And the day might yet come when a future Republic of Australia will feel sufficiently confident to chart her own course in the Pacific.

Notes

Abbreviations

AA	Australian Archives, Canberra
BL	British Library, London
CC	Churchill College Archives Centre, Cambridge
CUL	Cambridge University Library
FUL	Flinders University Library, Adelaide
HLRO	House of Lords Record Office, London
IOL	India Office Library, London
IWM	Imperial War Museum, London
KC	Liddell Hart Centre for Military Archives, King's College, London
LSE	London School of Economics
NLA	National Library of Australia, Canberra
NMM	National Maritime Museum, Greenwich
PRO	Public Record Office, London
RUL	Reading University Library, Reading
UMA	University of Melbourne Archives

CHAPTER 1

1 G. H. Gill, *Royal Australian Navy, 1942–5*, pp. 111–12.
2 War Cabinet Conclusions, 1 June 1942, CAB 65/26, WM (42)70; memorandum by Eden, 3 June 1942, CAB 66/25, WP (42)236: PRO.
3 R. Sherwood, *The White House papers of Harry L. Hopkins*, vol. 1, p. 478.
4 Conversation between Churchill and the Chinese Ambassador, 3 June 1942, PREM 3/158/6: PRO.
5 Minute, Churchill to Chiefs of Staff, 1 June 1942, PREM 3/143/9: PRO.
6 Cable, Wavell to Churchill, 3 June 1942, PREM 3/143/9: PRO.
7 Extract from COS (42) 51st Meeting (O), 8 June 1942, PREM 3/143/9: PRO.

[8] Letter, Somerville to H. MacQuarrie, 1 June 1942, ADD MS 50143, Somerville Papers: BL.
[9] Cable, Churchill to Wavell, 12 June 1942, PREM 3/143/9: PRO.
[10] Cable, Wavell to Churchill, 14 June 1942, PREM 3 143/9: PRO.
[11] Letter, Grigg to Lord Astor, 12 June 1942, MS 1066/1/823, Astor Papers: RUL.
[12] M. Gilbert, *Road to victory*, p. 122.
[13] Letter, Kirke to Pound, 10 June 1942, PREM 3/163/8; cable, Little to Pound, 12 June 1942, PREM 3/163/4: PRO.
[14] Cable, Pound to Little, 18 June 1942, PREM 3/163/4: PRO. cable, Churchill to Wavell, 5 June 1942, ISMAY VI/2, Ismay Papers: KC.
[15] Minutes of Prime Minister's War Conference, 11 June 1942, CRS A5954/1: AA.
[16] Cable, Little to Pound, 12 June 1942, PREM 3/163/4: PRO.
[17] ISMAY VI/2, Ismay Papers: KC.
[18] David Day, 'Anzacs on the run', May 1986.
[19] Talk with Sir James Grigg, 16 June 1942, CRS M100: AA.
[20] Letter, Croft to Percival, 6 August 1942, CRFT 1/17, Croft Papers: CC.
[21] Letter, Cross to Attlee, 16 September 1942, Rc/4/12, Cross Papers: IWM. *See also*, Dalton diary 11–14 September 1942, I/27/66, Dalton Papers: LSE.
[22] Letter, Norman Douglas to Bruce, 25 June 1942, and letter, Official Secretary to Douglas, 3 July 1942, AA 1970/559/2, 'High Commissioner Bruce—Miscellaneous Papers—1939–1945': AA
[23] Cable, Bruce to Curtin, 26 June 1942, CRS M100: AA.
[24] *See* f.n. 15.
[25] Minute, Churchill to Chiefs of Staff, 12 June 1942, PREM 3/151/4: PRO.
[26] Day, *The great betrayal*.
[27] Report by the Joint Planning Staff, 17 June 1942, PREM 3/158/3: PRO.
[28] War Cabinet conclusions/Confidential Annex, 22 June 1942, CAB 65/30, WM (42)79: PRO. Talk with Attlee, 28 June 1942, CRS M100: AA. Day, 'An undiplomatic incident', November 1986.
[29] Cable, Dominions Office to Curtin, 23 June 1942, PREM 3/150/7: PRO.
[30] Letter, Curtin to Evatt, 23 June 1942, CRS A5954/474: AA.
[31] Memorandum by Shedden, 23 June 1942, CRS A5954/474: AA. *See also*, Day, *The great betrayal*, ch. 13–14.
[32] Cable, Attlee to Churchill, 25 June 1942, PREM 3/150/7: PRO.
[33] Cable, Commonwealth Government to Dominions Office, 25 June 1942, PREM 3/150/7: PRO.
[34] Cable, Evatt to Bracken, 25 June 1942, PREM 3/150/7: PRO.
[35] Minute, Attlee to Churchill, 26 June 1942, PREM 3/150/7: PRO.
[36] Cable, Churchill to Curtin, 27 June 1942, VI/2, Ismay Papers: KC.
[37] Cable, Evatt to Bracken, 30 June 1942, PREM 3/150/7: PRO.
[38] Cable, Churchill to Curtin, 3 September 1942, CRS A5954/229: AA.
[39] 1941 trip diary, p. 197, Menzies Papers, MS 4936/13/3: NLA.
[40] The *Age*, Melbourne, 14 November 1970.
[41] War journal, 15 April 1943, Wilkinson papers, WILK 1/2: CC.
[42] Letters, one undated but probably late 1945 and the other 17 September 1945, Cranborne to Emrys-Evans, Emrys-Evans Papers, ADD MS 58263: BL.
[43] Shedden's draft memoirs, ch. 57, pp. 6–7, CRS A5954/771: AA.

[44] Cable, Evatt to Bruce, 15 November 1941, CRS M100: AA.
[45] Advisory War Council minutes, 17 June 1942, CRS A2682/5/967: AA.
[46] Appreciation by Willis, 19 June 1942, Willis Papers, WLLS 5/5: CC.
[47] Rev. J. W. C. Wand, *Has Britain let us down?*.
[48] Letter, Emrys-Evans to Cross, 19 June 1942, Cross Papers, Rc/4/9: IWM.
[49] Letter, Wakehurst to Emrys-Evans, 16 June 1942, Emrys-Evans Papers, ADD MS 58243: BL.
[50] Day, *The great betrayal*, pp. 338–9.
[51] Cable, Hughes to Churchill, 20 June 1942, PREM 3/150/7: PRO.
[52] Letter, Cunningham to Willis, 21 June 1942, Cunningham Papers, CUNN 5/9: CC.
[53] Day, 'Promise and performance', September 1986.
[54] Gilbert, *Road to victory*, p. 131.
[55] C. King, *With malice toward none*, pp. 181–2, extract from his diary for 9 July 1942. Paper by Waldorf Astor (?), 23 June 1942, canvassing possibility of replacing Churchill as Minister for Defence, Astor Papers, MS 1066/1/823: RUL. Day, 'An Undiplomatic Incident'.
[56] Gilbert, *Road to victory*, pp. 137–40.
[57] Gilbert, *Road to victory*, p. 137.
[58] Dalton diary, 27 August 1942, Dalton Papers: LSE.
[59] Harvey diary, 25 August 1942, ADD MS 536398, Harvey Papers: BL. Lockhart diary, 28 August 1942, in K. Young (ed.), *The diaries of Sir Robert Bruce Lockhart*, vol. 2, p. 191; W. J. Hudson, *Casey*, ch. 7.
[60] Shedden's draft memoirs, ch. 57, p. 6, CRS A5954/771: AA.
[61] Background briefing, 1 July 1942, Smith Papers, MS 4675: NLA.

CHAPTER 2

[1] C. King, pp. 183–4, extract from his diary for 17 and 24 July 1942.
[2] Report by Shedden, presented to Curtin on 10 July 1942 and to MacArthur on 20 July 1942, CRS A5954/587: AA.
[3] Report by Shedden.
[4] Day, *The great betrayal*, Ch. 10–14.
[5] *See* f.n. 2.
[6] Letter, Blamey to Curtin, 29 June 1942, CRS A5954/261; Cabinet minutes, 9 June 1942, CRS A2703/1[C]: AA.
[7] D. Horner, *High command*, pp. 353–7.
[8] Note by Shedden, 6 July 1942, CRS A5954/229: AA.
[9] War Cabinet minute, 7 July 1942, cited in Gilbert, *Road to victory*, p. 143.
[10] Minute, Churchill to Alexander and Pound, 4 July 1942, PREM 3/163/4: PRO.
[11] A. Bryant, *Turn of the tide*, p. 421.
[12] *See* f.n. 10.
[13] Minute, Alexander to Churchill, 14 July 1942, PREM 3/163/4: PRO.
[14] Cable, Pound to Little, 12 July 1942, PREM 3/163/5: PRO.
[15] Minute, Churchill to Alexander and Pound, 13 July 1942, PREM 3/163/5: PRO.
[16] *See* PREM 3/143/9: PRO.
[17] Gilbert, p. 149.
[18] Cable, Curtin to Churchill, 16 July 1942, PREM 3/63/10: PRO.
[19] Cable, Curtin to Churchill, 16 July 1942.
[20] Minute, Churchill to Ismay, 18 July 1942, PREM 3/63/10: PRO.

[21] Cable, Churchill to Curtin, 24 July 1942, PREM 3/63/10: PRO.
[22] Cabinet Minutes, 7 and 8 July 1942, CRS A2703/1[C]: AA.
[23] Letter, Curtin to his Ministers, 18 July 1942, CRS A5954/375: AA. *See also* background briefing, 29 July 1942, Smith Papers, MS 4675: NLA.
[24] Preliminary report by Tariff Board, 23 July 1942, CRS A2700/4/318: AA.
[25] Draft cable, Bruce to Curtin, 8 July 1942, AA 1970/559/2: AA.
[26] Gowrie Papers, MS 2852/4/21/24: NLA.
[27] Letter, Emrys-Evans to Cross, 24 July 1942, Emrys-Evans Papers, ADD MS 58243: BL.
[28] Letter, Emrys-Evans to Wakehurst, 23 July 1942.
[29] AWC minutes, 1 July 1942, CRS A2682/V/978: AA. For Menzies's report *see* Day, *Menzies and Churchill at war*, pp. 193–4.
[30] Letter, Robinson to Earl of Drogheda, 21 February 1963, Robinson Papers, 'Interesting Letters' folder: UMA.
[31] Day, *The great betrayal*, p. 331.
[32] Memorandum, Robinson to Evatt, 20 July 1942, Evatt Papers, 'Robinson, W. S., 1942–45 (a)' folder: FUL.
[33] Draft cable, Robinson to Bracken, undated but, according to a note, written between 20 July 1942 and January 1943, Robinson Papers, 'Wars' folder: UMA.
[34] War journal, 15 April 1943, Wilkinson Papers, WILK 1/2: CC. *See also* Lord Chandos, *The memoirs of Lord Chandos*.
[35] Talk with Robinson, 6 July 1943, CRS M100. Talk with Spender, 19 November 1942, CRS M100: AA.
[36] Day, 'P. G. Taylor and the alternative Pacific air route, 1939–45'. *See also* P. G. Taylor, *Forgotten Island*.
[37] Letter, Robinson to Evatt, 13 July 1942, Evatt Papers, 'Robinson, W. S., 1942–45 (a)' folder: FUL.
[38] Minutes of Prime Minister's War Conference, 17 July 1942, CRS A5954/1: AA.
[39] Background briefing, 23 July 1942, Smith Papers, MS 4675: NLA.
[40] CRS A5954/231: AA. PREM 3/150/7: PRO.
[41] AWC Minutes, 30 July 1942, CRS A2682/5/1009: AA.
[42] Cable, Curtin to Churchill, 31 July 1942, PREM 3/63/10: PRO.
[43] Cable, Dominions Office to Australian Government, 6 August 1942, PREM 3/150/9: PRO.
[44] Gilbert, *Road to victory*, p. 160.

CHAPTER 3

[1] Note by Astor, August 1942, MS 1066/1/823, Waldorf Astor Papers: RUL.
[2] Note by Astor, August 1942. *See also*, Day, *Menzies and Churchill at war*, ch. 11–13; D. Carlton, *Anthony Eden*, pp. 204–6; R. R. James, *Anthony Eden*, pp. 267–8.
[3] Background briefing, 4 August 1942, MS 4675, Smith Papers: NLA.
[4] P. Hasluck, *The government and the people 1939–1941*, p. 225.
[5] Cable, Attlee to Curtin, 6 August 1942, in *DAFP*, vol. 6, doc. 18.
[6] Gilbert, *Road to victory*, pp. 168–9.
[7] Letters, Blamey to Shedden, 29 July 1942, and Curtin to Cross, 6 August 1942, CRS A5954/463: AA.
[8] Cable, Bruce to Curtin, 20 August 1942, CRS A5954/463: AA. *See also* other documents in this box.
[9] Letters, Blamey to Shedden, 30 November 1942, and Shedden to

Blamey, 22 December 1942, CRS A5954/463: AA.
[10] Cable, Bruce to Curtin, 2 August, 1942, CRS M100: AA.
[11] Secret memorandum, Jones to Astor, 7 August 1942. *See also* letters, Thomas to Jones, 13 August 1942, and Astor to H. Brooke MP, 17 August 1942, MS 1066/1/823, Waldorf Astor Papers: RUL.
[12] War Cabinet Conclusions/Confidential Annex, 12 August 1942, CAB 65/31, WM (42)111: PRO.
[13] Talk with Cripps, 17 August 1942, CRS M100: AA.
[14] Cable, Churchill to Attlee, 17 August 1942, PREM 3/76A/11: PRO. *See also* G. Ross, 'Operation Bracelet: Churchill in Moscow, 1942', in D. Dilks (ed.), *Retreat from power*, vol. 2.
[15] *See* f.n. 13; *see also* P. Addison, *The road to 1945*, ch. 7.
[16] Note by Bruce, 6 October 1942, CRS M100: AA.
[17] Bruce received some support for his struggle when the former Secretary of the Committee of Imperial Defence, Lord Hankey, argued in a memorandum sent to Ministers and officials that Bruce should have access to 'all meetings, proceedings and documents, including those of the Defence Committee, of which he ought to be a member . . .' Memorandum by Hankey, 3 September 1942, HNKY 11/7, Hankey Papers: CC.
[18] Cabinet minutes, 3 August 1942, CRS A2703/1 (c): AA.
[19] Minutes of Prime Minister's War Conference, 10 August 1942, CRS A5954/1: AA
[20] Minutes of Prime Minister's War Conference, 10 August 1942.
[21] The *Sydney Morning Herald*, 10 August 1942.
[22] Day, *The great betrayal*, ch. 13–14.
[23] Horner, p. 218.
[24] Cable, Bruce to Curtin, 8 August 1942, *DAFP*, vol. 6, doc. 20.
[25] Cable, Evatt to Dixon, 11 August 1942, CRS A5954/229: AA.
[26] Cable, Evatt to King, 14 August 1942, *DAFP*, vol. 6, doc. 24.
[27] Letters, Cunningham to Pound, 31 July and 12 August 1942, ADD MSS 52561, Cunningham Papers: BL. As leader of the British Admiralty Delegation in Washington, Admiral Cunningham described the agreement allowing for the diversion to the Pacific of some American Air Force units as being 'a most poisonous document . . .'.
[28] Cable, Dixon to Evatt, 20 August 1942, CRS A5954/229: AA.
[29] Cable, Evatt to Dixon, 25 August 1942, CRS A5954/229: AA.
[30] Background briefing, 25 August 1942, MS 4675, Smith Papers: NLA.
[31] Cable, Dixon to Evatt, 25 August 1942, CRS A5954/229: AA.
[32] *See* f.n. 29.
[33] Letter, Watt to Hood, 10 August 1942, MS 3788/1/1, Watt Papers: NLA.
[34] J. Robertson and J. McCarthy (eds.), *Australian war strategy 1939–1945*, Doc. 327.
[35] Cables, Curtin to Churchill, 25 August 1942, *DAFP*, vol. 6, docs. 27 and 28.
[36] 'I consider that it might have a lasting effect on Australian sentiment if His Majesty's Government gave [a cruiser] freely and outright to the Royal Australian Navy . . .' Note by Churchill to War Cabinet, 28 August 1942, CAB 66/28, WP (42)384: PRO.
[37] D. Horner, p. 219.

CHAPTER 4

[1] Letter, Rowell to Vasey, 1 September 1942, in J. Robertson and J. Mc-Carthy (eds.), Doc. 328.

[2] Report by Hankey, 3 September 1942, sent to British and Dominion Ministers and officials including Bruce and Shedden, HNKY 11/7, Hankey Papers: CC.
[3] Report by Cross, 12 August 1942, Cross Papers, Rc/4/8: IWM.
[4] Cable, Curtin to Churchill, 31 August 1942, PREM 3/163/4: PRO.
[5] Cable, Churchill to Curtin, 1 September 1942, ISMAY VI/2, Ismay Papers: KC.
[6] Minute, Portal to Churchill, 28 August 1942, PREM 3/150/9: PRO.
[7] Cables, Bruce to Curtin, 1 and 2 September 1942, *DAFP*, vol. 6, docs. 36 and 38.
[8] Gilbert, *Road to victory*, p. 222.
[9] *See* f.n. 7.
[10] Cable, Curtin to Bruce, 5 September 1942, CRS A5954/229: AA.
[11] Letter, Jones to Shedden, 3 September 1942, CRS A5954/231: AA.
[12] Letter, Dewing to Ismay, 18 August 1943, ISMAY IV/Dew/1b, Ismay Papers: KC.
[13] Statement by Evatt, 3 September 1942, CRS A5954/474: AA.
[14] A. Bryant, *The turn of the tide*, ch. 10.
[15] Cable, Churchill to Curtin, 8 September 1942, PREM 3/163/6: PRO. G. H. Gill, *Royal Australian Navy 1942–1945*, pp. 185–7. S. W. Roskill, *The war at sea*, vol. 2, pp. 236–7.
[16] 'Operations in New Guinea—Review of 23rd November, 1942', CRS A5954/587: AA.
[17] Press conference, 9 September 1942, MS 4675, Smith Papers: NLA.
[18] Minutes of Prime Minister's War Conference, 7 September 1942, CRS A5954/1: AA.
[19] Background briefing, 23 July 1942, MS 4675, Smith Papers: NLA. Memoirs of Lord Wakehurst, ch. 18, p. 9, Wakehurst Papers: HLRO.
[20] Talk with Irvine Douglas, 8 September 1942, CRS M100: AA.
[21] Memo, Evatt to Robinson, undated, 'Dr H. V. Evatt Folder', Robinson Papers: UMA.
[22] Memorandum by Robinson, 7 September 1942, 'Robinson, W. S., 1942–45 (a) Folder', Evatt Papers: FUL.
[23] Memorandum by Robinson, 7 September 1942.
[24] Statement by Evatt, 3 September 1942, CRS A5954/474: AA.
[25] K. Buckley and K. Klugman, *The Australian presence in the Pacific*.
[26] At the end of August 1942, Bruce predicted that the issue of colonial trusteeship would be one where Australia and America would have to combine to force a 'realistic' attitude upon Whitehall. Cable, Bruce to Curtin, 28 August 1942, and talk with Lord Cranborne, 28 August 1942, CRS M100: AA.
[27] Cable, Curtin to Roosevelt, 11 September 1942, *DAFP*, vol. 6, doc. 43.
[28] Cable, Curtin to Churchill, 11 September 1942, PREM 3/163/6: PRO.
[29] Cable, Curtin to Churchill, 11 September 1942.
[30] Cable, Dill to Chiefs of Staff, 12 September 1942, PREM 3/163/4: PRO.
[31] Note of Secraphone Conversation between MacArthur and Curtin, 17 September 1942, *DAFP*, vol. 6, doc. 47.
[32] Cable, Page to Bruce, 14 September 1942, *DAFP*, vol. 6, doc. 45.
[33] Advisory War Council minute, 24 September 1942, CRS A5954/261: AA.
[34] Letters, Bruce to Churchill, 16 and 17 September 1942, PREM 3/163/6: PRO.

[35] Note, Cranborne to Churchill, 18 September 1942, PREM 3/163/6: PRO.
[36] Note by Bruce, 21 September 1942, *DAFP*, vol. 6, doc. 51.
[37] Note by Bruce, 21 September 1942.
[38] Talk with Austin Hopkinson, 17 September 1942, CRS M100: AA. For the disquiet in Westminster, *see* King, pp. 189–90; Lord Moran, *Winston Churchill: the struggle for survival*, ch. 9.
[39] *See* f.n. 36.
[40] Background briefing, 21 September 1942, MS 4675, Smith Papers: NLA.
[41] Press statement by Curtin, 23 September 1942, CRS A5954/305A: AA.
[42] J. Robertson, *Australia at war 1939–1945*, p. 143.

CHAPTER 5

[1] Lord Moran, p. 90.
[2] Harvey diary, 2 October 1942, cited in M. Gilbert, *Road to victory*, p. 237.
[3] R. R. James, *Victor Cazalet*, p. 281.
[4] Gilbert, *Road to victory*, p. 236.
[5] A good account of the Russian campaign is given in B. Liddell Hart, *History of the Second World War*, ch. 18.
[6] Shedden's draft memoirs, ch. 57, p. 6, CRS A5954/771: AA.
[7] Letter, Cross to Attlee, 16 September 1942, RC/4/12, Cross Papers: IWM.
[8] Teleprinter message, Polglaze to Shedden, 1 October 1942, CRS A5954/229: AA.
[9] Survey by the Minister of Production, 3 September 1942, CAB 66/28, WP (42)393: PRO.
[10] Cabinet agenda no. 335, 22 September 1942, CRS A2700/4: AA.
[11] Broadcast by Menzies, 9 October 1942, MS 3668, Folder S4, Tonkin Papers: NLA.
[12] Letter, Evatt to Hughes, 29 September 1942, 'Hughes, W. M. Folder', Evatt Papers: FUL.
[13] Talk with Sir Henry Self, 19 October 1942, CRS M100: AA.
[14] Cable, Chiefs of Staff to Wavell, 23 September 1942, PREM 3/143/9: PRO.
[15] Diary, 31 October 1942, ADD MS 56399, Harvey Papers: BL.
[16] *DAFP*, vol. 6, doc. 49.
[17] War journal, 19 October 1942, WILK 1/1, Wilkinson Papers: CC.
[18] War journal, 22 September 1942.
[19] Letter, Watt to Hood, 12 October 1942, MS 3788/1/1, Watt Papers: NLA.
[20] The report on air operations in Malaya was modified for publication 'to avoid unnecessarily wounding the susceptibilities of the Australians and the Dutch'. Letter, Air Commodore McEvoy to Brooke Popham, 31 October 1946, V/8/13, Brooke Popham Papers; draft report by Admiral Layton, 1946. The report included the following passage: 'the evidence is too overwhelming that many [of the Australian troops], and I am afraid the majority, threw their hands in as soon as things looked black, and spread the canker of their panic the further by the previous reputation of their country's fighting men'. Brooke Popham, the Commander-in-Chief in Malaya, recommended that the passage be excised from the published account. V/8/47/3, Brooke Popham Papers: KC.

[21] Note, Martin to Bridges, 4 October 1942, PREM 3/163/6: PRO. *See also* various documents in CRS M100, 'September 1942' and 'October 1942': AA.
[22] *See* f.n. 21.
[23] Minute, Peck to Churchill, 3 October 1942, PREM 3/163/6: PRO.
[24] Memorandum by Pound, 9 October 1942, PREM 3/163/6: PRO.
[25] Letter, Gowrie to Colonel Bankie, 20 October 1942, MS 2852/4/22/28, Gowrie Papers: NLA. Letter, Roosevelt to Churchill, 19 October 1942, in W. Kimball (ed.), *Churchill and Roosevelt*, vol. 1, p. 633.
[26] Background briefing, 6 October 1942, MS 4675, Smith Papers: NLA.
[27] AWC minutes, 15 October 1942, CRS A2682/6/1087: AA. WC minutes, 14 and 15 October 1942, CRS A2673/XII/2428 and 2446: AA. Cable, Curtin to Churchill, 17 October 1942, PREM 3/63/10: PRO.
[28] Minute, Churchill to Ismay, 18 October 1942, PREM 3/63/10: PRO.
[29] Letter, Robinson to Evatt, 12 August 1942, 'Robinson, W. S. 1942–45 (a)' Folder, Evatt Papers: FUL.
[30] Letter, Robinson to Evatt, 16 October 1942.
[31] Talk with Attlee, 20 October 1942; and cable, Bruce to Curtin, 20 October 1942, CRS M100: AA.
[32] Cable, Curtin to Bruce, 22 October 1942, *DAFP*, vol. 6, doc. 64; cable, Bruce to Curtin, 23 October 1942, CRS M100: AA.
[33] PREM 3/63/10: PRO.
[34] Liddell Hart, p. 316.
[35] Cable, Churchill to Curtin, 27 October 1942, PREM 3/63/10: PRO.
[36] *DAFP*, vol. 6, doc. 66.
[37] Cable, Alexander to Brooke, 28 October 1942, PREM 3/63/10: PRO.
[38] Minute, Churchill to Ismay, 28 October 1942, PREM 3/63/10: PRO.
[39] *DAFP*, vol. 6, doc. 68.
[40] Cable, Churchill to Roosevelt, 29 October 1942, PREM 3/63/10: PRO.
[41] Horner, p. 221.
[42] Note, Hollis to Churchill, 29 October 1942, PREM 3/63/10: PRO.
[43] War Cabinet paper, 30 October 1942, CAB 66/30, WP (42)491: PRO.
[44] Roskill, *The war at sea*, vol. 2, p. 229.
[45] Cadogan diary, 21 October 1942, in Dilks (ed.), *The diaries of Sir Alexander Cadogan*, p. 485.
[46] Cable, Curtin to Churchill, 29 October 1942, with note, Churchill to Attlee, 30 October 1942, PREM 3/63/10: PRO.
[47] Note, Attlee to Churchill, 30 October 1942, and draft cable, Churchill to Curtin, PREM 3/63/10: PRO.
[48] Cable, Dixon to External Affairs, 20 November 1942, CRS A5954/229: AA.

CHAPTER 6

[1] Cited in R. H. Spector, *Eagle against the sun*, p. 211.
[2] Bryant, *The turn of the tide*, p. 516.
[3] Cable, Curtin to Bruce, 4 November 1942, CRS M100: AA.
[4] Minute, Churchill to Brooke, 5 November 1942, PREM 3/63/10: PRO. Author's emphasis.
[5] Cable, Casey to Churchill, 3 November 1942, PREM 3/63/10: PRO.
[6] Gilbert, *Road to victory*, p. 254.
[7] W. S. Churchill, *The Second World War*, vol. 4, p. 583.
[8] Defence Committee (Operations) minutes, 16 November 1942, CAB 69/4, DO (42)17: PRO.

[9] Cables, Curtin to Bruce, 16 November 1942, CRS M100: AA.
[10] Cable, Curtin to Churchill, 16 November 1942, PREM 3/63/10: PRO.
[11] Cable, Curtin to Churchill, 16 November 1942, and attached note by Churchill, 18 November 1942, PREM 3/63/10: PRO.
[12] Minute, Dominions Office to J. Martin, 18 November 1942, and other documents in this file, PREM 3/63/10: PRO.
[13] Cable, Roosevelt to Churchill, 20 November 1942, PREM 3/63/10: PRO.
[14] Cable, Chiefs of Staff to Joint Staff Mission, 20 November 1942, PREM 3/63/10: PRO.
[15] Chiefs of Staff meeting minutes, 20 November 1942; cable, Fraser to Churchill, 20 November 1942, PREM 3/63/10: PRO.
[16] Cable, Joint Staff Mission to Chiefs of Staff, 21 November 1942, PREM 3/63/10: PRO.
[17] Cable, Churchill to Roosevelt, 23 November 1942, PREM 3/63/10: PRO.
[18] Note, Ismay to Churchill, 23 November 1942, PREM 3/63/10: PRO.
[19] Background briefing, 23 November 1942, MS 4675, Smith papers: NLA.
[20] Letter, Blamey to Curtin, 4 December 1942, CRS A5954/262: AA.
[21] Note by Bruce, 23 November 1942; *see also*, 'Talk with Spender', 17 November 1942, CRS M100: AA.
[22] Cable, Churchill to Roosevelt, 24 November 1942, PREM 3/63/10: PRO.
[23] Cables, Churchill to Curtin, and Churchill to Fraser, 24 November 1942, PREM 3/63/10: PRO.
[24] Cable, Bruce to Curtin, 25 November 1942, CRS M100: AA.
[25] Cable, Churchill to Alexander, 24 November 1942; cable, War Office to Alexander, 23 November 1942, PREM 3/63/10: PRO.
[26] Cable, Alexander to War Office, 24 November 1942, PREM 3/63/10: PRO.
[27] Minute, Churchill to Brooke, 25 November 1942, PREM 3/63/10: PRO.
[28] Note, Brooke to Churchill, 25 November 1942, with attached note by Churchill to Brooke, 27 November 1942, PREM 3/63/10: PRO.
[29] Memorandum by Brooke, 26 November 1942, PREM 3/63/10: PRO.
[30] Cable, Curtin to Churchill, 30 November 1942, PREM 3/63/10: PRO.
[31] Note, Attlee to Churchill, 1 December 1942; cable, Churchill to Roosevelt, 1 December 1942; cable, Churchill to Curtin, 2 December 1942; cable, Fraser to Churchill, 5 December 1942, PREM 3/63/10: PRO.
[32] Cable, Churchill to Fraser, 6 December 1942; cable, Roosevelt to Churchill, 6 December 1942, PREM 3/63/10: PRO.
[33] *See* Day, 'P. G. Taylor and the alternative Pacific air route, 1939–45'.
[34] Memo, W. P. Hildred to Balfour, 27 October 1942, BBK D/208, Beaverbrook Papers: HLRO.
[35] 'Note of Captain P. G. Taylor's Interview', 4 November 1942, BBK D/208, Beaverbrook Papers: HLRO.
[36] Memo, Balfour to Vice Chief of Air Staff, 28 October 1942, BBK D/208, Beaverbrook Papers: HLRO.
[37] Cable, Curtin to Bruce, 30 October 1942, CRS A2676/2200: AA.
[38] Memo, Assistant Chief of Air Staff (Plans) to Balfour, 31 October 1942, BBK D/208, Beaverbrook Papers: HLRO.
[39] Cable, Wavell to Churchill, 31 October 1942, PREM 3/143/9: PRO.

[40] Memo, Pound to Churchill, 5 November 1942, PREM 3/163/1: PRO.
[41] Cable, Churchill to Curtin, 2 December 1942, VI/2, Ismay Papers: KC.
[42] PREM 3/163/1: PRO.
[43] Bryant, *The turn of the tide*, pp. 525–35.
[44] Letter, Watt to Hood, 20 July 1942, MS 3788/1/1, Watt Papers: NLA.
[45] Shedden's draft memoirs, ch. 57, CRS A5954/771: AA.
[46] Shedden, ch. 56.
[47] CRS A2670/461/1942: AA.
[48] Minute, Bracken to Churchill, 16 December 1942, PREM 3/150/7: PRO.

CHAPTER 7

[1] 'Operations in New Guinea', 30 January 1943, in Robertson and McCarthy (eds.), Doc. 351.
[2] Gilbert, *Road to victory*, p. 272.
[3] Gilbert, *Road to victory*, pp. 281–2; Bryant, *The turn of the tide*, p. 535.
[4] *See* D. Reynolds, *The creation of the Anglo-American alliance 1937–41*; W. R. Louis, *Imperialism at bay*, and C. Thorne, *Allies of a kind*.
[5] Note by Grigg, 9 November 1942, CAB 66/30, WP (42)515: PRO. *See also* diary, 1 December 1942, ADD MS 56399, Harvey Papers: BL.
[6] Note by Lyttelton, 16 December 1942, CAB 66/32, WP (42)591: PRO.
[7] Letter, Bedell Smith to Ismay, 15 December 1942, IV/Smi/1a, Ismay Papers: KC. Letter, Dill to Cunningham, 1 December 1942, ADD MS 52570, Cunningham Papers: BL.
[8] Cable, Churchill to Attlee, 20 January 1943, CAB 65/37, WM (43)12: PRO. *See also* Gilbert, *Road to victory*, ch. 18; Bryant, *The turn of the tide*, ch. 11.
[9] Press conference, 19 January 1943, MS 4675, Smith Papers: NLA. *DAFP*, vol. 6, doc. 105.
[10] F. W. Perry, *The Commonwealth armies*, pp. 168–9.
[11] War journal, 18 and 19 October 1942, WILK 1/1, Wilkinson Papers: CC.
[12] Press statement by Curtin, 7 December 1942, CRS A5954/305A: AA.
[13] Background briefing, 30 December 1942, MS 4675, Smith Papers: NLA.
[14] *See* Thorne, p. 259; talk with Cripps, 29 December 1942, CRS M100: AA. *See also* memorandum by Attlee, 4 January 1943, CAB 66/33, WP (43)6: PRO.
[15] Letter, Robinson to Evatt, enclosing letter, Robinson to Bracken, 11 February 1943, 'Dr H. V. Evatt' folder, Robinson Papers: UMA. *See also* Day, *Menzies and Churchill at war*, ch. 15; war journal, 29 October 1942, WILK 1/1, Wilkinson Papers: CC.
[16] Background briefing, 30 December 1942, MS 4675, Smith Papers: NLA.
[17] *See* Day, 'An undiplomatic incident' and *Menzies and Churchill at war*.
[18] Background briefing, 30 December 1944, MS 4675, Smith Papers: NLA.
[19] Background briefings, 30 December 1942 and 1 January 1943, MS 4675, Smith Papers: NLA.
[20] *See* f.n. 19.
[21] AWC Minutes, 9 February 1943, CRS A2682/6/1138: AA. Talk with Lord Brabazon, 6 January 1943, CRS M100: AA. Background briefing, 30 December 1942, MS 4675, Smith Papers: NLA. Cable, Commonwealth Government to Bruce, 22 January 1943, CRS M100: AA.

[22] CRS A2670/42/1943: AA.
[23] Letter, Gowrie to Linlithgow, 14 January 1943, MS 2852/4/21/24, Gowrie Papers: NLA.
[24] Submission by Curtin, 1 January 1943, CRS A2700/5/412: AA.
[25] Carlton, p. 211. CAB 66/33, WP (43)33: PRO. Cable, Dixon to External Affairs Department, 9 December 1942, CRS A3300/229: AA. Memorandum by Bruce, 5 January 1943, CRS M100: AA.
[26] Memorandum by Amery, 12 November 1942, CAB 66/31, WP (42)524; memorandum by Amery, 25 January 1943, CAB 66/33, WP (43)39: PRO.
[27] Letter, Amery to Dorman-Smith, 29 December 1942, MSS Eur E 215/2, Dorman-Smith Papers: IOL.
[28] War journal, 22 February 1943, WILK 1/2, Wilkinson Papers: CC.
[29] War journal, 10 and 24 October, and 1 November 1942, WILK 1/1, Wilkinson Papers: CC. Letter, Dewing to Gowrie, 31 January 1943, MS 2852/4/22/28, Gowrie Papers: NLA. For Churchill's attempts to have Wilkinson returned to MacArthur's headquarters, *see* PREM 3/158/5: PRO.
[30] Roskill, *The war at sea*, vol. 1, p. 349.
[31] War Cabinet minutes, 22 March 1941, CRS A2673/6/909: AA; War Cabinet minutes, 26 November 1940, CRS A2673/4/632: AA.
[32] Minute, Hollis to Churchill, 29 October 1942, PREM 3/63/10: PRO.
[33] Minutes of Chiefs of Staff meeting, 20 November 1942, PREM 3/63/10: PRO.
[34] Memorandum by Brooke, 27 November 1942, PREM 3/63/10: PRO.
[35] Roskill, *The war at sea*, vol. 2, p. 433.
[36] Roskill, *The war at sea*, vol. 2, p. 481.
[37] Gill, pp. 295–6, 383–4.
[38] Roskill, *The war at sea*, vol. 2, p. 425.
[39] Diary, 7 January 1943, REDW 1/5, Edwards Papers: CC.
[40] Letter, Somerville to H. MacQuarrie, 3 January 1943, ADD MS 50143, Somerville Papers: BL.
[41] Diary, January–February 1943, REDW 1/5, Edwards Papers: CC.
[42] *See* f.n. 41.
[43] Letter, Somerville to Harold Guard, 6 February 1948, Misc. Box 30/557, Somerville Papers: IWM.
[44] Letter, Somerville to Pound, 11 March 1943, SMVL 8/1, Somerville Papers: CC. Diary, 8–11 February 1943, REDW 1/5, Edwards Papers: CC.
[45] Press conference, 24 February 1943, MS 4675, Smith Papers: NLA. AWC minutes, 17 February 1943, CRS A2682/6/1143: AA.
[46] Letter, Somerville to Pound, 11 March 1943, SMVL 8/1, Somerville Papers: CC.

CHAPTER 8

[1] Background briefing, 2 February 1943, MS 4675, Smith Papers: NLA.
[2] *See* f.n. 1.
[3] *See* f.n. 1.
[4] AWC minutes, 2 February 1943, CRS A2682/6/1129: AA.
[5] Press conference, 2 March 1943, MS 4675, Smith Papers: NLA.
[6] *DAFP*, vol. 6, docs. 119 and 120; Moran, p. 108.
[7] War journal, 15 March 1943, WILK 1/2, Wilkinson Papers: CC.
[8] *DAFP*, vol. 6, doc. 138.
[9] Notes of telephone conversation, 16 March 1943, CRS A5954/229: AA.

[10] Among the Allied aircraft attacking the Japanese convoy in the Bismarck Sea were a number of British-built Beaufighters which allowed the official British historian to record that, 'for all our acute needs for them at home and in the Mediterranean, [Beaufighters] had been sent out to Australia on the British Government's orders'. In fact, they had only been sent after a tough, and largely unsuccessful struggle by Menzies and after Britain had advised Australia against producing the aircraft herself. Roskill, *The war at sea*, vol. 2, p. 422; D. Gillison, *Royal Australian Air Force 1939–1942*, pp. 139–40; Day, *The great betrayal*, ch. 6–7.

[11] Cable, Curtin to Churchill, 18 March 1943, PREM 3/142/7: PRO. Cable, Curtin to Roosevelt, 18 March 1943, CRS A5954/229: AA. *See also* background briefing, 23 March 1943, MS 4675, Smith Papers: NLA.

[12] Spector, pp. 223–6.

[13] Talk with Ismay, 20 March 1943; cable, Bruce to Curtin, 23 March 1943, CRS M100: AA.

[14] *DAFP*, vol. 6, doc. 145; note, Attlee to Churchill, 25 March 1943, PREM 3/142/7: PRO.

[15] Note by Bruce, 22 March 1943, CRS M100: AA.

[16] Canada was treated with some respect because of its special position *vis-à-vis* the United States; South Africa was led by the venerable General Smuts, the former Boer rebel, whom Churchill and the rest of Whitehall came to regard as a fount of Imperial and strategic wisdom. *See* J. L. Granatstein, *Canada's war*, p. 119; for one of the innumerable and favourable references to Smuts, *see* 4 August and 14 October 1942, in Dilks (ed.), *The diaries of Sir Alexander Cadogan*, pp. 467, 483.

[17] C. Edwards, *Bruce of Melbourne*, pp. 4, 87.

[18] Letter, Colvin to Gowrie, 19 January 1943, MS 2852/4/21/24, Gowrie Papers: NLA. *See also* talks with Attlee, 13 January and 15 February 1943, CRS M100; letter, Bruce to Curtin, 5 March 1943, CRS M100: AA.

[19] Letter, Colvin to Gowrie, 19 January 1943, MS 2852/4/21/24, Gowrie Papers: NLA.

[20] Cable, Dixon to External Affairs Department, 25 March 1943, and cable, Evatt to Dixon, 31 March 1943, CRS A3300/262: AA. Evatt informed Dixon that the war had 'demonstrated that control or supervision of control over neighbouring territories will be vital to the security of Australia as well as to other Pacific countries; and we must have a full say in the determination of these questions'.

[21] Letter, Cranborne to Emrys-Evans, 18 March 1943, ADD MS 58240, Emrys-Evans Papers: BL.

[22] Letter, Dewing to Ismay, 18 November 1943, enclosing a report on 'Australia in Relation to the War', 20 April 1943, IV/Dew/3/2b, Ismay Papers: KC.

[23] Cable, Curtin to Churchill, 30 March 1943, PREM 3/142/7: PRO. *See also* cable, Curtin to Bruce, 29 March 1943, CRS M100: AA. Roosevelt had also rejected the Australian request, claiming that while the Japanese had the 'capability of massing 1,500 to 2,000 aircraft in the South West Pacific theatre . . . the United Nations have even greater capabilities'. Cable, Roosevelt to Curtin and Evatt, 29 March 1943, CRS A5954/229: AA.

[24] Talk with Sir Campbell Stuart, 31 March 1943, CRS M100: AA. Letter, Lord Wakehurst to Emrys-Evans, 22 March 1943, ADD MS 58243, Emrys-Evans Papers: BL.

[25] Broadcast by Evatt, 7 March 1943, 'War—Speeches by Evatt 43 (a)' folder, Evatt Papers: FUL.
[26] Talk with Colonel Wardell, 17 March 1943, CRS A5954/14: AA.
[27] Cable, Churchill to Wavell, 3 February 1943, PREM 3/143/10: PRO.
[28] Cable, Wavell to Churchill, 4 February 1943, PREM 3/143/10: PRO.
[29] Minute, Churchill to Ismay, 11 February 1943, PREM 3/143/3/4: PRO.
[30] PREM 3/143/10: PRO.
[31] War Cabinet Conclusions, Confidential Annex, 29 April 1943, CAB 65/38, WM (43)61: PRO.
[32] PREM 3/143/10: PRO.
[33] PREM 3/143/10: PRO. Gilbert, *Road to victory*, ch. 21.
[34] Chiefs of Staff meeting, 22 April 1943, PREM 3/143/7: PRO.
[35] Minute, Ismay to Churchill, 28 April 1943, PREM 3/143/10: PRO.
[36] War Cabinet conclusions, 29 April 1943, CAB 65/34, WM (43)62: PRO.
[37] Carlton, p. 212.
[38] Letter, Curtin to Evatt, 1 April 1943; teleprinter message, MacArthur to Shedden, 25 March 1943; and cable, Roosevelt to Curtin, 6 April 1943, CRS A5954/474: AA.
[39] Cable, Curtin to Roosevelt, 13 April 1943, CRS A5954/474: AA.

CHAPTER 9

[1] War journal, 3 May 1943, WILK 1/2, Wilkinson Papers: CC.
[2] War Cabinet Conclusions, 19 April 1943, CAB 65/34, WM (43)56: PRO.
[3] Aide memoire, 6 May 1943, PREM 3/151/4: PRO.
[4] Gilbert, *Road to victory*, p. 396.
[5] Talk with McVey, 6 April 1943, CRS M100: AA.
[6] S. J. Butlin and C. B. Schedvin, *War economy*, pp. 410–13; CRS A5954/219: AA. Press conference, 14 June 1943, MS 4675, Smith Papers: NLA.
[7] Day, *Menzies and Churchill at war*.
[8] Letter, Dalton to Jowitt, 6 January 1943, II/7/5, Dalton Papers: LSE.
[9] War Cabinet Conclusions, 25 February 1943, CAB 65/33, WM (43)35; Memorandum by Cripps and Sinclair, 24 February 1943, CAB 66/34, WP (43)83: PRO.
[10] Diary, 19 January 1943, ADD MS 56399, Harvey Papers: BL.
[11] Letter, Fysh to Forde, 5 November 1942, CRS A461, T314/1/4/2: AA.
[12] Letter, M. E. Antrobus to Secretary, Prime Minister's Department, 23 February 1943, CRS A461, T314/1/4/2: AA.
[13] *See* 'Evatt—Overseas Trips—1943', Evatt Papers: FUL. *See also* H. Fysh, *Qantas at war*, pp. 176–7; talk with Fysh, 6 May 1943, CRS M100: AA. Cable, Dominions Secretary to Commonwealth Government, 28 April 1943, CRS A5954/343: AA.
[14] Submission by Drakeford, 13 April 1943, CRS A2670/183/1943: AA.
[15] *See* f.n. 14.
[16] Note by Shedden, and notes of discussions with MacArthur, 25–31 May 1943, CRS A5954/2: AA. *See also* Shedden's memoirs, ch. 57, pp. 4–5, CRS A5954/771: AA.
[17] *See* f.n. 16.
[18] Talk with McVey, 6 April 1943, CRS M100: AA.
[19] Talk with Lord Burghley, 20 April 1943; cable, Bruce to Curtin, 21 April 1943; cable, Evatt to Bruce, 24 April 1943; cable, Bruce to Evatt, 24 April 1943, CRS M100: AA.

[20] Memorandum to Evatt, 8 June 1943, 'Aviation—Post War' folder, Evatt Papers: FUL. Although the memorandum is not signed, it is most likely that it was written by Robinson, although it may have been by Fysh who was also in London at the time.
[21] Memorandum to Evatt, 8 June 1943. For details of the British opposition, *see* J. McCarthy, *Australia and imperial defence 1918–39.*
[22] Memorandum by the Society of British Aircraft Constructors, June 1943, BBK D/228, Beaverbrook Papers: HLRO.
[23] Note by Churchill, 22 June 1943, CAB 66/38, WP (43)257: PRO.
[24] Letter, Curtin to Evatt, 1 April 1943, CRS A5954/474: AA.
[25] Note, Evatt to Churchill, 12 May 1943, PREM 3/150/8: PRO; *DAFP*, vol. 6, doc. 191.
[26] *DAFP*, vol. 6, doc. 191.
[27] Note, Evatt to Churchill, 12 May 1943, PREM 3/150/8: PRO.
[28] AWC minutes, 13 May 1943, CRS A2682/6/1188: AA.
[29] Note, Churchill to Portal, 13 May 1943, PREM 3/150/8: PRO.
[30] Note, Somerville to Pound, 17 May 1943, PREM 3/163/6: PRO.
[31] Even Bruce, who seemed to fit in to London life so well, complained of British officials having a 'superiority complex'. Talk with Attlee, 24 May 1943, CRS M100: AA.
[32] Note, Somerville to Pound, 17 May 1943, PREM 3/163/6: PRO.
[33] *DAFP*, vol. 6, doc. 207.
[34] Minute, Portal to Churchill, 18 May 1943, PREM 3/150/8: PRO.
[35] Letter, Evatt to Churchill, 18 May 1943, PREM 3/150/8: PRO.
[36] Letter, Evatt to Churchill, 20 May 1943, PREM 3/150/8: PRO.
[37] Minute, Churchill to Portal, 21 May 1943, PREM 3/150/8: PRO.
[38] Letter, Watt to Hodgson, 24 May 1943, MS 3788/1/1, Watt Papers: NLA.
[39] *See* f.n. 38.
[40] Cable, Evatt to Curtin, 22 May 1943, 'Washington' Folder, Evatt Papers: FUL.
[41] Letter, Thompson to Beaverbrook, 23 May 1943, enclosing précis of Churchill's press conference, BBK D/182, Beaverbrook Papers: HLRO.
[42] Day, 'Promise and Performance'. For a contrary view, *see* R. Quinault, 'Churchill and Australia: The Military Relationship 1899–1945', May 1988.
[43] Minute, Churchill to Attlee, 8 April 1943, PREM 3/63/5: PRO.
[44] Cable, Batterbee to Dominions Office, 9 June 1943, PREM 3/63/5: PRO. *See also DAFP*, vol. 6, docs 198 and 206.
[45] Letter, Evatt to Churchill, 23 May 1943, PREM 3/150/8: PRO.
[46] Letter, Evatt to Churchill, 23 May 1943, PREM 3/150/8: PRO.
[47] *DAFP*, vol. 6, docs 192–3; War Cabinet conclusions, 15 June 1943, CAB 65/34, WM (43)85: PRO.
[48] Note by Portal, 24 May 1943, PREM 3/150/8: PRO.
[49] Letter, Robinson to Churchill, 25 May 1943, PREM 3/150/8: PRO.
[50] Cable, Churchill to Evatt, 26 May 1943, PREM 3/150/8: PRO.
[51] Gilbert, *Road to victory*, ch. 24; Bryant, *The turn of the tide*, ch. 12.
[52] III/4/9/4, Ismay Papers: KC.
[53] Cable, Evatt to Churchill, 4 June 1943, PREM 3/150/8: PRO.
[54] War Cabinet conclusions, 5 June 1943, CAB 65/34, WM (43)81: PRO.
[55] *See* f.n. 54.
[56] Cable, Churchill to Dill, 7 June 1943, and cable, Dill to Churchill, 9 June 1943, PREM 3/150/8: PRO.
[57] *DAFP*, vol. 6, doc. 210; cable, Dill to Churchill, 13 June 1943, PREM 3/150/8: PRO.

[58] *DAFP*, vol. 6, doc. 222.
[59] *DAFP*, vol. 6, doc. 223.
[60] Background briefing, 1 June 1943, MS 4675, Smith Papers: NLA.
[61] Cable, Curtin to Evatt, 3 June 1943, 'Washington' folder, Evatt Papers: FUL. *See also* notes by Shedden, 25–31 May 1943, CRS A5954/2: AA.
[62] Background briefing, 1 June 1943, MS 4675, Smith Papers: NLA.
[63] Prime Minister's War Conference minute, 7 June 1943, CRS A5954/2: AA.
[64] Background briefing, 9 June 1943, MS 4675, Smith Papers: NLA.
[65] Cable, Evatt to Churchill, 12 June 1943, PREM 3/150/8: PRO.
[66] Cable, Courtney to Portal, 14 June 1943, PREM 3/150/8: PRO.

CHAPTER 10

[1] Minute, Churchill to Brooke, 21 May 1943, PREM 3/143/10: PRO.
[2] Cable, Churchill to Roosevelt, 16 June 1943, PREM 3/150/8: PRO.
[3] Robertson, appendices 2 and 3.
[4] Minute, Portal to Churchill, 16 June 1943, PREM 3/150/8: PRO.
[5] Minute, Portal to Churchill, 16 June 1943, PREM 3/150/8: PRO.
[6] Letter, Evatt to Churchill, 2 July 1943, PREM 3/150/8: PRO.
[7] Minute, Portal to Churchill, 6 July 1943, PREM 3/150/8: PRO.
[8] Cable, Evatt to Curtin, 8 July 1943, CRS A5954/474: AA.
[9] Cable, Evatt to Churchill, 14 July 1943, PREM 3/150/8: PRO.
[10] Letters, Williams to Evatt, both 11 July 1943, CRS A3300/258: AA.
[11] *See* f.n. 10.
[12] Minute, Churchill to Sinclair and Portal, 12 July 1943, PREM 3/150/8: PRO.
[13] *See* f.n. 12.
[14] Minute, Sinclair to Churchill, 19 July 1943, PREM 3/150/8: PRO.
[15] Minute, Hollis to Churchill, 21 July 1943; draft cable, Churchill to Evatt, undated; and minute, Hollis to Churchill, 22 July 1943, PREM 3/150/8: PRO.
[16] *DAFP*, vol. 6, doc. 244.
[17] Cable, Evatt to Churchill, 23 July 1943, PREM 3/150/8: PRO.
[18] Minute, Churchill to Sinclair, 23 July 1943, PREM 3/150/8: PRO.
[19] Minute, Sinclair to Churchill, 26 July 1943, PREM 3/150/8: PRO.
[20] Minute, Cherwell to Churchill, 29 July 1943, PREM 3/150/8: PRO.
[21] Letter, Eggleston to Evatt, 21 November 1944, CRS A5954/293: AA.
[22] Cable, Evatt to Churchill, and note, K. H. to Churchill, 1 August 1943, PREM 3/150/8: PRO.
[23] Minute, Churchill to Sinclair and Portal, 1 August 1943, and cable, Churchill to Evatt, 2 August 1943, PREM 3/150/8: PRO.
[24] *Sydney Morning Herald*, 2 August 1943.
[25] *See* f.n. 24.
[26] Letter, Latham to Eggleston, 2 May 1943, MS 423/1/181a, Eggleston Papers: NLA.
[27] L. F. Fitzhardinge, *The little digger 1914–1952*, p. 661.
[28] In thanking Churchill, Evatt requested that he ensure that Portal provide the latest type of Spitfire 'tropicalised' for conditions in the Pacific. *See* PREM 3/150/8: PRO.
[29] Letter, Curtin to MacArthur, 4 September 1943, CRS A5954/231: AA.
[30] Message, Evatt to Curtin, 28 August 1943, CRS A5954/231: AA.
[31] Note by Churchill, 10 June 1943, CAB 66/37, WP (43)233: PRO.
[32] War journal, 31 March 1943, WILK 1/2, Wilkinson Papers: CC.

[33] War Cabinet Conclusions, Confidential Annex, 13 April 1943, CAB 65/38, WM (43)53: PRO.
[34] Letter, Emrys-Evans to Wakehurst, 4 June 1943, ADD MS 58243, Emrys-Evans Papers: BL.
[35] Note by Robinson sent to various British ministers in July 1943, BBK D/214, Beaverbrook Papers: HLRO.
[36] *See* f.n. 35.
[37] Background briefing, 5 July 1943, MS 4675, Smith Papers: NLA.
[38] Report by Dewing, 20 April 1943, enclosed in letter to Ismay, 18 November 1943, IV/Dew/3/2d, Ismay Papers: KC.
[39] *See* f.n. 35.
[40] *See* f.n. 35.
[41] Interview with Hudson Fysh, 15 July 1943, BBK D/214, Beaverbrook Papers: HLRO.
[42] Draft War Cabinet paper by Bevin, 21 June 1943, BEVN 2/4, Bevin Papers: CC.
[43] Memorandum by Amery, 5 June 1943, CAB 66/37, WP (43)232: PRO.
[44] Memorandum by Grigg, 14 June 1943, CAB 66/37, WP (43)239: PRO.
[45] Memorandum by Churchill, 21 July 1943, CAB 66/39, WP (43)327: PRO. Diary, 7 June 1943, ADD MS 56399, Harvey Papers: BL.
[46] Diary, 16 July 1943, ADD MS 56399, Harvey Papers: BL.
[47] Letter, Churchill to MacArthur, 20 July 1943, PREM 3/158/5: PRO.
[48] Day, 'Promise and Performance'.
[49] Minute, Churchill to Ismay, 19 July 1943, PREM 3/143/8: PRO.
[50] Letter, Somerville to MacQuarrie, 22 July 1943, ADD MS 50143, Somerville Papers: BL.
[51] 'The P.M. is determined to oust Wavell from his command in India and Burma. He has never thought highly of W. and now he says he is shocked at the lifelessness of our command there. He wants to put Wavell on the shelf as Governor General of Australia . . .' Diary, 30 May 1943, ADD MS 56399, Harvey Papers: BL.
[52] Cables, Churchill to Attlee, 29 and 31 May 1943, and Attlee to Churchill, 30 May 1943, PREM 3/53/5: PRO.
[53] British intelligence agent, Gerald Wilkinson, had been attached to MacArthur's headquarters when he first arrived in Australia, despite Curtin's opposition to the appointment, MacArthur's apparent personal regard for Wilkinson did not, however, outweigh the paranoia that prompted him to expel Wilkinson from his entourage. When Wilkinson left to visit London, MacArthur refused to allow his return despite pleas from Evatt and Churchill. Evatt reportedly assured Wilkinson that he and Curtin would act to allow Wilkinson to return to Australia to 'propound Churchill's policy', but to no avail. General Dewing, the head of the British military mission to Australia, fared little better, with MacArthur according only the most limited access to his headquarters. *See* war journal, 23 and 29 July 1943, WILK 1/2, Wilkinson Papers: CC. *See also* talk with Dewing, 30 June 1943, CRS M100: AA; *and* documents in CRS Box 463: AA.
[54] Cable, Curtin to Churchill, 7 June 1943, MS 2852/4/22/28, Gowrie Papers: NLA.
[55] Minute, Amery to Churchill, 27 April 1943, PREM 3/53/1: PRO. Edwards diary, 22 May 1943, REDW 1/5, Edwards Papers: CC. *See also* PREM 3/154/1: PRO.
[56] Minute, Churchill to Ismay, 26 July 1943, PREM 3/143/8: PRO.

[57] Minutes of Defence Committee (Operations), 28 July 1943, PREM 3/143/8: PRO.
[58] Cable, Bruce to Curtin, 6 August 1943, CRS M100: PRO.

CHAPTER 11

[1] Cable, Churchill to Attlee, 11 August 1943, PREM 3/53/4: PRO.
[2] Chiefs of Staff Committee minutes, 7 August 1943, PREM 3/147/2: PRO.
[3] Alanbrooke diary, 6 August 1943, in Bryant, *The turn of the tide*, p. 693; Pownall diary, 14 September 1943, in B. Bond (ed.), *Chief of Staff*, vol. 2, p. 108; for a defence of Mountbatten, *see* P. Ziegler, *Mountbatten*.
[4] Gilbert, *Road to victory*, p. 467.
[5] Minute, Churchill to Ismay, 7 August 1943, PREM 3/147/3: PRO. Bryant, *The turn of the tide*, p. 694; Gilbert, *Road to victory*, pp. 460–1, 465.
[6] *See* f.n. 5.
[7] Cable, Churchill to Chiang Kai-shek, 11 August 1943, PREM 3/143/10: PRO. *See also* documents in PREM 3/147/3: PRO; *and* Bryant, *Triumph in the West*.
[8] Letter, Ismay to Pownall, undated but most likely 27 May 1944, IV/Pow/4/2a, Ismay Papers: KC.
[9] Letter, Robinson to Evatt, 13 August 1943, 'Dr H. V. Evatt' folder, Robinson Papers: UMA.
[10] Background briefing, 7 September 1943, MS 4675, Smith Papers: NLA.
[11] Press conference, 9 June 1943, MS 4675, Smith Papers: NLA.
[12] *DAFP*, vol. 5, doc. 358.
[13] Undated note, CRS A5954/287: AA.
[14] Letter, Curtin to MacArthur, 5 August 1943, CRS A5954/306: AA.
[15] *See* f.n. 14. *See also* other documents this file; Day, *The great betrayal*, pp. 278–9.
[16] *Sydney Morning Herald*, 16 August 1943.
[17] *Sydney Morning Herald*, 7 September 1943.
[18] Background briefing, 6 September 1943, MS 4675, Smith Papers: NLA.
[19] *Sydney Morning Herald*, 7 September 1943; Day, *Menzies and Churchill at war*, pp. 112, 176; C. Hazlehurst, *Menzies observed*, pp. 257–8.
[20] *Sydney Morning Herald*, 7 September 1943.
[21] Letter, Churchill to Bruce, 6 October 1943; talk with Cranborne, 19 October 1943; letter, Churchill to Bruce, 21 October 1943; letter, Bruce to Churchill, 24 October 1943; letter, Churchill to Bruce, 3 November 1943; letter, Bruce to Churchill, 8 November 1943; all in CRS M100: AA. *See also* letter, Bruce to Curtin, 15 November 1943, CRS A1608/H33/1/2: AA.
[22] Background briefing, 6 September 1943, MS 4675, Smith Papers: NLA. Talk with Evatt, 16 June 1943, CRS M100: AA. Note by Bruce, 22 February 1943, CRS A1970/559/2: AA. P. G. Edwards, 'The rise and fall of the High Commissioner', in A. F. Madden and W. H. Morris-Jones (eds.), *Australia and Britain: studies in a changing relationship*, pp. 48–9, 55; letter, Bruce to Officer, 6 August 1943, MS 2629/1, Officer Papers: NLA. Talks with Attlee, 22 April and 24 May 1943, CRS M100: AA.
[23] *DAFP*, vol. 6, docs 260–1.

[24] Memorandum by Attlee, 22 September 1943, CAB 66/41. WP (43)412: PRO. War Cabinet conclusions, 24 September 1943, CAB 65/35, WM (43)131: PRO.
[25] Memorandum, Robinson to Evatt, 13 August 1943, ibid.; radio broadcast by Evatt, 21 March 1943, 'War—Speeches by Evatt 43(a)' folder, Evatt Papers: FUL.
[26] Minute, Churchill to Ismay, 17 August 1943, PREM 3/147/3: PRO.
[27] Memorandum by Combined Staff Planners, 18 August 1943, PREM 3/147/1: PRO.
[28] Minute, Churchill to Ismay, 20 August 1943, PREM 3/147/3: PRO. *See also* other documents in this file.
[29] Cable, Churchill to Attlee, undated, PREM 3/366/10: PRO.
[30] Gilbert, *Road to victory*, p. 478.
[31] Roskill, *The war at sea*, vol. 2, pp. 420–1; P. Ziegler, *Mountbatten* p. 221.
[32] Letter, Cross to Churchill, 11 August 1943, and cable, Churchill to Cross, 2 September 1943, PREM 3/151/4: PRO.
[33] Talk with Sir Campbell Stuart, 31 March 1943, CRS M100: AA.
[34] Note by Churchill, 5 October 1943, CAB 66/41, WP (43)430: PRO. Cable, Dixon to Evatt and Curtin, 4 September 1943, CRS A3300/258: AA.
[35] Cable, Churchill to Attlee, 9 September 1943, PREM 3/163/7: PRO. *See also* other documents in this file.
[36] Minute, Martin to Churchill, 29 September 1943, PREM 3/147/10: PRO. Minute, Churchill to Ismay, 2 October 1943, PREM 3/53/16: PRO.
[37] Minute, Churchill to Ismay, 2 October 1943, PREM 3/53/16: PRO.
[38] Ziegler, *Mountbatten*, p. 223.
[39] In June 1943 Attlee had argued in a memorandum to the War Cabinet that: 'If we are to carry our full weight in the post-war world with the United States and U.S.S.R., it can only be as a united British Commonwealth'. Memorandum by Attlee, 15 June 1943, CAB 66/37, WP (43)244: PRO.
[40] Memorandum by Attlee, 19 July 1943, CAB 66/39, WP (43)321: PRO.
[41] Memorandum by Attlee, 17 September 1943, CAB 66/41, WP (43)404: PRO.
[42] *See* f.n. 41.
[43] Memorandum by Amery, 2 September 1943, CAB 66/40, WP (43)388: PRO.
[44] Letter, Grigg to his father, 9 September 1943, PJGG 9/6/24, Grigg Papers: CC.
[45] War Cabinet conclusions, 8 April 1943, CAB 65/34, WM (43)50: PRO. Talk with Coombs, 23 June 1943, CRS M100: AA. Dalton diary, 6 July 1943, I/29/7, Dalton Papers: LSE. Note by Dalton, 23 July 1943, CAB 66/39, WP (43)334: PRO.
[46] *See* f.n. 42.
[47] War Cabinet conclusions, 22 September 1943, CAB 65/35, WM (43)130: PRO.

CHAPTER 12

[1] War Cabinet minutes, 1 October 1943, CRS A2673/XIII/3065: AA. Press release, undated, CRS A5954/305A: AA. Cabinet Minutes, 21 October 1943, CRS A2703/1(D): AA.
[2] Memorandum by Shedden, 1 July 1943, CRS A5954/301: AA.

[3] Dalton diary, 9 December 1943, I/29/154, Dalton Papers: LSE.
[4] Cabinet minutes, 22 September 1943, CRS A2703/1(D): AA.
[5] Pownall diary, 2 October 1943, in Bond, vol. 2, p. 109; minute, Lambe to Cunningham, 3 November 1943, and Admiralty Board minutes, 4 November 1943, ADM 205/33: PRO.
[6] Notes of discussion with MacArthur, 25–31 May 1943, CRS A5954/2: AA.
[7] Letters, Blamey to Curtin, 2 June and 5 October 1943, CRS A5954/306: AA. *See also* letter, Forde to Curtin, 31 July 1943, same file.
[8] Notes on discussions with MacArthur, 16–20 January 1943, CRS A5954/2: AA. Note by Sir Walter Layton, October 1943, PREM 3/159/2: PRO. Spector, pp. 232–3.
[9] Notes on talk with Curtin, 22 October 1943, PREM 3/159/2: PRO. *See also* other documents this file.
[10] *See* Perry.
[11] *See* Liddell Hart, pp. 507–8.
[12] Cable, Curtin to Churchill, 8 October 1943, PREM 3/63/8: PRO.
[13] Note, Ismay to Churchill, 23 January 1944, and note, Cranborne to Churchill, 21 January 1944, PREM 3/63/8: PRO.
[14] Minute, Churchill to Cranborne, 25 January 1944, PREM 3/63/8: PRO.
[15] Note by Chiefs of Staff, 1 May 1944, PREM 3/63/8: PRO.
[16] Letter, MacArthur to Curtin, 8 September 1943, CRS A5954/218: AA.
[17] Letters, Storey to Curtin, 4 October 1943, and Curtin to Storey, 11 October 1943; War Cabinet minute, 11 November 1943, CRS A5954/218: AA. War Cabinet agendum by Curtin, 23 September 1943, CRS A5954/345: AA.
[18] *See* f.n. 17.
[19] AWC minute, 11 November 1943, CRS A5954/218: AA.
[20] The *Age*, Melbourne, 16 October 1943.
[21] The *Age*, Melbourne, 15 October 1943; *see also* cable, Commonwealth Government to Bruce, 8 October 1943, CRS A5954/345: AA.
[22] Letter, Dewing to Ismay, 18 November 1943, IV/Dew/3/3c, Ismay Papers: KC.
[23] Cabinet minutes, 23 November 1943, CRS A2703/1(D): AA.
[24] Cited in The *Argus*, Melbourne, 17 November 1943; *see also* letter, Campbell Stuart to Gowrie, 17 November 1943, MS 2852/4/21/24, Gowrie Papers: NLA.
[25] Background briefing, 8 December 1943, MS 4675, Smith Papers: NLA.
[26] Cable, Churchill to War Cabinet office, 25 August 1943, PREM 3/150/8: PRO. Message, Shedden to MacArthur, 27 August 1943; letter, Blamey to Brooke, 29 November 1943, and other documents in this file, CRS A5954/463: AA.
[27] Prime Minister's War Conference, 29 November–1 December 1943, CRS A5954/2: AA. Letter, Air Marshal Williams to Drakeford, 24 November 1943, CRS A3300/258: AA. The *Argus*, Melbourne, 27 October 1943.
[28] Prime Minister's War Conference, 29 November–1 December 1943, CRS A5954/2: AA. Letter, McVey to Colonel Wilson, 29 November 1943, CRS A5954/218: AA.
[29] The *Herald*, Melbourne, 1 October 1943; Kimball (ed.), vol. 2, pp. 527–30.
[30] Cable, Law to Eden and Churchill, 8 October 1943, and minute, Churchill to Eden, 9 October 1943, PREM 3/158/7: PRO.
[31] Cable, Dominions Secretary to Commonwealth Government, 3 July 1943, CRS A5954/345: AA.

[32] 'Summary of Air Ministry Memorandum', by British High Commission, Canberra, 16 September 1943, CRS A5954/345: AA. For an extensive file on British concern about the American development of air facilities in Fiji and their fear that the Americans would try to retain residual rights, *see* BBK D/214, Beaverbrook Papers: HLRO.
[33] Letter, Cross to Churchill, 11 August 1943, PREM 3/151/4: PRO.
[34] For the Anglo-American view of the future United Nations organization, *see* War Cabinet conclusions/Confidential Annex, 13 April 1943, CAB 65/38, WM (43)53, and 'United Nations Plan for Organising Peace', memorandum by Eden, 7 July 1943, CAB 66/38, WP (43)300: PRO.
[35] PREM 3/366/8: PRO.
[36] Memorandum by Beaverbrook, 3 December 1943, CAB 66/43, WP (43)537: PRO.
[37] Letters, Amery to Beaverbrook and Beaverbrook to Amery, 5 and 6 October 1943, BBK D/228, Beaverbrook Papers: HLRO.
[38] Letter, J. E. Stephenson to Batterbee, 28 October 1943, BBK D/274, Beaverbrook Papers: HLRO.
[39] Cable, Bruce to Commonwealth Government, 27 October 1943, CRS A5954/345: AA. War Cabinet conclusions, 27 October 1943, CAB 65/36, WM (43)146: PRO.
[40] Talk with Beaverbrook, 22 December 1943, CRS M100: AA. Cable, New Zealand Government to Dominions Office, 25 November 1943, and cable, Dominions Office to New Zealand Government, 9 December 1943, BBK D/274, Beaverbrook Papers: HLRO.
[41] Extract from cable, Campbell to Foreign Office, 12 August 1943, BBK D/221, Beaverbrook Papers: HLRO. Cable, Dixon to Curtin, 11 August 1943, CRS A3300/264: AA. Louis, pp. 269–73.
[42] Letter, Taylor to Bowhill, 11 November 1943, MS 2852/4/22/28, Gowrie Papers: NLA. P. G. Taylor, *The sky beyond*, pp. 159–61.
[43] AWC minute, 14 October 1943, CRS A5954/261: AA.
[44] War Cabinet minute, 13 July 1943, and War Cabinet agendum, CRS A2670/299/1943: AA. Butlin and Schedvin, pp. 70–9; Day, *The great betrayal*, pp. 56–8.
[45] *See* documents in CRS A5954/306, 309 and 843: AA.
[46] Press conference, 17 November 1943, MS 4675, Smith Papers: NLA.
[47] Draft press statement, 28 November 1943, CRS A5954/306: AA.
[48] *See* f.n. 47.
[49] Minute, Shedden to Curtin, 1 December 1943, and message, Rodgers to Diller, 3 December 1943, CRS A5954/306: AA.
[50] War Cabinet conclusions/Confidential Annex, 13 December 1943, CAB 65/40, WM (43)169: PRO.
[51] Moran diary, 5 December 1943, in Moran, p. 145.
[52] Gilbert, *Road to victory*, pp. 596–8.
[53] The quick British capture and equally quick fall of the Italian-held islands of Leros and Samos off the coast of Turkey at the end of 1943 was an example of Churchill's eagerness to take wild gambles with forces in the Mediterranean in pursuit of distant and doubtful objectives. Harvey diary, 18 November 1943, ADD MS 56400, Harvey Papers: BL. Bryant, *Triumph in the West*, pp. 48–58.
[54] Press conference, 6 December 1943, MS 4675, Smith Papers: NLA.
[55] *DAFP*, vol. 6, doc. 340.
[56] War Cabinet conclusion, 13 December 1943, CAB 65/40, WM (43)169: PRO. *DAFP*, vol. 6, p. 606, f.n. 1.
[57] *See DAFP*, vol. 6, docs 330, 334, 344, 347–9.

CHAPTER 13

[1] Letter, Ismay to Somerville, 20 December 1943, IV/Som/4b, Ismay Papers: KC.
[2] Letter, Cunningham to Somerville, 19 December 1943, ADD MS 52563: BL.
[3] Gill, p. 358.
[4] Cable, Hollis to Jacob, 31 December 1943, PREM 3/164/5: PRO. Gilbert, *Road to Victory*, p. 599.
[5] Minute, Hollis to Churchill, 1 January 1944; cable, Chiefs of Staff to Churchill, 13 January 1944, PREM 3/160/7: PRO. *See also* letter, Dewing to Ismay, 8 February 1944, IV/Dew/6/1a, Ismay Papers: KC.
[6] Cable, Churchill to Chiefs of Staff, 14 January 1944, PREM 3/160/7: PRO.
[7] Letter, Royle to Somerville, 30 December 1943, SMVL 8/7, Somerville Papers: CC. War Cabinet minutes, 4 February 1944, CRS A2673/XIV: AA. Letter, Royle to Shedden, 6 January 1944, CRS A5954/294: AA. Cable, Bruce to Curtin, 10 February 1944, in which Bruce backed up British pressure to appoint a British officer to replace Royle, CRS A5954/509: AA.
[8] Minute, Shedden to Curtin, 10 December 1942, CRS A5954/393: AA.
[9] Speech by Curtin, Canberra, 14 December 1943, CRS A5954/294: AA.
[10] For an extended discussion of this point, *see* Day, 'Aliens in a hostile land'.
[11] Notes by a British press delegation, submitted to Churchill, 3 February 1944, PREM 3/159/2: PRO.
[12] Background briefing, 25 November 1943, MS 4675, Smith Papers: NLA.
[13] C. Thorne's *Allies of a kind* and J. Dower's *War without mercy* are notable exceptions.
[14] Moran, p. 131.
[15] Somerville journal, 20 November 1943, SMVL 2/2: CC.
[16] Minute, Harvey to Eden with note appended by Eden, 7 September 1941, ADD MS 56402: BL.
[17] Harvey diary, 21 April 1943, ADD MS 56399: BL.
[18] Letters, ex-Senator R. D. Elliott to Beaverbrook, 21 September 1939 and Beaverbrook to Elliott, 25 October 1939, BBK C/130: HLRO.
[19] Shedden's emphasis; Minute, Shedden to Curtin, 21 December 1943, CRS A5954/305: AA. *See also* letter, W. J. Scully to Curtin, 7 December 1943, same file.
[20] Letter, MacArthur to Curtin, 4 February 1944, CRS A5954/218: AA. Letter, Dewing to Ismay, 31 December 1943, IV/Dew/4c, Ismay Papers: KC. Letter, Gowrie to Cranborne, 6 January 1944, MS 2852/4/21/25: NLA. Defence Committee minutes, 23 February 1944, CRS A5954/306: AA.
[21] *DAFP*, vol. 7, doc. 26.
[22] *See* f.n. 210.
[23] Memorandum by Cranborne, 2 February 1944, CAB 66/46, VP (44)70: PRO.
[24] War Cabinet conclusions, 11 February 1944, CAB 65/41, WN (44)18: PRO.
[25] Memorandum by Cranborne, 15 February 1944, CAB 66/47, WP (44)106: PRO.
[26] Letter, Curtin to Evatt, 5 February 1944, enclosing message, Hull to Curtin, 3 February 1944, CRS A1608/Y41/1/1: AA.

[27] Thorne, *Allies of a kind*, p. 486.
[28] Letter, Evatt to Johnson, 24 February 1944, and initialled by Curtin, CRS A1608/Y41/1/1: AA. For details of the Pacific War Council meeting on 12 January, *see* cable, Dixon to External Affairs Department, 12 January 1944, CRS A3300/265: AA. *See also DAFP*, vol. 7, doc. 8.
[29] *See* f.n. 28.
[30] Report by Defence Committee, 5 January 1944, CRS A2031/12/2/1944: AA. *See also* letter, Royle to Shedden, 6 January 1944, CRS A5954/294: AA.
[31] Paper by Admiral Parry, early 1944, 71/19/4, Parry Papers: IWM.
[32] Letter, Blamey to Shedden, 15 January 1944, CRS A5954/294: AA.
[33] *See* f.n. 32.
[34] Statement by Curtin, 18 January 1944, CRS A 5954/294: AA.
[35] *DAFP*, vol. 7, doc. 21.
[36] Letters, Noble to Cunningham, 12, 24 and 30 January 1944, ADD MS 52571, Cunningham Papers: BL.
[37] Memorandum by Lambe, 5 February 1944, ADD MS 52571, Cunningham Papers: BL.
[38] AWC minutes, 20 January 1944, CRS A2682/7/1284: AA.
[39] Cable, Curtin to Churchill, 5 February 1944, PREM 3/164/5: PRO.
[40] Defence Committee (Operations) minutes, 19 January 1944, CAB 69/6, D.O. (44)3: PRO.
[41] *See* f.n. 40; letter, Churchill to de Wiart, 18 January 1944, PREM 3/159/14: PRO.
[42] *See* f.n. 40.
[43] Note, Attlee to Churchill, 20 January 1944, and minute, Churchill to Ismay, 24 January 1944, PREM 3/160/7: PRO.
[44] *See* f.n. 43.
[45] Minutes, Ismay to Churchill, 25 January 1944, and minute, Churchill to Ismay, 31 January 1944, PREM 3/160/7: PRO.
[46] Pownall diary, 12 January and 5 February 1944, in Bond (ed.), pp. 132, 139; letter, Somerville to Tennant, 22 January 1944, TEN 25: NMM. *See also* letter, Somerville to H. MacQuarrie, 27 January 1944, ADD MS 50143: BL.
[47] Letter, Mountbatten to Beaverbrook, 4 February 1944, BBK D/141, Beaverbrook Papers: HLRO.
[48] Letter, Layton to Cunningham, 7 February 1944, ADD MS 52571: BL. *See also* letter, Ismay to Auchinleck, 2 February 1944, IV/Con/1/1Fa, Ismay Papers: KC.
[49] Minute, Churchill to Ismay, 13 February 1944, PREM 3/159/14; minute, Churchill to Alexander and Cunningham, 14 February 1944, PREM 3/164/5: PRO.
[50] Draft memorandum by Churchill, 14 February 1944, and minutes of a Staff Conference, 14 February 1944, PREM 3/160/7: PRO. C.O.S. (44)161(O), 14 February 1944, PREM 3/148/10: PRO. Brooke diary, 14 February 1944, in Bryant, *Triumph in the West*, p. 148.
[51] PREM 3/148/10: PRO. Note by Churchill, 17 February 1944, PREM 3/159/14: PRO. Note, Alexander to Churchill, 20 February 1944, PREM 3/164/5: PRO.
[52] Brooke diary, 21 and 22 February 1944, in Bryant, *Triumph in the West*, p. 152; report by Chiefs of Staff, 23 February 1944, PREM 3/148/2: PRO.
[53] Memorandum by Dening, 17 February 1944, and minute, Eden to Churchill, 21 February 1944, PREM 3/160/7: PRO.
[54] Minute, Churchill to Ismay, 23 February 1944, PREM 3/160/7: PRO.

[55] Letter, Somerville to Mountbatten, 23 February 1944, SMVL 8/3: CC.
[56] Edwards diary, 24 February 1944, REDW 1/6: CC.
[57] Pownall diary, 25 February 1944, in Bond (ed.), pp. 145–6; *see also* cable, Mountbatten to Chiefs of Staff, 25 February 1944, PREM 3/164/1: PRO. Gill, p. 388.

CHAPTER 14

[1] Background briefing, 2 March 1944, MS 4675, Smith Papers: NLA.
[2] Somerville diary, 2 March 1944, ADD MS 52564, Cunningham Papers: BL.
[3] Cable, Curtin to Churchill, 4 March 1944, PREM 3/160/1: PRO.
[4] Minute, Cranborne to Churchill, 28 February 1944, and cable, Churchill to Curtin, 3 March 1944, PREM 3/164/1: PRO.
[5] *See* f.n. 4; cable, Mountbatten to Air Ministry, 25 February 1944, PREM 3/164/1: PRO.
[6] Background briefing, 7 March 1944, MS 4675, Smith Papers: NLA.
[7] Report by Captain Hillgarth, PREM 3/159/10: PRO.
[8] *See* f.n. 7; Somerville diary, 8 and 11 March 1944, ADD MS 52564, Cunningham Papers: BL.
[9] Churchill instructed that his favoured operation against Sumatra must be postponed because of the Japanese move, but he ordered that planning for it should continue. Minute, Churchill to Ismay, 7 March 1944, PREM 3/164/1: PRO.
[10] Spector, p. 361.
[11] Letter, Ismay to Casey, 14 March 1944, IV/Cas/2d, Ismay Papers: KC.
[12] For details of these discussions *see* Reynolds, *The Creation of the Anglo–American Alliance* and Gilbert, *Finest hour.*
[13] Letter, Amery to Eden, 11 February 1944, CRFT 1/2: CC.
[14] War Cabinet conclusions/Confidential Annex, 11 February 1944, CAB 65/45, WM (44)18: PRO.
[15] Letter, Cranborne to Beaverbrook, 16 February 1944, BBK D/131: HLRO.
[16] Talk with Lord Croft, 18 February 1944, CRS M100: AA.
[17] Cabinet submission by Evatt, 18 January 1944, CRS A2700/8/594: AA.
[18] Letter, Beaverbrook to Churchill, 8 February 1944, BBK D/183: HLRO.
[19] 'Synopsis of World Oil Reserve Situation', unsigned and undated paper, BBK D/189: HLRO.
[20] War Cabinet conclusions/Confidential Annex, 11 February 1944, CAB 65/45, WM (44)18: PRO.
[21] Memorandum by Beaverbrook, 21 February 1944, BBK D/421: HLRO.
[22] Cable, Bruce to Curtin, 5 January 1944, enclosing a copy of Halifax's cable, CRS A5954/345: AA. Report by Attlee, 3 February 1944, CAB 66/46, WP (44)73: PRO.
[23] Letter, Taylor to Gowrie, 1 March 1944, MS 2852/4/22/28, Gowrie Papers: NLA. Letter, Taylor to J. G. Beohm, 27 March 1944, MS 2594/86, Taylor Papers: NLA.
[24] Report by Hillgarth, PREM 3/159/10: PRO. Note by Beamish, 29 February 1944, BEAM 3/5: CC.
[25] Report by Hillgarth, PREM 3/159/10: PRO.
[26] AWC Minutes, 21 March 1944, CRS A2682/7/1322: AA. Gill, pp. 471–2; message, Colonel Wilson to Royle, 22 March 1944, CRS A5954/305: AA. *See also* AWC Minute, 7 March 1944, CRS A5954/510: AA.

[27] Memorandum by Shedden, 9 February 1944, CRS A5954/468: AA.
[28] Memorandum by Shedden for Curtin, 23 March 1944, CRS A5958/305: AA.
[29] Letter, Shedden to Wilson, 1 April 1944, CRS A5954/305: AA.
[30] War Cabinet conclusions, 17 February 1944, CAB 65/41, WM (44)22: PRO. Dalton diary, 15 December 1943, I/29/160, Dalton Papers: LSE.
[31] P. Smith, *Task force 57*, p. 112; note by Lyttelton, 29 March 1944, CAB 66/48, WP (44)173: PRO.
[32] Minute, Churchill to Alexander and Cunningham, 9 April 1944, PREM 3/164/5: PRO. *See also* minute, Churchill to Alexander and Cunningham, 10 March 1944, PREM 3/164/5: PRO.
[33] Dalton diary, 13 April 1944, I/30/100, Dalton Papers: LSE.
[34] Cable, Churchill to Mountbatten, 25 February 1944, and cable, Roosevelt to Churchill, 25 February 1944, PREM 3/148/10: PRO. Alanbrooke diary, 24–25 February 1944, in Bryant, *Triumph in the West*, pp. 154–6; note by Churchill, 28 February 1944, PREM 3/160/7: PRO.
[35] Bryant, *Triumph in the West*, pp. 161–71; PREM 3/160/1–8 and PREM 3/148/1–12: PRO. Minute, Churchill to Ismay, 15 April 1944, and cable, Churchill to Roosevelt, 14 April 1944, PREM 3/260/12: PRO.
[36] *See* f.n. 35.
[37] Cable, Churchill to Roosevelt, 10 March 1944; minute, Colville to Private Office, 4 March 1944; minute, Ismay to Churchill, 6 March 1944; and cable, Roosevelt to Churchill, 13 March 1944, PREM 3/160/8: PRO.
[38] Note, Leathers to Churchill, 5 March 1944, PREM 3/160/7; cable, Churchill to Dill, 10 March 1944, PREM 3/160/8; note by Churchill, 7 March 1944, and minute, Churchill to Leathers and Ismay, 14 March 1944, PREM 3/160/3: PRO. Somerville diary, 15 March 1944, ADD MSS 52564, Cunningham Papers: BL.
[39] Cable, Churchill to Curtin, 11 March 1944, PREM 3/160/1: PRO. *See also* letter, Churchill to MacArthur, 12 March 1944, PREM 3/159/14: PRO.
[40] Cable, Churchill to Mountbatten, 18 March 1944, PREM 3/160/1; memorandum, Chiefs of Staff to Churchill, March 1944, PREM 3/160/7: PRO.
[41] Defence Committee minute, 16 March 1944, CRS A2031/12/87/1944: AA. Shedden's memoirs, ch. 46, pp. 4–5, CRS A5954/771: AA.
[42] Minute, Hollis to Churchill, 25 March 1944, and minute, Churchill to Ismay, 27 March 1944, PREM 3/160/1: PRO. Somerville diary, 3 April 1944, ADD MS 52564, and letter, Cunningham to Noble, 8 April 1944, ADD MS 52571: BL.
[43] Somerville diary, 26 March 1944, ADD MSS 52564, Cunningham Papers: BL.
[44] Minutes of a Staff Conference, 8 April 1944, PREM 3/160/8; report by Hillgarth, PREM 3/159/10; cable, Dill to Churchill, 30 March 1944, PREM 3/148/10: PRO. 'Summary of Work carried out by Australian Chemical Warfare Research and Experimental Section November 1943–May 1944', CRS A5954/362: AA.
[45] Shedden memoirs, ch. 57, pp. 1–3, CRS A5954/771: AA.
[46] Letter, Cross to Cranborne, 13 April 1944, RC/4/24, Cross Papers: IWM.

CHAPTER 15

[1] Shedden's memoirs, ch. 57, CRS A5954/771: AA. Horner, pp. 313–14, 507–8.

[2] Letter, Watt to Hood, 4 April 1944, MS 3788/1/1, Watt Papers: NLA. *DAFP*, vol. 7, pp. 133–8.
[3] *DAFP*, vol. 7, pp. 197–200, 218–20; P. G. Edwards, *Prime ministers and diplomats*, pp. 162–3.
[4] Horner, pp. 308–13.
[5] *Daily Telegraph*, Sydney, 20 April 1944.
[6] Thorne, *Allies of a kind*, p. 486.
[7] Horner, pp. 314, 507–8.
[8] Critique of Curtin by Cross, 2 April 1944, enclosed in a letter to Cranborne, 13 April 1944, Rc/4/13, Cross Papers: IWM.
[9] Memorandum by Cranborne, 18 April 1944, CAB 66/49, WP (44)210: PRO. *See also* V. Massey, *What's past is prologue*, p. 417.
[10] Granatstein, pp. 319–20.
[11] War Cabinet conclusions, 27 April 1944, CAB 65/42, WM (44)58: PRO.
[12] *See* f.n. 11.
[13] *See* f.n. 11; letter, Beaverbrook to Croft, 2 May 1944, BBK D/131, Beaverbrook Papers: HLRO.
[14] AWC Minutes, 20 April 1944; cable, Forde to Curtin, 20 April 1944, CRS A5954/305: AA.
[15] Message, Forde to Fraser, 8 April 1944; cable, Curtin to Forde, 19 April 1944, CRS A5954/305: AA.
[16] Minute, Churchill to Ismay, 16 April 1944, PREM 3/63/8: PRO.
[17] Note, Ismay to Churchill, 1 May 1944, PREM 3/63/8: PRO.
[18] Minute, Colville to Jacob, 2 May 1944, PREM 3/63/8: PRO. Letter, Hankey to his son, 7 May 1944, HNKY 3/46: CC.
[19] Minute, Churchill to Ismay, 4 May 1944, and memorandum, Ismay to Churchill, 6 May 1944, PREM 3/63/8: PRO.
[20] J. W. Pickersgill (ed.), *The Mackenzie King record*, vol. 1, p. 667.
[21] Note by Shedden, 5 April 1944, CRS A5954/305: AA.
[22] War Cabinet minutes, 1 May 1944, CRS A2673/XIV: AA. Press conference, 28 April 1944, MS 4675, Smith Papers: NLA.
[23] AWC Minutes, 2 May 1944, CRS A2682/7: AA. Statement by Royle, 3 May 1944, CRS A5954/305: AA.
[24] Cable, Forde to Curtin, 4 May 1944, CRS A5954/305: AA.
[25] Cable, Curtin to Forde, 19 May 1944, same file.
[26] Letter, Churchill to Curtin, 12 May 1944; minute, Alexander to Churchill, 19 May 1944; PREM 3/160/1: PRO.
[27] Note, Alexander to Churchill, 23 May 1944, PREM 3/63/8: PRO.
[28] Letter, Churchill to Curtin, 27 May 1944, PREM 3/160/2: PRO.
[29] Cable, Curtin to Forde, 30 May 1944, CRS A5954/305: AA.
[30] Shedden's memoirs, ch. 46, pp. 5–7, CRS A5954/771: AA. Summary of observations by Curtin, 3 May 1944, PREM 3/160/1; PMM (44)5th Meeting, Confidential Annex, PREM 3/160/2: PRO. Cunningham diary, 3 May 1944, ADD MS 52577: BL.
[31] Cunningham diary, 5 May 1944.
[32] Minutes, Churchill to Ismay, 5 and 14 May 1944, PREM 3/148/9 and 3/160/2: PRO.
[33] Pickersgill, p. 685.
[34] Letter, Amery to Churchill, 4 May 1944, PREM 3/160/1: PRO. *See also* note, Cherwell to Churchill, 3 May 1944, and other documents in this file, PREM 3/160/3: PRO.
[35] Minute, Churchill to Hollis, 7 May 1944; extract from COS(44)148, Staff Conference, 8 May 1944, PREM 3/148/9: PRO.
[36] Letter, Ismay to Pownall, undated but most likely May 1944, Ismay IV/Pow/4/2a, Ismay Papers: KC.

[37] Cunningham diary, 18 May 1944, ADD MS 52577: BL.
[38] Cable, Curtin to Forde, 19 May 1944, CRS A2679/16/1944: AA.
[39] *See* f.n. 38.
[40] Letter, Curtin to Churchill, 17 May 1944, PREM 3/63/8: PRO.
[41] Talk with Curtin, 25 May 1944, CRS M100: AA.
[42] Minutes of a Conference at Chequers, 21 May 1944, and letter, Curtin to Churchill, 23 May 1944, PREM 3/63/8: PRO.
[43] *See* f.n. 42; War Cabinet conclusions, 16 and 18 May 1944, CAB 65/42, WM (44)64 and 65: PRO.
[44] Defence Committee minute, 28 and 29 March 1944, CRS A2031/12/91/1944: AA.
[45] Address by Curtin, 17 May 1944, BEVN 6/21: CC.
[46] Talk with McVey, 18 April 1944, BBK D/214: HLRO.
[47] Letters, Beaverbrook to Cranborne, 25 April 1944, and Cranborne to Beaverbrook, 26 April 1944, BBK D/254: HLRO.
[48] Message, Secretary, External Affairs to Director General, Civil Aviation, 15 May 1944, CRS A5954/345: AA. Documents in BBK D/208: HLRO. *DAFP*, vol. 7, p. 207.
[49] *DAFP*, vol. 7, pp. 270–1; memorandum, McVey to Curtin, 15 May 1944, BBK D/214: HLRO.
[50] Cable, Curtin to Forde, 19 May 1944, CRS A5954/345: AA.
[51] Report by Taylor, June 1944; letter, Taylor to Burghley, 18 June 1944, MS 2594/86, Taylor Papers: NLA.
[52] *See* f.n. 51 and other documents in this box.
[53] Talk with McVey, 31 May 1944, CRS M104: AA. For Beaverbrook's version of the conversation, *see* 'McVey on Commonwealth Airline', 31 May 1944, BBK D/214: HLRO.
[54] Talk with Beaverbrook, 2 June 1944, and talk with Curtin, 25 May 1944, CRS M100: AA. *DAFP*, vol. 7, p. 390.
[55] Cunningham diary, 5 June 1944, ADD MS 52577: BL.

CHAPTER 16

[1] *DAFP*, vol. 7, docs 154, 200–1, 210 and 284. *See also* background briefing, 3 July 1944, MS 4675, Smith Papers: NLA.
[2] Cabinet minutes, 4 July 1944, CRS A2703, vol. II: AA.
[3] Background briefing, 3 July 1944, MS 4675, Smith Papers: NLA.
[4] *DAFP*, vol. 7 doc. 143; background briefing, 3 July 1944, MS 4675, Smith Papers: NLA.
[5] Shedden's memoirs, ch. 52, CRS A5954/771: AA. Horner, pp. 327–31.
[6] Minute, Chiefs of Staff to Churchill, undated, PREM 3/160/5: PRO.
[7] Horner, pp. 335–7.
[8] Cunningham diary, 11 and 14 June 1944, ADD MS 52577: BL. Brooke diary, 14 June 1944, in Bryant, *Triumph in the West*, p. 217.
[9] Note, Eden to Churchill, 12 June 1944, PREM 3/160/4: PRO.
[10] Pownall diary, 17 and 19 June 1944, in Bond, pp. 175–7.
[11] Minute, Churchill to Ismay, 24 June 1944; COS (44)225th Meeting (0), 6 July 1944, PREM 3/160/5: PRO. Brooke diary, 6 July 1944, in Bryant, *Triumph in the West*, p. 230.
[12] Report by the Chiefs of Staff, 7 July 1944, CAB 66/52, WP (44)380: PRO.
[13] Memorandum by Anderson, 14 July 1944, CAB 66/52, WP (44)381: PRO.
[14] Cunningham diary, 26 July 1944, ADD MS 52577: BL.
[15] Cunningham diary, 30 July 1944.

[16] Draft directive by Churchill, 3 August 1944, CAB 66/53, WP (44)431; memorandum by Chiefs of Staff, 18 August 1944, CAB 66/54, WP(44)452: PRO. Cunningham diary, 4 August 1944, ADD MS 52577: BL.
[17] War Cabinet minute, 11 May 1944, CRS A5954/305A: AA.
[18] Letter, MacArthur to Curtin, 12 July 1944, CRS A5954/306: AA.
[19] Report by Senator R. V. Keane, 18 August 1944, CRS A2670/415/1944: AA.
[20] War Cabinet minutes, 4 August 1944, CRS A2673/15/3691: AA. *See also* background briefing, 21 August 1944, MS 4675, Smith Papers: NLA.
[21] Letter, Alexander to Curtin, 26 May 1944; letter, Curtin to Makin, 13 July 1944; letter, Shedden to Curtin, 31 July 1944; letter, Shedden to Northcott, 18 April 1945, and other documents in this file, CRS A5954/509; War Cabinet minutes, 4 August 1944, CRS A2673/15/3692: AA. Cunningham diary, 9–14 October 1944, ADD MS 52577: BL.
[22] Brooke diary, 14 July 1944, in Bryant, *Triumph in the West*, p. 232; Cunningham diary, 14 July 1944, ADD MS 52577: BL.
[23] Report by the Chiefs of Staff, 13 July 1944; minutes of Staff Conference, 14 July 1944, PREM 3/160/5: PRO.
[24] Pownall diary, 20 July 1944, in Bond, pp. 181–2; letter, Ismay to Pownall, undated, IV/Pow/4/2a, Ismay Papers: KC. Letter, Cunningham to Layton, 19 July 1944, ADD MS 52571: BL.
[25] Minute, Chiefs of Staff to Churchill, 20 July 1944, PREM 3/160/5: PRO. Minute, Churchill to Chiefs of Staff, 24 July 1944, PREM 3/148/9: PRO.
[26] Brooke diary, 7 and 8 August 1944, in Bryant, *Triumph in the West*, p. 248; Cunningham diary, 7 and 8 August 1944, ADD MS 52577: BL.
[27] Minutes of Staff Conference, 9 August 1944, PREM 3/149/7: PRO. Brooke diary, 9 August 1944, in Bryant, *Triumph in the West*, pp. 249–51; Cunningham diary, 9 August 1944, ADD MS 52577: BL. Pownall diary, 29 August 1944, in Bond, pp. 184–5.
[28] Minutes of Staff Conference, 9 August 1944, PREM 3/149/7: PRO. Cunningham diary, 10 August 1944, ADD MS 52577: BL. *See also* Somerville journal, 22 and 24 August 1944, SMVL 2/2: CC.
[29] Cable, Chiefs of Staff to Joint Staff Mission, 12 August 1944, PREM 3/149/7: PRO.
[30] This description of Churchill was reported by Ismay during a closed meeting of the Chiefs of Staff. Cunningham diary, 11 August 1944, ADD MS 52577: BL.
[31] Cable, Curtin to Churchill, 12 August 1944, and Cable 5, Churchill to Curtin, 23 August 1944, CAB 69/6, DO (44)13: PRO.
[32] Note by General Lumsden, PREM 3/159/4; cable, Curtin to Churchill, 3 September 1944, CAB 69/6, DO (44)13: PRO.
[33] Note, Cross to Ismay, 4 September 1944, PREM 3/159/4: PRO.
[34] Account of conversation between Cross and MacArthur on 16 August 1944, submitted to Churchill by Cross on 30 August 1944, Rc/4/19, Cross Papers: IWM.
[35] Cable, Churchill to Lumsden, 4 September 1944, and cable, Churchill to Curtin, 10 September 1944, CAB 69/6, DO (44)13: PRO.
[36] *See* f.n. 35.
[37] Cable, Curtin to Churchill, 14 September 1944, and cable, Churchill to Curtin, 18 September 1944, CAB 69/6, DO(44)13: PRO. *See also* *Daily Telegraph*, Sydney, 18 September 1944.
[38] Gilbert, *Road to victory*, ch. 50; Brooke diary, 5–10 September 1944, in

Bryant, *Triumph in the West*, pp. 268–70; Cunningham diary, 8 September 1944, ADD MS 52577: BL.

[39] Cunningham's draft memoirs, p. 122, ADD MS 52581B: BL. Brooke diary, 14 September 1944, in Bryant, *Triumph in the West*, p. 274; minute, Churchill to Ismay, 12 September 1944, PREM 3/160/6: PRO.

[40] Brooke diary, 12 September 1944, in Bryant, *Triumph in the West*, p. 272.

[41] Minute, Churchill to Ismay, 12 September 1944, PREM 3/160/6: PRO.

[42] *See* f.n. 41; Canadian Cabinet War Committee minutes, 14 September 1944, PREM 3/329/6: PRO. *See also* note by Hollis, 24 June 1944, CAB 69/6, DO (44)12: PRO.

[43] Draft statement by Churchill, 2 September 1944, PREM 3/149/11; note, Ismay to Churchill, 10 September 1944, PREM 3/159/14: PRO.

[44] Gilbert, *Road to victory*, p. 967.

[45] *Times-Herald*, Washington, 16 September 1944.

[46] Gilbert, *Road to victory*, p. 980.

[47] Cable, Churchill to Curtin, 18 September 1944, and cable, Curtin to Churchill, 22 September 1944; minute, Alexander to Churchill, 26 September 1944, PREM 3/159/4: PRO. Shedden's memoirs, ch. 52, CRS A5954/771: AA.

[48] Report by Production Executive, 21 September 1944, CRS A2670/473/1944: AA. Minute, Churchill to Ismay, 12 September 1944, PREM 3/160/6: PRO.

[49] AWC minutes, 28 September 1944, CRS A2682/8/1431: AA.

[50] Notes of discussions with MacArthur, 30 September 1944, CRS A5954/843: AA.

[51] Pownall diary, 24 September 1944, in Bond, p. 187.

CHAPTER 17

[1] Gilbert, *Road to victory*, pp. 1020, 1038–9.

[2] P. Charlton, *The unnecessary war*.

[3] AWC minutes, 28 September 1944, CRS A2682/8/1430: AA.

[4] Memorandum by Shedden, 28 October 1944, CRS A5954/312: AA.

[5] Letter, Robinson to Evatt, 15 October 1944, 'Robinson, WS, 1942–45(b)' folder, Evatt Collection: FUL.

[6] *See* f.n. 4.

[7] *Province*, Vancouver, 3 January 1945.

[8] For instance, Churchill was looking to Indian troops to garrison British territories as they were recaptured from the Japanese. Minute, Churchill to Ismay, 25 May 1944, PREM 3/159/14: PRO.

[9] Taylor Papers, MS 2594/86: NLA.

[10] Letter, Air Marshal Welsh to Admiral King, 7 August 1944, MS 2594/86: NLA.

[11] Draft letter (not sent), Taylor to Air Marshal Welsh, MS 2594/46: NLA.

[12] Note by Taylor, MS 2594/46: NLA.

[13] Paper by Colyer, 3 September 1944, BBK D/208: HLRO.

[14] P. G. Taylor, *The sky beyond*, p. 184.

[15] Message, RAF Dorval to Taylor, 1 November 1944, MS 2594/86: NLA.

[16] P. G. Taylor, *The sky beyond*, pp. 222–3.

[17] Thorne, *Allies of a kind*, p. 666.

[18] Cable, Roosevelt to Churchill, 27 November 1944, BBK D/422, Beaverbrook Papers: HLRO.

[19] Minute, Churchill to Ismay, 28 November 1944, and minute,

Beaverbrook to Churchill, 4 December 1944, BBK D/422, Beaverbrook Papers: HLRO. Minute, Eden to Churchill, 4 December 1944, PREM 3/95: PRO.

20 Report by Chiefs of Staff, 1 December 1944, PREM 3/95: PRO.

21 Cable, Halifax to Foreign Office, 4 January 1945, PREM 3/95: PRO.

22 Minute, Churchill to Chiefs of Staff and Sinclair, 8 January 1945; cable, Mexico to Foreign Office, 4 January 1945; minute, Churchill to Eden, 6 January 1945, PREM 3/95: PRO. Cable, Churchill to Roosevelt, 10 January 1945, BBK D/423: HLRO.

23 Note by Taylor, undated, and letter (draft), Taylor to Churchill, undated, MS 2594/86: NLA. Thorne, *Allies of a kind*, p. 667.

24 Message, 25 August 1944, and other documents in this file, CRS A5954/345; Cabinet minutes, 25 September 1944, CRS A2703/II: AA.

25 War Cabinet conclusions, 1 September 1944, CAB 65/43, WM (44)114; War Cabinet conclusions, 29 September 1944, CAB 65/47, WM (44)129; CAB 66/57, WP (44)628: PRO. Cable, Cranborne to Commonwealth Government, 22 November 1944, CRS A5954/343; cable, Drakeford to Forde, 9 November 1944, CRS A5954/345: AA.

26 Smith, p. 71; Lord Halifax, *Fulness of days*, p. 259.

27 Cunningham diary, 5, 24 October and 13 November 1944, ADD MS 52577: BL.

28 Directive by Churchill, and directive to Viceroy by Churchill, 20 November 1944, CAB 66/58, WP (44)670 and 671: PRO.

29 *See* J. M. Lee, *The Churchill Coalition*, pp. 174–5.

30 *See* Dower.

31 Letter, Mountbatten to Beaverbrook, 27 October 1944, BBK D/141: HLRO.

32 As Adolf Berle, the former American Assistant Secretary of State, assured Beaverbrook, America 'wants the British Commonwealth a strong, solid, flourishing, prosperous, going concern, just as I should imagine England wants a strong, solid, and prosperous America—not merely for sentimental reasons but because both conditions are to the solid interest of both'. Letter, Berle to Beaverbrook, 30 December 1944, BBK D/151: HLRO. For the description of Hurley, *see* letter, Henderson to Colville, 17 March 1945, PREM 3/159/12: PRO.

33 Memorandum by Eden, October 1944, PREM 3/180/7: PRO.

34 Letter, Mountbatten to Somerville, 14 November 1944, SMVL 9/2: CC.

35 *See* f.n. 27.

36 *DAFP*, vol. 7, doc. 337.

37 Memorandum by Cranborne, 10 November 1944, CAB 66/57, WP (44)641: PRO.

38 War Cabinet conclusions, 13 November 1944, CAB 65/44, WM (44)149: PRO; *DAFP*, vii, Doc. 347.

39 Letter, Eggleston to Evatt, 21 November 1944, CRS A5954/293: AA.

40 *See DAFP*, vol. 7, docs 349, 352–4; for Curtin's account of the May meeting on colonial policy, *see* CRS A2679/16/1944: AA.

41 War Cabinet conclusions, 24 November 1944, CAB 65/44, WM (44)155: PRO.

42 Cable, Bruce to Evatt, 27 November 1944, CRS M100: AA. *DAFP*, vol. 7, doc. 344; Louis, pp. 416–7.

43 Letter, Croft to Cranborne, 28 November 1944, CRFT 1/18: CC.

44 War Cabinet conclusions, 20 December 1944, CAB 65/44, WM (44)172: PRO.

45 Carlton, p. 251; Louis, ch. 29–30; War Cabinet conclusions, 19 March

1945, CAB 65/49, WM (45)33: PRO.
[46] Dower, p. 71.

CHAPTER 18

[1] R. Rhodes, *The making of the atomic bomb*, p. 560.
[2] Robertson, p. 174.
[3] In early October Curtin had expressed his disappointment to the acting British High Commissioner in Canberra at the lack of any British naval component in the American attack on the Philippines. In this and other instances, Curtin was forever reminding London of the need to restore the prestige of the British Empire in Asia and the Pacific. Cable, Acting High Commissioner to Dominions Office, 3 October 1944, PREM 3/164/6: PRO.
[4] Letter, Curtin to Hankinson, 7 September 1944, and letter, Strahan to Hankinson, 14 September 1944, CRS A1608/S41/1/9 Pt 1: PRO.
[5] Speech by Curtin, 22 February 1945, CRS A5954/1605: AA.
[6] Letters, Shedden to MacArthur, 15 and 27 February 1945, CRS A5954/75: AA.
[7] Discussion with General MacArthur, Tokyo, May 1946, CRS A5954/3: AA.
[8] Letters, Mountbatten to Murdoch, 26 June 1944, Murdoch to Mountbatten, 24 July 1944, and Mountbatten to Murdoch, 2 September 1944, Murdoch Papers, MS 2823/44: NLA.
[9] Minutes, Hollis to Churchill, 20 February and 6 March 1945, Churchill to Ismay, 7 and 10 March 1945, and Ismay to Churchill, 9 March 1945, PREM 3/53/14: PRO.
[10] Letter, Cross to Cranborne, 23 April 1945, and minute, Ismay to Churchill, 2 June 1945, PREM 3/53/14: PRO.
[11] Minute by Churchill, 1 June 1945; minutes, Peck to Ismay and Ismay to Peck, and Ismay to Churchill, 2 June 1945, PREM 3/53/14: PRO.
[12] Cunningham diary, 13 December 1944, ADD MS 52577: BL.
[13] *See* Day, *Menzies and Churchill at war; The great betrayal*; and 'Anzacs on the run', May 1986.
[14] Note by Anderson with memorandum by Keynes, 3 April 1945, CAB 66/65, WP (45)301: PRO.
[15] Submission by W. J. Scully, 30 January 1945, CRS A2700/14/1/794: AA. *See also* cable, Bruce to Forde, 25 May 1945, CRS M100: AA; *and* memorandum by Minister of Food, 11 June 1945, CAB 66/66, CP (45)29: PRO.
[16] Smith, p. 108.
[17] Although Curtin had explained Australia's problem during his visit to London in May 1944, Bruce was still having to explain it to the Admiralty six months later. *See* cable, Bruce to Curtin, 27 November 1944, CRS M100: AA.
[18] Shedden's memoirs, ch. 52, p. 8, CRS A5954/771; note by Shedden, 15 November 1944, CRS A5954/309; documents in CRS A5954/306: AA.
[19] A. Bullock, *Ernest Bevin*, p. 294.
[20] Report by Keynes and Sinclair, 2 February 1945, CAB 66/61, WP (45)77: PRO.
[21] Letter, Cunningham to Fraser, 19 January 1945, ADD MS 52572; Cunningham diary, 22 January 1945, ADD MS 52578; letter, Cunningham to Power, 23 January 1945, ADD MS 52562: BL.
[22] Minute, Cherwell to Churchill, 22 January 1945, PREM 3/164/5;

minutes, Brooke to Churchill, 19 January 1945, and Churchill to Brooke, 21 January 1945, PREM 3/63/7: PRO. Cunningham diary, 26 January 1945, ADD MS 52578: BL.

[23] Cable, Bruce to Curtin, 26 January 1945, CRS M100: AA. J. C. W. Reith, *Into the wind*, p. 505.

[24] Gilbert, *Road to victory*, p. 1162.

[25] CCS 184th Meeting, 1 February 1945, PREM 3/51/8: PRO.

[26] Minutes of First Plenary Session, 2 February 1945, PREM 3/51/4: PRO.

[27] *See*, e.g., Cunningham diary, 28 December 1944, ADD MS 52577: BL; *and* note by Officer, 24 February 1945, MS 2629/1, Officer Papers: NLA. *See also* A. Shai, *Britain and China, 1941–47*; Thorne, *Allies of a kind*, ch. 26.

[28] Minutes of Second Plenary Session, 9 February 1945, PREM 3/51/4: PRO.

[29] Gilbert, *Road to victory*, p. 1207; minute, Ismay to Churchill, 25 February 1945, PREM 3/51/10: PRO.

[30] Memorandum by Anderson, 12 February 1945, CAB 66/61, WP (45)87: PRO.

[31] Cunningham diary, 23 February 1945, ADD MS 52578: BL.

[32] Directive by Churchill, 26 February 1945, CAB 66/62, WP (45)117: PRO.

[33] Cunningham diary, 1 March 1945, ADD MS 52578: BL.

[34] For delays to Mountbatten's plans in February 1945, *see* PREM 3/149/3: PRO; *and* letters, Power to Cunningham, 1 March 1945 and Cunningham to Power, 13 March 1945, ADD MS 52562; Cunningham diary, 6 March 1945, ADD MS 52578: BL; *and* Mountbatten diary, 17 February 1945, in P. Ziegler, (ed.), *Personal diary of Admiral the Lord Louis Mountbatten*, p. 182.

[35] Cable, Chiefs of Staff to Joint Staff Mission, 7 March 1945, and minute, Ismay to Churchill, 7 March 1945, PREM 3/149/3: PRO. Somerville journal, 28 February and 7 March 1945, SMVL 2/3: CC. Gill, pp. 616–7; minute, Ismay to Churchill, 7 March 1945, PREM 3/164/6: PRO.

[36] Cable, Somerville to Cunningham, 9 March 1945, PREM 3/164/6: PRO.

[37] Letter, Fraser to Cunningham, 14 March 1945, ADD MS 52572: BL.

[38] AWC Minutes, 7 March 1945, CRS A2682/8/988: AA.

[39] Minute, Sinclair to Churchill, 27 February 1945, PREM 3/150/9: PRO. Cable, Cranborne to Commonwealth Government, 20 April 1945, CRS A5954/231: AA.

[40] Memorandum, Chiefs of Staff to Churchill, 19 March 1945, and other documents in PREM 3/142/5: PRO. *See also* Cunningham diary, 19 March 1945, ADD MS 52578: BL.

[41] Cable, Curtin to Churchill, 23 March 1945, and cable, Churchill to Curtin, 20 April 1945, PREM 3/63/8: PRO.

[42] Report by Attlee, 20 March 1945, CAB 66/63, WP (45)197: PRO.

[43] Minute, Lyttelton to Churchill, 22 March 1945, and other documents in PREM 3/159/14: PRO.

[44] Memorandum by Attlee, 24 March 1945, CAB 66/63, WP (45)192: PRO.

[45] Directive by Churchill, 14 April 1945, CAB 66/64, WP (45)250: PRO.

[46] Note by Churchill, 16 April 1945, CAB 66/65, WP (45)252: PRO; Bryant, *Triumph in the West*, p. 460.

[47] Cunningham diary, 3 April 1945, ADD MS 52578: BL. Edwards diary, 21 April 1945, REDW 1/7: CC.

CHAPTER 19

[1] Letter, Penney to Sinclair, 2 May 1945, PENNEY 5/1: KC.
[2] Cited in Rhodes, p. 649.
[3] Appreciation by Blamey, 18 May 1945, CRS A2670/209/1945: AA.
[4] Rhodes,· p. 618.
[5] L. Ross, *John Curtin*, p. 378.
[6] Memorandum by Evatt, 8 January 1945, CRS A2670/23/1945: AA.
[7] War Cabinet conclusions, 11 January 1945, CAB 65/49, WM (45)4; memorandum by Cranborne, 16 February 1945, CAB 66/62, WP (45)99: PRO.
[8] Talk with Evatt and talk with Forde, both on 2 April 1945, CRS M100: AA.
[9] Talk with Sir Eric Machtig, and talk with Evatt, both on 3 April 1945, CRS M100: AA.
[10] P. Hasluck, *Diplomatic witness*, p. 173.
[11] Speech by Evatt, 9 April 1945, BEVN 6/29: CC.
[12] Draft policy document discussed on 5 April 1944, BEVN 2/12: CC. C. Attlee, *As it happened*, p. 156.
[13] C. Hartley Grattan, 'The Southwest Pacific since the First World War', in W. S. Livingston and W. R. Louis (eds.), *Australia, New Zealand, and the Pacific Islands since the First World War*, p. 215.
[14] 'The International Post-War Settlement', BEVN 2/12: CC.
[15] Cable, Chiefs of Staff to Joint Staff Mission, 21 April 1945, PREM 3/180/7: PRO. *See also* minute, Eden to Churchill, 2 June 1945, PREM 3/221/7: PRO.
[16] Note by the Foreign Office, 28 March 1945, PREM 3/159/12: PRO.
[17] Note by Churchill, 11 April 1945, PREM 3/159/12: PRO.
[18] War Cabinet conclusions, 3 April 1945, CAB 65/50, WM (45)38: PRO.
[19] Cadogan diary, 4 to 13 April 1945, ACAD 1/15, Cadogan papers: CC. Louis, p. 505.
[20] Memorandum by Cranborne, 10 April 1945, CAB 66/64, WP (45)228: PRO.
[21] *See* f.n. 20; War Cabinet conclusions, 12 April 1945, CAB 65/50, WM (45)42: PRO. Louis, p. 510.
[22] Letter, Cranborne to Emrys-Evans, 29 April 1945, ADD MS 58263, Emrys-Evans Papers: BL.
[23] Cadogan diary, 23 May 1945, in Dilks (ed.), p. 745.
[24] Letter, Cranborne to Emrys-Evans, 29 May 1945, ADD MS 58263, Emrys-Evans Papers: BL.
[25] Louis, ch. 33; War Cabinet conclusions, 14 May 1945, CAB 65/50, WM (45)61: PRO.
[26] Letter, Cranborne to Emrys-Evans, undated, ADD MS 58263, Emrys-Evans Papers: BL.
[27] *See* f.n. 26.
[28] Extract from cable, UK Delegation to the Foreign Office, 23 June 1945, CRS M100: AA.
[29] Memorandum by Chifley, 9 May 1945, CRS A2679/15/1945: AA. *See also* AWC Minutes, 10 May 1945, CRS A2682/8/1528: AA.
[30] Minute, Ismay to Churchill, 2 May 1945, and minute, Churchill to Ismay, 3 May 1945, PREM 3/142/7: PRO.
[31] Minute, Churchill to Leathers, 10 May 1945, and minute, Leathers to Churchill, 16 May 1945, PREM 3/142/7: PRO.
[32] Talk by Cross, 8 May 1945, RC/2/47, Cross Papers: IWM.
[33] Memorandum by Alexander, 23 May 1945, CAB 66/65, WP (45)323; note by Cherwell, 31 May 1945, PREM 3/164/5: PRO.

[34] Cabinet conclusions, 5 June 1945, CAB 65/53, CM (45)5: PRO.
[35] Mountbatten's attempts to retrieve ships from the British Pacific Fleet angered officials at the Admiralty; Somerville agreed with Cunningham that Britain's name 'would certainly stink if we took ships away from the B.P.F. for mopping up in S.E.A.C.'. Letter, Somerville to Cunningham, 15 May 1945, ADD MS 52563: BL. Letter, Cunningham to Somerville, 9 May 1945, SMVL 9/3: CC. Cable, Mountbatten to Chiefs of Staff, 4 May 1945, PREM 3/149/10: PRO.
[36] Cables, Joint Staff Mission to Chiefs of Staff, 6 and 30 April and 12 May 1945, and Chiefs of Staff to Mountbatten, 16 May 1945, PREM 3/149/11, 3/149/10 and 3/160/8: PRO.
[37] Cunningham diary, 11 May 1945, ADD MS 52578: BL; cable, Eden to Churchill, 14 May 1945, PREM 3/484; memorandum by Grigg, 10 May 1945, CAB 66/65, WP (45)292: PRO.
[38] Cable, Churchill to Eden, 11 May 1945, PREM 3/484: PRO.
[39] Cables, Churchill to Truman, 12 May 1944, PREM 3/484: PRO.
[40] H. Thomas, *Armed truce*, ch. 23.
[41] Cable, Wilson to Chiefs of Staff, 16 May 1945, and minute, Churchill to Ismay, 9 June 1945, PREM 3/484: PRO.
[42] Cunningham diary, 9 June 1945, ADD MS 52578: BL. *See also* Alanbrooke diary, 11, 19 and 20 June 1945, in Bryant, *Triumph in the West*, p. 464; minute, Churchill to Ismay, 16 June 1945, PREM 3/149/10: PRO.
[43] Alanbrooke diary, 20, 21 and 28 June 1945, in Bryant, *Triumph in the West*, pp. 464–5; Cunningham diary, 28 June 1945, ADD MS 52578: BL.
[44] Memorandum by Anderson, 29 June 1945, CAB 66/67, CP (45)53: PRO. Letter, Nicolson to his son, 27 May 1945, in Sir H. Nicholson, *Diaries and letters*, vol. 2, p. 466.
[45] Paper by Shedden, 9 February 1945, CRS A5954/309; speech by Curtin, 28 February 1945, CRS A5954/1605: AA.
[46] Cunningham diary, 16, 19 and 26 April 1945, ADD MS 52578: BL. *See also* cable, Chiefs of Staff to Joint Staff Mission, 12 April 1945, and cable, Wilson to Chiefs of Staff, 13 April 1945, PREM 3/159/7: PRO.
[47] Cable, Gairdner to Churchill, 12 May 1945, PREM 3/159/7: PRO. War Cabinet Minutes, 28 May 1945, CRS A2673/XVI/4216: AA.
[48] Cable, Forde and Evatt to Chifley, 5 June 1945, 'United Nations—Cables' folder, Evatt Papers: FUL. War Cabinet minutes, 28 June 1945, CRS A2673/XVI/4291 and 4292: AA.
[49] AWC Minutes, 6 June 1945, CRS A2682/8/1550: AA.
[50] Memorandum by Chifley, 31 May 1945, CRS A2679/24/1945; War Cabinet minutes, CRS A2673/XVI/4241: AA; *and* War Cabinet conclusions, 20 April 1945, CAB 65/50, WM (45)48: PRO.

CHAPTER 20

[1] Gilbert, *Never despair*, p. 62.
[2] Cited in Rhodes, p. 676.
[3] Note by Churchill, 3 July 1945, CAB 66/67, CP (45)61: PRO.
[4] Gilbert, *Never despair*, p. 60.
[5] Memorandum by Anderson, 11 July 1945, CAB 66/67, CP (45)72: PRO.
[6] Minute, Ismay to Churchill, 5 June 1945, PREM 3/142/7: PRO.
[7] Memorandum by Macmillan, 11 July 1945, CAB 66/67, CP (45)75. For documents on Canada's reluctance to contribute to the VLR force, *see*

PREM 3/142/5: PRO.

[8] Cunningham diary, 16 July 1945, ADD MS 52578: BL.

[9] Gilbert, *Never despair*, pp. 63–4; 68–9.

[10] W. S. Churchill, *The Second World War*, vol. 6, pp. 552–3.

[11] *See* e.g., talk with M. E. Dening, 18 June 1945, CRS M100: AA. *See also* letter, Penney to Sinclair, 2 May 1945, 5/1, Penney Papers: KC.

[12] Cunningham diary, 20 and 23 July 1945, ADD MS 52578: BL.

[13] Cited in Rhodes, p. 689.

[14] Cunningham diary, 18 and 23 July 1945, ADD MS 52578: BL. *See also* Brooke diary, 17 and 18 July 1945, in Bryant, *Triumph in the West*, p. 475.

[15] Cited in Rhodes, p. 692.

[16] Rhodes, p. 688.

[17] Cunningham's draft memoirs, p. 136, ADD MS 52581B; BL.

[18] Cited in Rhodes, p. 688.

[19] Gilbert, p. 109.

[20] Cable, Dominions Office to Chifley, 17 July 1945, CRS A5954/453: AA.

[21] Cable, Bruce to Chifley, 18 July 1945, CRS M100: AA.

[22] Cable, Evatt to Bruce, 27 July 1945, CRS M100: AA.

[23] Cable, Evatt to Bruce, 30 July 1945, CRS M100: AA.

[24] *See* f.n. 23.

[25] Talk with Ernest Bevin, 3 August 1945, CRS M100: AA.

[26] Letter, Rourke to Shedden, 26 June 1945, CRS A5954/1615: AA.

[27] Brooke diary, 4 July 1945, in Bryant, *Triumph in the West*, p. 465.

[28] Cable, Bruce to Forde, 6 July 1945, CRS M100: AA.

[29] AWC Minutes, 19 July 1945, CRS A2682/8/1583: AA.

[30] AWC Agendum, CRS A2679/35/1945: AA.

[31] Draft cable, Chifley to Dominions Office, CRS A2679/35/1945: AA.

[32] Memorandum by Fraser, 13 June 1945, CRS A2670/259/1945: AA.

[33] Letters, Cunningham to Fraser, 5 July 1945, and Fraser to Cunningham, 17 July 1945, ADD MS 52572: BL. War Cabinet minutes, 11 July 1945, CRS A2673/XVI/4328; cable, Chifley to Bruce, 17 July 1945, CRS M100: AA.

[34] Cable, Mountbatten to Chiefs of Staff, 10 July 1945, PREM 3/149/9: PRO. Letter, Cunningham to Fraser, 5 July 1945, ADD MS 52572: BL.

[35] Memorandum by Chiefs of Staff, 7 August 1945, CAB 69/7, DO (45)2: PRO. Gill, p. 665.

[36] Memorandum by Bevin, 8 August 1945, CAB 69/7, DO (45)3: PRO.

[37] Defence Committee Minutes, 8 August 1945, CAB 69/7, DO (45)2: PRO.

[38] *See* f.n. 37.

[39] Cunningham diary, 10 August 1945, ADD MS 52578: BL.

[40] Brooke diary, 10 August 1945, in Bryant, *Triumph in the West*, p. 484.

[41] Cable, External Affairs to Australian Legation, Washington, 10 August 1945, CRS A3300/290: AA.

[42] Cables, Commonwealth Government to Dominions Secretary, 11 and 12 August 1945, CRS A5954/453: AA.

[43] Cables, Dominions Secretary to Commonwealth Government, 11 August 1945, CRS A5954/453; cable, Evatt to Bruce, 10 August 1945, CRS M100: AA.

[44] Cadogan diary, 11 and 14 August 1945, in Dilks, p. 781; cable, External Affairs to Bruce, 11 August 1945, CRS M100; cable, Dominions

Secretary to Commonwealth Government, 11 August 1945, CRS A5954/453: AA.

[45] Cited in Rhodes, pp. 745–6.

EPILOGUE

[1] Halifax, pp. 249–50; note by Churchill, 11 April 1945, PREM 3/159/12: PRO.

[2] Cabinet conclusions, 10 August 1945, CAB 128/1, CM (45)20: PRO. Brooke diary, 10 August 1945, in Bryant, *Triumph in the West*, p. 485.

[3] Cable, Attlee to Chifley, 11 August 1945, CRS A5954/453: AA. Memorandum by Chiefs of Staff, 13 August 1945, CAB 69/7, DO (45)5: PRO. Memoir of Allied Disarmament Mission in Saigon 1945, Cheshire Papers: CC.

[4] Message, Shedden to Chifley, 12 August 1945; paper by Shedden, 17 August 1945; *see also* cable, Rourke to Shedden, 12 August 1945, CRS A5954/453; War Cabinet minute, 17 August 1945, CRS A2670/379/1945: AA.

[5] Cunningham diary, 15–17, 22–23 August 1945, ADD MS 52578, and letter, Cunningham to Fraser, 20 August 1945, ADD MS 52572: BL. Gill, p. 683; cable, Dominions Secretary to Commonwealth Government, 18 August and 3 September 1945, CRS A5954/453: AA. Defence Committee minutes, 31 August 1945, CAB 69/7, DO (45)4: PRO. Paper by Penney after visit to MacArthur, 23 August 1945, 5/13, Penney Papers: KC.

[6] Cable, Dominions Secretary to Commonwealth Government, 23 August 1945, CRS A5954/437: AA.

[7] Note by Hollis, 4 September 1945, CAB 69/7, DO (45)10; Defence Committee minutes, 7 and 14 September 1945, CAB 69/7, DO (45)5 and 6: PRO. Cunningham diary, 20 August, 7, 14, 17, 21 and 22 September, 5 October 1945, ADD MS 52578, and letter, Cunningham to Power, 17 September 1945, ADD MS 52562: BL.

[8] Message, Secretary, Department of the Army to Shedden, 13 August 1945, CRS A5954/453: AA.

[9] War Cabinet minute, 17 August 1945, CRS A2670/379/1945: AA. Defence Committee minutes, 13 August 1945, CAB 69/7, DO (45)3: PRO. *See also* note by Mr Quealy, 14 August 1945, and 'Surrender Terms for Japan', unsigned note for Shedden, 14 August 1945, CRS A5954/453: AA.

[10] Press statement by Chifley, 17 August 1945, and cable, Dominions Secretary to Commonwealth Government, 20 August 1945, CRS A5954/453: AA.

[11] Cable, Bruce to Chifley, and draft cable, Attlee to Chifley, 25 August 1945, CRS M100: AA.

[12] Defence Committee minute, 23 August 1945, CRS A5954/453: AA.

[13] Defence Committee minute, 22 August 1945, CRS A5954/453: AA.

[14] Note by Hollis, 30 August 1945, CAB 69/7, DO (45)9; Defence Committee minutes, 31 August 1945, CAB 69/7, DO (45)4: PRO. Alanbrooke diary, 30 August 1945, in Bryant, *Triumph in the West*, p. 490.

[15] Cunningham diary, 27 August and 14 September 1945, ADD MS 52578: BL. Defence Committee minutes, 14 September 1945, CAB 69/7, DO (45)6: PRO.

[16] War Cabinet minutes, 19 September 1945, CRS A2673/XVI/4400: AA.

Alanbrooke diary, 27 September 1945, in Bryant, *Triumph in the West*, p. 490. *See also* letter, Rourke to Shedden, 10 September 1945, CRS A5954/1615: AA.

[17] Cables, Dominions Secretary to Commonwealth Government, 23 and 30 August 1945, CRS A5954/437: AA. Letter, Dening to Sterndale-Bennett, 31 August 1945, 5/18, Penney Papers: KC.

[18] Paper by Penney, 23 August 1945, and letter, Mountbatten to MacArthur, 16 August 1945, 5/11 and 13, Penney Papers: KC.

[19] Speech by Curtin, 28 February 1945, CRS A5954/1605: AA.

[20] *Herald*, Melbourne, 17 October 1939.

CONCLUSION

[1] Cited in J. D. B. Miller, *Britain and the old dominions*, p. 39.

Select Bibliography

OFFICIAL DOCUMENTS

Australian Archives, Canberra

CRS CP 290/7, Cables from Britain 1939–1943.
CRS CP 290/16, Papers relating to Wartime Policy 1940–1945.
CRS M100, S. M. Bruce, Monthly War Files.
CRS M103, S. M. Bruce, Supplementary War Files.
CRS M104, S. M. Bruce, Folders of Annual Correspondence.
AA 1970/559, S. M. Bruce, Miscellaneous Papers, 1939–45.
CRS A1608, Prime Minister's Department, Correspondence, Secret and Confidential War Series, 1939–1945.
CRS A2031, Defence Committee Minutes 1939–45.
CRS A2670, War Cabinet Agenda, 1939–1946.
CRS A2673, War Cabinet Minutes, 1939–1946.
CRS A2682, Advisory War Council Minutes, 1940–5.
CRS A2700, Cabinet Agenda 1941–1949.
CRS A2703, Cabinet Minutes 1941–1949.
CRS A3300, Australian Legation to USA, Correspondence, 1939–48.
CRS A5954, Sir Frederick Shedden, Papers.

OFFICIAL DOCUMENTS

Public Record Office, London

CAB 65, War Cabinet Conclusions and Confidential Annexes.
CAB 66, War Cabinet Memoranda.
CAB 69, Defence Committee (Operations), Minutes and Memoranda.
PREM 1, 3, 4, 7, and 10, Prime Minister's Papers.

PRIVATE PAPERS

Australia

Lord Casey: NLA.
J. J. Dedman: NLA.
Sir Frederick Eggleston: NLA.
Dr H. V. Evatt: FUL.
Lord Gowrie: NLA.
Henry B. S. Gullett: NLA.
R. V. Keane: NLA.
Sir John Latham: NLA.
Norman Makin: NLA.
Sir Robert Menzies: NLA.
Sir Keith Murdoch: NLA.
Sir Keith Officer: NLA.
Sir Earle Page: NLA.
W. S. Robinson: UMA.
Sir Percy Spender: NLA.
F. T. Smith: NLA.
Sir Gordon Taylor: NLA.
Sir Alan Watt: NLA.

PRIVATE PAPERS

Great Britain

Lord Alanbrooke: KC.
A. V. Alexander: CC.
Waldorf Astor: RUL.
Rear Admiral T. P. H. Beamish: CC.
Lord Beaverbrook: HLRO.
Ernest Bevin: CC.
Sir Alexander Cadogan: CC.
Lord Chandos: CC.
Admiral Sir John Crace: IWM.
Lord Croft: CC.
Sir Ronald Cross, Papers: IWM.
Admiral Sir A. B. Cunningham: BL and CC.
Hugh Dalton: LSE.
Admiral Sir William Davis: CC.
Sir Reginald Dorman-Smith: IOL.
Vice Admiral J. W. Durnford: IWM.
Admiral Sir Ralph Edwards: CC.
Paul Emrys-Evans: BL.
Captain Godfrey French: CC.
David Lloyd George: HLRO.
Admiral J. H. Godfrey: CC.
Sir Percy James Grigg: CC.
Lord Halifax, Papers: CC.
Lord Hankey: CC.
Captain Liddell Hart: KC.
Oliver Harvey: BL.
Lord Ismay: KC.
Admiral Kelly: NMM.
W. Mackenzie King: CUL.

Admiral Sir Edward Parry: IWM.
Major-General W. R. C. Penney: KC.
Admiral Sir James Somerville: BL, CC and IWM.
Admiral Sir William Tennant: NMM.
Lord Wakehurst: HLRO.
Gerald Wilkinson: CC.
Admiral Sir Algernon Willis: CC.

MEMOIRS, COLLECTED LETTERS, DIARIES, SPEECHES, ETC.

ATTLEE, C., *As it happened*, London, 1954.
AVON, EARL OF, *The Eden memoirs: the reckoning*, London, 1965.
BARNES, J. and NICHOLSON, D. (eds.), *The Empire at bay: the Leo Amery diaries 1929–1945*, London, 1988.
BLAINEY, G. (ed.), *The memoirs of W. S. Robinson, 1876–1963*, Melbourne, 1967.
BOND, B. (ed.), *Chief of staff*, vol. 2, London, 1974.
BRYANT, A., *The turn of the tide, 1939–1943*, London, 1957.
——, *Triumph in the West*, London, 1959.
CASEY, R. G., *Personal experience 1939–46*, London, 1962.
CHANDOS, LORD, *The memoirs of Lord Chandos*, London, 1962.
COLVILLE, J., *The Churchillians*, London, 1981.
——, *Footprints in time*, London, 1976.
——, *The fringes of power*, London, 1985.
CROWLEY, F. K. (ed.), *Modern Australia in documents*, vol. 2, Melbourne, 1973.
DILKS, D. (ed.), *The diaries of Sir Alexander Cadogan*, London, 1971.
EDWARDS, P. G. (ed.), *Australia through American eyes*, Brisbane, 1979.
FADDEN, A., *They called me Artie*, Melbourne, 1969.
FYSH, H., *Qantas at war*, Sydney, 1968.
HALIFAX, LORD, *Fulness of days*, London, 1957.
HARRIMAN, W. A. and ABEL, E., *Special envoy to Churchill and Stalin*, New York, 1975.
HASLUCK, P., *Diplomatic witness*, Melbourne, 1980.
HULL, C., *The memoirs of Cordell Hull*, New York, 1948.
JAMES, R. R. (ed.), *Chips: the diaries of Sir Henry Channon*, London, 1967.
——, *Victor Cazalet*, London, 1976.
JONES, T., *A diary with letters 1931–1950*, London, 1954.
KIMBALL, W., *Churchill and Roosevelt*, vols 1, 2 and 3, London, 1984.
KING, C., *With malice toward none*, London, 1970.
MACMILLAN, H., *War diaries: the Mediterranean, 1943–1945*, London, 1984.
MANSERGH, N. (ed.), *Documents and speeches on British Commonwealth affairs 1931–1952*, vol. 1, London, 1953.
MASSEY, V., *What's past is prologue*, Toronto, 1963.
MENZIES, (SIR) R. G., *Afternoon light*, London, 1967.
——, *The measure of the years*, London, 1970.
——, *Speech is of time*, London, 1958.
MORAN, LORD, *Winston Churchill: the struggle for survival*, London, 1968.
NEALE, R. G. *et al* (eds.), *Documents on Australian foreign policy 1937–49*, vols 2–7, Canberra, 1976–88.
NICHOLSON, SIR H., *Diaries and letters*, vol. 2, London, 1967.
PAGE, E., *Truant surgeon*, Sydney, 1963.
PICKERSGILL, J., *The Mackenzie King record*, vols 1 and 2, Toronto, 1960, 1968.
REITH, J. C. W., *Into the wind*, London, 1949.

ROBERTSON, J. and McCARTHY, J. (eds), *Australian war strategy 1939–1945*, Brisbane, 1985.
SHERWOOD, R., *The White House papers of Harry L. Hopkins*, London, 1948.
TAYLOR, P. G., *Forgotten island*, New York, 1946.
——, *The sky beyond*, Melbourne, 1963.
VAN DER POEL, J. (ed.), *Selections from the Smuts papers*, Cambridge, 1973.
WAND, REV. J. W. C, *Has Britain let us down?*, Sydney, 1942.
WATT, (SIR) A., *Australian diplomat*, Sydney, 1972.
WILLIAMS, SIR R., *These are facts*, Canberra, 1977.
WINANT, J. G., *A letter from Grosvenor Square*, London, 1947.
YOUNG, K., *The diaries of Sir Robert Bruce Lockhart*, vol. 2, London, 1980.
ZIEGLER, P., *Personal diary of Admiral the Lord Louis Mountbatten 1943–1946*, London, 1988.

SECONDARY WORKS

ADDISON, P., *The road to 1945*, London, 1975.
BARKER, E., *Churchill and Eden at war*, London, 1978.
BARNETT, C., *The audit of war*, London, 1986.
BELL, R. J., *Unequal allies: Australian–American relations and the Pacific War*, Melbourne, 1977.
BULLOCK, A., *Ernest Bevin*, Oxford, 1985.
BUTLIN, S. J. and SCHEDVIN, C. B., *War economy 1942–1945*, Canberra, 1977.
CALLAHAN, R. A., *Churchill: retreat from empire*, Delaware, 1984.
CARLTON, D., *Anthony Eden*, London, 1981.
CHARLTON, P., *The unnecessary war*, Melbourne, 1983.
CHURCHILL, W. S., *The Second World War*, vols 1–6, 1948–54.
CRISP, L. F., *Ben Chifley*, Melbourne, 1961.
DAY, D. A., *Menzies and Churchill at war*, Sydney, 1986.
——, *The great betrayal: Britain, Australia and the onset of the Pacific War 1939–42*, New York, 1989.
DENNIS, P., *Troubled days of peace*, Manchester, 1987.
DILKS, D. (ed.), *Retreat from power*, vol. 2, London, 1981.
DOWER, J., *War without mercy*, London, 1986.
EDWARDS, C., *Bruce of Melbourne*, London, 1965.
EDWARDS, P. G., *Prime ministers and diplomats*, Melbourne, 1983.
ESTHUS, R. A., *From enmity to alliance*, Seattle, 1964.
FITZHARDINGE, L. F., *The little digger 1914–1952*, Sydney, 1979.
GILBERT, M., *Finest hour: Winston Churchill 1939–1941*, London, 1983.
——, *Road to victory: Winston Churchill 1941–1945*, London, 1986.
——, *Never despair: Winston Churchill 1945–1965*, London, 1988.
GILL, G. H., *Royal Australian Navy*, vols 1 and 2, Canberra, 1957–8.
GILLISON, D., *Royal Australian Air Force 1939–1942*, Canberra, 1962.
GRANATSTEIN, J. L., *Canada's war: the politics of the Mackenzie King Government, 1939–1945*, Toronto, 1975.
HANCOCK, W. K., *Smuts*, vol. 2, Cambridge, 1968.
HARPER, N. (ed.), *Australia and the United States*, Melbourne, 1971.
HASLUCK, P., *The government and the people 1939–1941*, Canberra, 1952.
HASTINGS, M., *Bomber command*, London, 1979.
HAZLEHURST, C., *Menzies observed*, Sydney, 1979.
HORNER, D., *High command*, Sydney, 1982.
HOYT, E. P., *Japan's war: The great Pacific conflict*, London, 1986.

HUDSON, W. J., *Casey*, Melbourne, 1986.
——, and SHARP, M. P., *Australian independence: colony to reluctant kingdom*, Melbourne, 1988.
IRIYA, A., *The origins of the Second World War in Asia and the Pacific*, London, 1987.
IRVING, D., *Churchill's war: the struggle for power*, Australia, 1987.
JAMES, R. R., *Anthony Eden*, London, 1986.
KENNEDY, P., *The rise and fall of the great powers*, London, 1988.
LARRABEE, E., *Commander in chief*, London, 1987.
LEE, J. M., *The Churchill coalition 1940–1945*, London, 1980.
LEWIN, R., *Slim: the standardbearer*, London, 1976.
——, *The chief*, London, 1980.
LIDDELL HART, B., *History of the Second World War*, London, 1973.
LIVINGSTON, W. S. and LOUIS, W. R. (eds.), *Australia, New Zealand, and the Pacific Islands since the First World War*, Austin, 1979.
LOUIS, W. R., *Imperialism at bay*, Oxford, 1977.
LOWE, P., *Britain in the Far East*, London, 1981.
MCCARTHY, J., *Australia and imperial defence 1918–39*, Brisbane, 1976.
——, *A last call of empire: Australian aircrew, Britain and the Empire Air Training Scheme*, Canberra, 1988.
MACINTYRE, CAPTAIN D., *The battle for the Pacific*, London, 1966.
——, *Fighting admiral*, London, 1961.
MCKERNAN, M. and BROWNE, M. (eds.), *Australia: Two centuries of war and peace*, Canberra, 1988.
MADDEN, A. F. and MORRIS-JONES, W. H. (eds.), *Australia and Britain: studies in a changing relationship*, London, 1980.
MANSERGH, N., *The Commonwealth experience*, London, 1969.
——, *Survey of British Commonwealth affairs*, London, 1958.
MARDER, A. J., *Old friends, New Enemies*, Oxford, 1981.
MILLAR, T. B., *Australia in peace and war*, London, 1978.
MILLER, J. D. B., *Britain and the old dominions*, London, 1966.
ODGERS, G., *Air war against Japan 1943–1945*, Canberra, 1957.
OVENDALE, R., *The English-speaking alliance*, London, 1985.
PELLING, H., *Winston Churchill*, London, 1974.
PERRY, F. W., *The Commonwealth armies*, Manchester, 1988.
PIMLOTT, B., *Hugh Dalton*, London, 1985.
POTTS, E. D. and POTTS A., *Yanks down under 1941–45*, Melbourne, 1985.
RENOUF, A., *Let justice be done*, Brisbane, 1983.
REYNOLDS, D., *The creation of the Anglo-American alliance 1937–41*, London, 1981.
RHODES, R., *The making of the atomic bomb*, London, 1988.
ROBERTSON, J., *Australia at war 1939–1945*, Melbourne, 1981.
ROSKILL, S., *Hankey, man of secrets*, vol. 3, London, 1974.
——, *Churchill and the admirals*, London, 1977.
——, *The war at sea 1939–1945*, vols 1 and 2, London, 1954 and 1956.
ROSS, L., *John Curtin*, Melbourne, 1977.
SHAI, A., *Britain and China, 1941–47*, Oxford, 1984.
SMITH, P., *Task force 57*, London.
SPECTOR, R. H., *Eagle against the sun*, London, 1987.
STIRLING, A., *Lord Bruce*, Melbourne, 1974.
TAYLOR, A. J. P., *Beaverbrook*, London, 1972.
——, *Churchill: four faces and the man*, London, 1969.
TENNANT, K., *Evatt*, Sydney, 1970.
THOMAS, H., *Armed truce*, London, 1986.

THORNE, C., *Allies of a kind*, London, 1978.
——, *The issue of war*, London, 1985.
WATT, (SIR) A., *The evolution of Australian foreign policy*, Cambridge, 1967.
WINGATE, SIR R., *Lord Ismay*, London, 1970.
WRINCH, P. M., *The military strategy of Winston Churchill*, Boston, 1961.
YOUNG, K., *Churchill and Beaverbrook*, London, 1966.
ZIEGLER, P., *Mountbatten*, London, 1985.

ARTICLES

DAY, D. A., 'Anzacs on the run: the view from Whitehall, 1941–2', *JICH*, May 1986.
——, 'H. V. Evatt and the "Beat Hitler First" strategy: scheming politician or an innocent abroad?', *HS*, October 1987.
——, 'P. G. Taylor and the alternative Pacific air route, 1939–45', *AJPH*, 32, 1 (1986).
——, 'Promise and performance: Britain's Pacific pledge, 1943–45', *WS*, September 1986.
——, 'An undiplomatic incident: S. M. Bruce and the moves to curb Churchill, February 1942', *JAS*, November 1986.
——, 'Aliens in a hostile land: a reappraisal of Australian history', *JAS*, November 1988.
DEDMAN, J. J., 'The return of the A.I.F. from the Middle East', *AO*, 1967.
MCCARTHY, J., 'Australia: a view from Whitehall 1939–45', *AO*, December 1974.
MEANEY, N. K., 'Australia's foreign policy: history and myth', *AO*, August 1969.
MILLAR, T. B., 'A rejoinder', *AO*, August 1969.
QUINAULT, R., 'Churchill and Australia: The Military Relationship 1899–1945', *WS*, May 1988.
ROBERTSON, J., 'Australian war policy 1939–1945', *HS*, October 1977.
——, 'Australia and the "Beat Hitler First" strategy, 1941–42: a problem in wartime consultation', *JICH*, May 1983.

INDEX